**WATER AND
WASTEWATER
TREATMENT**

McGRAW-HILL SERIES IN WATER RESOURCES AND ENVIRONMENTAL ENGINEERING

Ven Te Chow, Rolf Eliassen, and Ray K. Linsley, Consulting Editors

BAILEY AND OLLIS: Biochemical Engineering Fundamentals
BOCKRATH: Environmental Law for Engineers, Scientists, and Managers
CHANLETT: Environmental Protection
GRAF: Hydraulics of Sediment Transport
HALL AND DRACUP: Water Resources Systems Engineering
JAMES AND LEE: Economics of Water Resources Planning
LINSLEY AND FRANZINI: Water Resources Engineering
LINSLEY, KOHLER, AND PAULHUS: Hydrology for Engineers
METCALF AND EDDY, INC.: Wastewater Engineering: Collection, Treatment, Disposal
NEMEROW: Scientific Stream Pollution Analysis
RICH: Environmental Systems Engineering
SCHROEDER: Water and Wastewater Treatment
TCHOBANOGLOUS, THEISEN, AND ELIASSEN: Solid Wastes: Engineering Principles and Management
WALTON: Groundwater Resource Evaluation
WIENER: The Role of Water in Development: An Analysis of Principles of Comprehensive Planning

**McGRAW-HILL
BOOK COMPANY**
New York
St. Louis
San Francisco
Auckland
Bogotá
Düsseldorf
Johannesburg
London
Madrid
Mexico
Montreal
New Delhi
Panama
Paris
São Paulo
Singapore
Sydney
Tokyo
Toronto

EDWARD D. SCHROEDER
Professor of Civil Engineering
University of California, Davis

Water and Wastewater Treatment

This book was set in Times New Roman.
The editors were B. J. Clark and Madelaine Eichberg;
the production supervisor was Charles Hess.
The drawings were done by J & R Services, Inc.
R. R. Donnelley & Sons Company was printer and binder.

Library of Congress Cataloging in Publication Data

Schroeder, Edward D
    Water and wastewater treatment.

    (McGraw-Hill series in water resources and environmental engineering)
    Includes bibliographical references and index.
    1. Sewage—Purification. 2. Water—Purification. I. Title.
TD645.S37   628.1′6   76-13213
ISBN 0-07-055643-1

**WATER AND
WASTEWATER
TREATMENT**

1 2 3 4 5 6 7 8 9 0 D O D O 7 8 3 2 1 0 9 8 7 6

CONTENTS

PREFACE

This book is intended for advanced undergraduate and graduate-level courses on water and wastewater treatment. A previous course or experience with the subject is helpful in that the processes and systems discussed will be familiar. In addition, some knowledge of the nomenclature of water and wastewater treatment is assumed. Students using this book for their first course in the area can usually solve any problems associated with a lack of background by taking field trips and consulting standard undergraduate textbooks.

The material is presented in a slightly modified unit operations–unit processes approach. A general description of reacting systems and methods of modeling treatment processes is given in Chap. 2. Following chapters utilize the format developed in Chap. 2. Each chapter deals with a treatment objective rather than a type of process. Thus Chap. 3 discusses physical-chemical removal of dissolved materials and includes sections on adsorption, ion exchange, precipitation, chemical oxidation, and membrane processes. Chapter 4 discusses gas transfer, including both aeration and stripping, and other chapters include similar topic groupings. Including all the methods used to meet a treatment objective in a single chapter allows the reader to associate the process used with the objective and to compare available treatment methods. This combines many of the strengths of the unit operations approach and traditional methods of presentation.

Because the book is intended for use in advanced level courses, the emphasis has been on conceptual and theoretical development rather than on design procedures. Limitations of theoretical expressions, development of experimentally derived coefficients, and relationships between theory and practice are discussed on a case-by-case basis. The objective of this approach is to help the reader understand how processes work and the theoretical and practical constraints involved in process design. Insight into how and why processes function as they do, limitations or constraints on process operation, and limitations of design equations is as useful to the practicing design engineer as to the student. Thus it is hoped that this book will be an aid to practicing engineers as well as students.

Emphasis on theory should not be taken as a slight against empiricism. Theoretical models help the reader gain understanding of the mechanisms involved in a process. In many cases, satisfactory theoretical models are not available and wholly empirical models must be used in process design. More commonly, a theoretical model is modified by including factors or coefficients based upon experience. The empirical nature of these relationships is noted where they are presented. An interesting corollary to the place of empirical relationships in design is that quite often new processes are developed with empirical design relationships long before theoretical descriptions are derived. When the theory is developed, it simply supports the empirical relationships.

Chapters 1 and 2 should be read first, and Chap. 6 should be read before Chaps. 7 to 10. Otherwise there is no specific order in which the book should be read. The order used is based on a graduate sequence of courses presented at the University of California, Davis. Others may find a completely different order suitable, however. Similarly, the relative emphasis on certain topics reflects our particular program interests and goals. Many instructors may wish to supplement the text areas in which they have a particular interest.

This book is a composite of the ideas of many individuals with whom I have been fortunate to work over the past 12 years. Virtually daily conversations with George Tchobanoglous have contributed enormously to my understanding of the relationship between process theory and process design. A. A. Friedman carefully read the entire manuscript; his suggestions covered the range of improved clarity to theoretical development and were virtually all incorporated into the text. J. H. Sherrard reviewed the manuscript and made many helpful suggestions, particularly on subject coverage. Discussions over the past 12 years with R. L. Irvine, G. J. Kehrberger, A. P. Jackman, and D. P. Y. Chang have been very helpful in developing the text and the course on which it is based. Graduate students I have been fortunate enough to work with have also contributed a great deal to my understanding of process kinetics and stoichiometry. Their work is noted in a number of the chapters on biological wastewater treatment. The original manuscript was written while I was a visiting staff member at University College of Swansea, Wales. Conversations with B. Atkinson and J. A. Howell during this period were very helpful, and I am grateful for their interest and support. Joyce Brown edited and typed the final manuscript. Her concern for detail has kept this book from being a grammatical embarrassment.

This book is dedicated to two close friends who have also served as my teachers, A. W. Busch and R. B. Krone.

EDWARD D. SCHROEDER

1
INTRODUCTION

Water quality management is a societal necessity rather than an optional undertaking. Where industries such as steel or gasoline production may be considered necessary to the formation or maintenance of a scale or form of society, water quality management is necessary if society is to operate at all. As the structure of society becomes more complex, water quality requirements, wastes produced, management processes, and environmental impact of the wastes become greater in subtlety, complexity, and magnitude. Water quality management and the subareas of water and wastewater treatment reflect this increasing complexity in a number of ways. Modern society is centered around the industrial city. The flow of wastewaters from these centers is generally large, and the area or volume into which the wastewaters are discharged is generally small in a relative sense (at the other end of the scale is the nomadic community, small in population, water usage, and wastewater production with large areas or volumes into which the wastes can be discharged). In addition, industrial processes often produce wastewaters that are toxic to many forms of living organisms. Examples are abundant and include acids and bases that normally change the environment rather than directly attacking organisms, poisons such as cyanide from metal-plating operations, toxic substances such as phenol from the petrochemical industry, and high-temperature cooling waters that can both alter the environment or cause thermal shock. Thus the increasing sophistication of society has resulted in the production of larger quantities of wastewater that generally are far more concentrated and potentially harmful to the receiving environment.

Even if toxic materials were not a problem in modern wastewaters, the increasing volume would place great pressure on the engineering community to develop improved methods of wastewater treatment. Most cities have been using the same receiving waters for many years even though they have grown considerably in population and, hence, discharge quantity. Arguments can be made that any discharge is damaging, but in any case, the effects are closer to being an exponential rather than a linear function. Only minor, if any, measurable effects are noted up to a particular waste loading above which receiving-water quality declines at a relatively high rate. As a result, many cities have found that the level of wastewater treatment necessary to protect receiving-water quality has increased much faster than the population, and, of course, the cost of treatment is roughly exponentially related to the extent of treatment also.

Water treatment for domestic and industrial use is directly related to the quality of the source. In many cases, the source is also a receiver of industrial and domestic wastewaters, and therefore water treatment and wastewater treatment are closely tied together. In recent years, there has been an increasing concern with asbestos, pesticides, and chlorinated hydrocarbons found in small quantities in water supplies. Many of these compounds are known or suspected carcinogens. Because threshold concentrations and mechanisms of infection are not known or understood, regulatory agencies are understandably conservative in setting requirements.

Both treatment level and cost are engineering problems. The general problem statement is to provide maximum treatment at minimum cost. However, certain constraints must also be considered. A minimum quality must be related to the use and local environment of sources or receiving waters. For example, a generally accepted societal value is that natural fisheries should not be damaged even if they are of limited direct commercial value. This means that many communities or industries must provide treatment systems that will remove the toxic materials, growth stimulants (e.g., nitrogen and phosphate), and residual organics which are not easily removed by conventional wastewater treatment processes. The need for these *advanced wastewater treatment* processes may often be limited to short periods of the year, and therefore the problem of cost control becomes increasingly difficult.

Engineering responsibility for water and wastewater treatment begins with the determination of the level of treatment necessary or desirable and extends to system design and operation. In most cases, the level of treatment required is set by a regulatory agency that, in the United States at least, is largely influenced by engineers. These agencies receive advice and other needed information from other disciplines and governmental bodies. For example, fisheries management, mines, water resources development, and agriculture specialists often are involved in regulatory-agency decision formulation, and in some cases, their inclusion is legally required. Most regulatory agencies have also been adding biologists, geologists, land-use planners, and attorneys to their staffs. Engineers remain the predominant group, and agency statements and decisions usually have an engineering flavor, however. While this fact is advantageous with respect to making communication between the agency and the designer simple, it places a responsibility on the engineering community to find methods of incorporating non-quantitative information into regulatory and design decisions.

1-1 STANDARDS AND REQUIREMENTS

Water quality standards have both qualitative and quantitative aspects. The qualitative aspects include the concepts of what is a standard and what type of standards should be set. For example, the question of what level of water quality should be maintained, or in the case of many rivers and lakes returned to, is basically a conceptual problem. Until recently standards were locally determined in the United States. People directly affected thus were the predominant decision makers with respect to water quality. Unfortunately, the involvement of taxes, industrial jobs, and desired community growth in the standard formulation process usually resulted in a decision to fish upstream. Since 1960 there has been an increasing movement of the authority to formulate and maintain water quality standards into state and federal regulatory agencies. This change has been primarily due to the development of a greater understanding by the public of the importance of environment quality on the overall quality of life and recognition of this increased understanding by Congress and the state legislatures. Unfortunately the engineering community has had little positive input into the process.

Standards are defined here as values of water quality parameters (e.g., dissolved-oxygen concentration, temperature, turbidity) which must be met in a stream or lake to maintain a specified environment. Thus there will be a specified minimum dissolved-oxygen value in a given stream for the maintenance of trout fishery. *Requirements* differ from standards, at least as defined here, in that they are set by regulatory agencies and take into account other constraints. For example, a given river may have year-round dissolved-oxygen values greater than 8 mg/l, well above the value needed to support game fish. Dischargers would be required to maintain this level rather than meet the actual water quality standard for game fish. Justification for this approach is based primarily on the complexity of the ecosystem. All the organisms in a stream interact, and it is difficult to determine how a decrease in dissolved oxygen from 8 mg/l to, say, 5 mg/l would affect the total system over a long period of time. In addition, the decrease in oxygen would normally mean the addition of substantial numbers of microorganisms to the stream, a generally undesirable change and one that would both modify the natural (i.e., predischarge) situation considerably and decrease the stability of the ecosystem.

Requirements can also be used to equalize treatment costs for industrial discharges. Arguments against discharge requirements have often been made on the basis of making a manufacturing site uncompetitive. When requirements are made uniform over a large area without regard to local receiving-water quality, the problem of competitive advantage is greatly decreased. The U.S. Environmental Protection Agency (EPA) has proceeded on this philosophy by setting guidelines for the treatment of wastewaters from various industries. Authority for this approach was given in the 1970 Environmental Protection Act,[1] which requires that all wastewaters be treated to a technologically feasible level. Feasibility is construed to include economic considerations, but the EPA has considerable latitude in the matter.

Another difference between standards and requirements is that standards are based upon the existing or desired environment and, therefore, are usually tied to these conditions. Examples are dissolved oxygen, turbidity, or pH standards in a stream. Requirements are related to the standard but may well be placed upon the wastewater rather than the stream. California's dissolved oxygen requirements are normally stated for the stream (e.g., the discharge shall not depress the dissolved-oxygen concentration in the stream more than n tenths mg/l or below m mg/l) while biochemical oxygen demand (BOD) requirements are placed on the discharge stream (e.g., the discharge shall not have a BOD concentration greater than p mg/l or a total daily mass greater than q grams). Requirements for other parameters are set in the same manner. Of *special* engineering significance is that this approach virtually eliminates the concept of assimilative capacity of a stream or lake.

1-2   WATER TREATMENT PROCESSES AND SYSTEMS

Water treatment involves the removal from water of constituents detrimental to a specific use. Domestic water supplies must be nearly sterile and turbidity-free and should have a low total-dissolved-solids (TDS) concentration. Specific chemical species such as the hardness ions, calcium and magnesium, or toxic materials such as lead and the pesticide dieldrin must be removed in some cases.

Requirements for industrial uses of water vary widely. Cooling is a major industrial water use and has relatively loose requirements. Corrosion, scale formation, and bacterial growth in pipes and cooling towers are the primary concerns. Boiler feedwater is necessary in many industries. Because of the high temperatures and pressures in boilers, scale formation is a major problem. Boiler feedwater must be low in turbidity, dissolved oxygen, and hardness. Silica is a particular problem in boilers. Water is often directly in contact with or incorporated into an industrial product. Quality is obviously an important factor in such cases, and the requirements are specific to the particular use. A summary of such uses and an excellent list of references is given in "Water Quality Criteria."[2]

The intended use of a water directly affects the choice of sources and treatment method. Groundwaters are normally higher in TDS than surface waters and are often relatively hard. Surface waters usually require turbidity removal. The quality of water from either source varies widely. For example, groundwaters in the Willamette Valley of Oregon often have TDS concentrations less that 100 mg/l, while Los Angeles utilizes surface sources such as the Colorado River with TDS concentrations well over 500 mg/l. Portland, Ore., utilizes a surface-water source that is nearly turbidity-free. This would be a rare situation in the Eastern United States where most streams carry a considerable turbidity load.

Processes commonly used to treat water are listed in Table 1-1. Ground-waters are often distributed without treatment or with disinfection only. Surface

Table 1-1 WATER TREATMENT PROCESSES IN COMMON USE

Purpose	Process	Comments
Turbidity removal	Coagulation Flocculation Sedimentation Filtration: Depth Precoat	Often used for domestic water treatment
Dissolved-solids removal	Precipitation ion exchange	Often used to remove hardness
	Distillation Reverse osmosis	Reduction in TDS
Dissolved-organic removal	Activated-carbon adsorption	Used for removal of color, tastes, and odors and low concentrations of toxic organics
	Chemical oxidation	Chlorine, ozone, permanganate or peroxide used to oxidize organics
Cooling	Cooling towers	
Disinfection	Chlorination	Most common method in United States
	Ozonation	Effective, leaves no residual products, more expensive than chlorination

waters usually require coagulation, flocculation, sedimentation, filtration and disinfection prior to distribution. The first four steps only remove turbidity, and therefore water quality characteristics related to dissolved materials are unchanged.

Dissolved inorganic materials are often removed by precipitation. The most common examples of the use of precipitation are for hardness removal (Ca^{2+}, Mg^{2+}) and iron and manganese removal. In the latter case, the ions are first oxidized to the insoluble trivalent form. Ion exchange is also used quite often for the removal of undesirable dissolved constituents of water. Home water softeners are usually simple ion-exchange units that exchange Na^+ for Ca^{2+} and Mg^{2+}.

Organic materials can be removed by adsorption on activated carbon. This material has a very large surface-to-mass ratio (1000 m^2/g is typical) and provides a great number of adsorption sites for nonpolar molecules. Organic contaminants of concern include those responsible for taste, odor, and color in water as well as low concentrations of pesticides and petrochemical wastes. In some cases, oxidation of the organics provides satisfactory organic removals or can be used to improve adsorption-process effectiveness.

The most important application of disinfection is the production of safe drinking water. Industrially, disinfection also serves to prevent or control slime growth in pipes, cooling towers, and mechanical systems. A number of disinfectants are technically possible to use, but chlorine and ozone are preferred for operation and cost reasons.

1-3 WASTEWATER TREATMENT PROCESSES AND SYSTEMS

Treatment processes and systems for wastewater renovation can be classified in a number of ways. The most common method is to characterize them by function, e.g., precipitation, biooxidation, or adsorption. In the case of wastewater treatment, the majority of treatment plants built are for municipalities and, until very recently, have been generally restricted to a narrow group of operations and processes. The terms *primary, secondary,* and *tertiary* treatment have become synonomous with sedimentation, biological treatment with sedimentation, and removal of residual or nonbiodegradable materials by any means, respectively. In the treatment of domestic wastes, primary treatment usually removes about 35 percent of the BOD and 60 percent of the suspended solids.† Secondary treatment can be expected to remove an additional 50 percent of the incoming BOD (77 percent of primary effluent BOD) and reduce the suspended-solids concentration by an additional 33 percent. These estimates are based on "typical" domestic sewages and are often quite misleading. For example, a correctly designed and operated treatment system should consistently produce an effluent

---

† Measured as residual dry weight by filtration.

with BOD *and* suspended-solids concentrations less than 20 mg/l regardless of the influent concentrations. Most domestic wastes are similar (see Table 1-2), and the percentages quoted are reasonable estimates if used wisely. Unfortunately, there has been a tendency to use them as if they were characteristics of the processes rather than artifacts of the classical treatment of domestic sewage, however. The result of this tendency has been the assumption, in many cases, that wastewaters with high organic concentrations cannot be treated to the same extent as domestic sewage. As will be explained in later chapters, biological processes designed on this basis (relatively high effluent BOD) are doomed to operational failure.

A recent modification in nomenclature has been encouraged by the EPA because of the development of nonbiological methods of organic removal. Current guidelines identify primary treatment as removing approximately 40 percent and secondary treatment as removing 80 percent or greater of the incoming BOD. The author understands the motivation for the definitions but believes that a far better approach would have been to drop classification systems altogether. Very little is gained except to categorize an additional number of treatment processes by percent-removal potential. A common result of this practice is the incorrect comparison of plants. For example, a system producing an unacceptable effluent of 100 mg/l BOD from a 2000 mg/l influent (95 percent removal) will be compared favorably with a plant producing a 20 mg/l BOD effluent from an influent of

Table 1-2 TYPICAL MUNICIPAL WASTEWATER CHARACTERISTICS

	Concentration, mg/l
Solids, total	700
Dissolved	500
Fixed	300
Volatile	200
Suspended	200
Fixed	50
Volatile	150
Ultimate biochemical oxygen demand (BOD_L)	300
Total organic carbon (TOC)	200
Chemical oxygen demand (COD)	400
Nitrogen (as N)	40
Organic	15
Free ammonia	25
Nitrites	0
Nitrates	0
Phosphorous (as P)	10
Organic	3
Inorganic	7
Grease	100

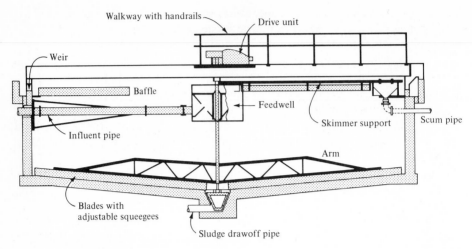

Walkway with handrails

Drive unit

Weir

Baffle

Feedwell

Skimmer support

Scum pipe

Influent pipe

Arm

Blades with
adjustable squeegees

Sludge drawoff pipe

FIGURE 1-1
Typical primary-clarifier cross section. (*Municipal Equipment Division, Envirotech
Corporation.*)

100 mg/l (80 percent removal). A corollary to the percent-removal characterization
is the concept of removable BOD. This concept has been discussed by Busch,[3]
and it will be left to the reader to investigate.

Primary-solids Removal

Primary-solids removal is used where the wastewater contains a high enough
concentration of large, dense suspended or settleable solids to make removal
economical. Two stages are often incorporated, the first removing dense inorganic
grit and the second separating large organic particles from the flow. Grit removal
has the advantage of protecting pumps and other equipment. Both processes are
usually designed using standard "rules of thumb" derived partly from consideration of Stokes' law settling theory (see Chap. 5) and experience.[4] Modern primary-
solids-removal processes utilize continuous sludge removal as depicted in Fig. 1-1.

It should be noted that many wastewaters do not require solids removal
at this point because they do not contain significant quantities of settleable
material. Other wastewaters contain little else, and some form of solids removal
is all that is needed to produce an acceptable effluent.

Biological Processes

Biological treatment can be accomplished in a number of ways, but the basic
characteristic of the system is the use of a mixed (heterogeneous) bacterial culture
for the conversion of pollutants. In most cases, organic materials are converted

FIGURE 1-2
Trickling filter at Woodburn, Ore. (*Clark and Groff Engineers, Salem, Ore.*)

to oxidized end products, mostly carbon dioxide and new bacterial cells. The organic material serves as both an energy source and a source of carbon for cell synthesis under these circumstances. In a few cases (e.g., nitrification), bacterial cultures are used for the conversion of inorganic materials, and in one process (algal stripping of nutrients), algae rather than bacteria are used for the conversion process. This latter process is part of the system which functions in oxidation ponds or sewage lagoons.

Biological processes are most commonly used for secondary treatment, i.e., conversion of organics remaining after a primary sedimentation process is applied to remove suspended solids. Two configurations are in common use: film flow and suspended (or fluidized cultures). The trickling filter (Fig. 1-2) is the most commonly used example of the film-flow-type process. Wastewater is sprayed over a bed of support media (gravel, rock, or formed plastic) covered with a biological slime and allowed to flow downward through the bed in the form of a liquid film. Organic material and inorganic nutrients are extracted from the liquid film by the bacteria in the slime. Oxygen needed for aerobic conversion processes is transferred from the gas phase in the pores through the liquid film to the biological slime. Microorganisms are active in the film as well as in the slime, and therefore the reaction system is both in the film and in the slime.

FIGURE 1-3
Rotating biological contactor (RBC). (*Bio-Systems Division, Autotrol Corporation.*)

Other configurations of film-flow reactors are in use, one of which is the rotating biological contactor (RBC). These systems utilize a lightweight material such as Styrofoam to support the slime, and they operate by moving the slime through the wastewater (Fig. 1-3). A film of wastewater is carried out of the sump by the rotating disk, and the system obviously functions much the same as a trickling filter.

Suspended-culture processes in use include activated sludge, aerated lagoons (actually a form of activated sludge), oxidation ponds, and anaerobic treatment processes. Anaerobic processes are most often used for the breakdown of concentrated organic sludges, while the aerobic processes are used in a similar manner to trickling filters. Activated sludge like the trickling filter is an aerobic process but utilizes a culture in which bacterial cells are agglomerated together in *flocs*. These flocs may be of the order of 0.1 mm in diameter or larger, whereas a single bacterium is of the order of 1 or 2 μm in diameter. A large fraction of the total floc volume is usually taken up by an extracellular slime produced by the organisms. This slime is assumed to be the binding agent in the floc formation. Flocs are maintained in suspension either by diffused air added at the bottom of the tank or by mechanical agitation (Figs. 1-4 and 1-5). In either case, the mixing device is usually the aeration system also. The only major deviation from this rule occurs in processes that use pure oxygen rather than air.

FIGURE 1-4
Diffused-aeration system for activated sludge at Hamilton, Ohio, wastewater treatment plant.

Flocculant cultures are necessary for activated sludge because of the need to separate the cells from the treated liquid. This is done by sedimentation for economic reasons, and consequently large particles are very desirable. Activated-sludge cultures are relatively concentrated with mass concentrations normally maintained in the range of 1000 to 5000 mg/l. These cultures do not settle as individual flocs but act together in what is commonly referred to as *zone* or *hindered settling*. Settled cells are recycled to the aeration-reaction tank, and the clarified supernatant is either discharged or treated further.

Other biological processes of current importance include lagoons, anaerobic digesters, and nitrogen-conversion systems. The term *lagoon* is used to cover a wide variety of systems, the simplest of which is nothing more than a holding basin that hopefully has been sized to allow for satisfactory levels of oxygen production by naturally generated algae populations. In this type of lagoon (usually referred to as *facultative*), influent wastewater solids and organic solids resulting from dying algae, bacteria, and plants settle to the bottom and are broken down by anaerobic organisms. Algae growing near the surface carry out photosynthesis and thus provide oxygen to the upper layer. In between (these ponds are usually about 3 to 5 ft deep), a range of oxygen tensions exist, and soluble and colloidal organics are oxidized by bacteria. The major design problem is

FIGURE 1-5
Surface aerators at Muskegon County, Mich. (*Aqua Aerobic Systems, Inc.*)

to make sure that organic uptake rates do not exceed oxygen production rates for any great length of time. If this does happen and the pond becomes completely anaerobic, odors are generated in sufficient quality and quantity to cause anguish for radii of several miles.

Lagoons were originally developed as inexpensive waste disposal methods for small rural communities with large amounts of relatively inexpensive land nearby. As design procedures became more firmly established, these systems were used for larger and larger communities and stronger wastes. The major difficulty has been that removal of algae grown in these lagoons is very difficult, and until recently there has been no requirement to do so. Organic concentrations of pond effluents with algae included often exceed the influent. Thus either algae removal or a return to the original idea of no discharge from oxidation lagoons is necessary if lagoons are to remain a viable treatment system in terms of current effluent requirements.

Modifications of the facultative lagoon concept have been developed, notably aerobic lagoons, anaerobic lagoons, and aerated lagoons. The aerated lagoon is really an activated-sludge process without recycle. Aerobic lagoons are very shallow and do not have a significant anaerobic zone. They are rarely used except as a secondary process behind anaerobic lagoons and even there do not produce a satisfactory effluent because of the large amount of algae grown and discharged. Anaerobic lagoons are really primitive anaerobic digesters. They were first used in the meat-processing industry because the waste is peculiarly well adapted to this form of treatment. Meat-processing wastes contain high concentrations of fats that form a scum on the pond surface. This scum often reaches a thickness of 2 cm or more and provides insulation and prevents the escape of large amounts of reduced-sulfide compounds. Insulation is important because methane fermentation, the critical step in anaerobic digestion or treatment, is extremely temperature sensitive.

Biological processes used for nitrogen conversion are of two types: the aerobic processes used for oxidizing ammonia (-3 valence) nitrogen to the nitrate form and the anaerobic processes used for reducing nitrate nitrogen to molecular (N_2) nitrogen. Ammonia-nitrogen oxidation (nitrification) is a chemoautotrophic process. The bacteria are obligate aerobes that use ammonia nitrogen (actually ammonium ion) both as an energy and a nitrogen source. Carbon dioxide serves as the carbon source. Nitrification can be carried out in conventionally designed trickling filters and activated-sludge units if desired or in special units, which to date have been physically designed to be very much like conventional process designs. Nitrate reduction or denitrification is carried out by heterotrophic bacteria that use nitrate as an alternative electron acceptor to oxygen. An organic material is needed to serve as a carbon and energy source. Virtually any biodegradable nitrogen-free or low-nitrogen organic is suitable. The physical systems used for denitrification can be packed beds, fluidized cultures, or ponds, depending on the designer's needs or desires.

Physical-Chemical Treatment

Removal of soluble or colloidal organic materials has until recently been accomplished only by biological processes. A number of treatment plants in the United States are now being designed as *physical-chemical* plants. The principal processes used are coagulation, filtration, and adsorption on activated carbon. None of these processes are new, but developments and changing economics have made them more competitive with biological processes. All three processes require frequent regeneration of one form or another. Coagulant requirements are quite high in most cases, making regeneration necessary. Filters used are similar to water treatment filters and must be backwashed frequently. Activated-carbon processes must include regeneration facilities in nearly all cases in order to be economically competitive with biological processes.

The same three processes are often considered for *tertiary* or *advanced* wastewater treatment, i.e., following biological treatment. In this case, concentrations of materials to be removed are much smaller, and the overall costs are less. Because of the volumes being treated and the requirements being met, costs are still quite high on a dollars per volume treated basis, however.

A number of other physical-chemical processes have been investigated and are in use to some extent in the removal of organic materials. Reverse osmosis, ultrafiltration, vacuum filtration, centrifugation, and distillation are all useful in special circumstances. They will be only briefly discussed in this text but are well treated elsewhere.[5, 6]

Many wastewaters contain little or no organic material, and hence some form of physical-chemical treatment is necessary. Among these are metal processing, mining, and inorganic chemical-manufacturing wastewaters. Neutralization, ion exchange, precipitation, and evaporation are the principal treatment methods. To a certain extent, treatment of inorganic industrial wastes is difficult to discuss because each industry, and in some cases each operation, is different. Processes commonly used are always based upon the same principles.

REFERENCES

1. U.S. CONGRESS: "1970 Environmental Protection Act," PL 92-500.
2. Report of the Commission on "Water Quality Criteria," Environmental Studies Board, National Academy of Engineering, published as Environmental Protection Agency Rept. No. EPA-R3-73-033, Washington, D.C., 1973.
3. BUSCH, A. W.: Conceptions and Misconceptions about Biological Waste Treatment, in E. F. Gloyna and W. W. Eckenfelder, Jr. (eds.), "Advances in Water Quality Improvement," University of Texas Press, Austin, 1968.
4. METCALF AND EDDY, INC.: "Wastewater Engineering," McGraw-Hill Book Company, New York, 1972.
5. COULSON, J. M., and J. F. RICHARDSON: "Chemical Engineering," Pergamon Press, New York, 1968.
6. WEBER, W.: "Physico-Chemical Treatment," Wiley-Interscience, New York, 1973.

2
REACTORS

All water and wastewater treatment operations take place in a tank of some sort, which will be given the general term *reactor*. The term is quite naturally applied to the tanks where conversion processes take place, such as the trickling filter, adsorption units, or activated-sludge aeration units. This term will also be used for describing sedimentation tanks and other less obvious reactors. Use of a general term to describe all the physical systems allows a general nomenclature and mathematical development. While this is not an original or particularly complicated concept, it is so useful that the rationale bears explaining.

Given that a reactor is some sort of a vessel or tank in which water or wastewater constituents are either removed or converted in some way, the next step is to consider how we might describe these processes. Clearly, a hydraulic description of the reactor will be necessary as well as a mathematical description of the conversion reactions. This latter description must include the effects of environmental variables such as temperature and/or pH in some manner. Finally, consideration must be given to the effect of mass-transfer limitations on the conversion rate. To understand this concept, it must be recognized that all reactions take place at specific sites. In some cases the site is a molecule; in others the surface of a colloidal or larger particle will serve as the reaction site. In every case the reactive materials must in some manner be transported to the site. The rate of transport may be less than or greater than the rate of reaction. Interaction of these two rates controls the conversion process and the manner in which the overall reaction system responds.

Rate expressions are important in process design because they directly affect the size of the reactor and, thus, the capital cost. The higher the rate of removal the smaller the reactor required. In some cases reactions interact or interfere with each other. Unless control is exerted over the reaction system, undesirable products occur. This situation exists in coagulation and in the combined aeration and biooxidation reaction system.

2.1 A SIMPLISTIC VIEW OF REACTIONS

The development presented here is not intended to be a rigorous explanation of chemical reactions. Instead, methods of describing conversion processes from an engineering viewpoint are given with the goal of providing useful tools for design, operation, and analysis. Reactions have two basic features that interest us: stoichiometry and rate. Stoichiometry relates the number of moles of reactants to the number of moles of products of a given reaction. If v_i is the stoichiometric number of the ith material, i.e., the relative number of moles entering into or produced by a reaction, and if v_i is negative for reactants and positive for products (a statement of conversation of mass), we can write

$$v_i A_i + v_2 A_2 + \cdots + v_n A_n = 0 \qquad (2\text{-}1)$$

where A_i is the mass per mole of material i. For a simple reaction such as $A_1 + A_2 \rightarrow 2A_3$, the stoichiometric numbers are $v_1 = v_2 = -1$ and $v_3 = 2$. The rate of conversion of A_1 and A_2 into A_3 is related to the stoichiometric numbers. For example, each time 1 mole of A_1 is consumed (or A_2), 2 moles of A_3 is produced. Thus the formation rate of A_3 is twice as fast as the conversion (disappearance) rate of the two reactants on a molar basis. Another way of stating this is to say that the rate of formation of A_3 is twice as great and opposite in sign to the rates of formation of A_1 and A_2.

$$r_{A_3} = -2r_{A_1} = -2r_{A_2} \qquad (2\text{-}2)$$

Stating Eq. (2-2) for a more general reaction gives

$$r_{A_m} = \frac{v_m}{v_i} r_{A_i} \qquad (2\text{-}3)$$

where i is any reactant or product. Further manipulation of Eq. (2-3) gives a general term for the *reaction rate* μ, which is independent of the chemical species being monitored.

$$\mu = \frac{r_{A_1}}{v_1} = \frac{r_{A_2}}{v_2} = \frac{r_{A_3}}{v_3} = \cdots = \frac{r_{A_n}}{v_n} \qquad (2\text{-}4)$$

Equations (2-3) and (2-4) are very important because they provide a method of relating the rate of formation of one chemical species to that of another species. Situations where all reacting constituents are easily measurable are virtually nonexistent, and these relationships provide a method of predicting nonmeasured values from those that are readily or economically monitored.

Reaction rates must be expressed as functional relationships in order to be useful in describing wastewater treatment processes. Usually the rates are related to the concentration of the reactants, and in some cases they are dependent upon product concentrations as well. An example is bacterial growth, where the increased product (bacteria) concentration increases conversion rate. As previously noted, environmental variables such as temperature and pH also affect both the stoichiometry and the rate of reaction. Conventional nomenclature refers to the order of reaction in describing the relationship between reaction rate and reactants. For example, in the reaction $A_1 + A_2 \rightarrow$ products the reaction rate μ might be found to be $kC_{A_1}C_{A_2}$, where k is a constant for a given set of environmental conditions. This reaction would be second order but first order (i.e., to the first power) with respect to reactant A_1 and first order with respect to reactant A_2, respectively. The reaction rate might be found to be independent of A_2 and dependent upon the concentration of A_1 squared. In this case, the reaction would be second order again but zero order with respect to A_2 and second order with respect to A_1.

Quite often reaction rate functions take more complex forms, and several of these will be discussed in later chapters. Many of these complex functions reduce to relatively simple functions for certain limiting conditions. For example, a rate expression of the form $\mu = k_1 C_{A_1} + (k_2 C_{A_2})/(K + k_2 C_{A_2})$ reduces to the simple first-order function $\mu = k_1 C_{A_1} + 1$ when $C_{A_2} \gg K$. Often the region of interest is narrow, or one of the reactants is present in such excess quantities that its effective concentration does not change and, hence, alter the overall reaction rate. In these cases, reaction rate functions usually reduce to simple forms. A possibly more significant point is that *all rate expressions must be experimentally verified.* Experimental data on rate relationships are difficult to obtain and are rarely extremely precise. For this reason real-world rate equations are usually quite simple and limited to use in narrow regions of reactant concentration. This situation is nearly universal in water and wastewater treatment.

In this book, rate equations will follow the general form

$$\mu = kf(C_{A_1}, C_{A_2}, C_{A_3}, \ldots, C_{A_n}) \qquad (2\text{-}5)$$

where k is a constant for a given set of environmental conditions and has units dependent upon the order of reaction. The effect of temperature on the reaction rate can be expected to follow a relationship similar to the Arrhenius relationship in most cases [Eq. (2-6)].

$$\mu(T) = He^{-E/RT}f(C_{A_1}, C_{A_2}, C_{A_3}, \ldots, C_{A_n}) \qquad (2\text{-}6)$$

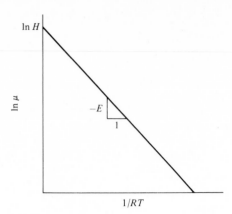

FIGURE 2-1
Typical rate-temperature relationship.

where H = frequency factor
$\quad\quad E$ = activation energy
$\quad\quad R$ = gas constant
$\quad\quad T$ = absolute temperature
Temperature effects are related by H and E through experiments in which other variables are eliminated.

EXAMPLE 2-1   The rate of the reaction $A_1 \rightarrow A_2$ has been determined to obey the expression

$$\mu = \frac{kA_1}{K + A_1}$$

at a constant temperature. Experiments under conditions where $A_1 \gg K$ were run at different temperatures, and the data were plotted as shown in Fig. 2-1. The frequency factor H is the ordinate intercept, and the activation energy E is the negative of the slope of the curve. An important fact to note is that the form of the rate expression must be known before temperature functions can be developed. ////

The effect of pH on reaction rate is a more complex situation. When pH is changed, it follows that ionic balances are also changed. Using ammonia $(NH_3) \rightleftharpoons$ ammonium ion (NH_4^+) as an example, we find that at pH 7 the percentage of ammonia is about 0.7, while at pH 12 the percentage of ammonia is about 99.9. In biological systems, there is usually a flat region over which pH changes have little measurable effect. This region is generally in the area of pH 6.0 to pH 8.5 for wastewater treatment cultures. Every organism has specific requirements, however, and consequently, generalization is hazardous.

2-2 HOMOGENEOUS REACTIONS

In many cases the reaction sites are dispersed throughout the liquid phase and are small enough to allow the approximation of homogeneity. This may be thought of as each point in the liquid phase having the same potential for a reaction, and thus we can still consider concentration gradients of the reactants and their effects. For example, consider a coagulation reaction in which a coagulant such as alum is added to a colloidal suspension. If the colloid is evenly dispersed through the liquid, the reaction between the alum and the colloid will appear to be a function of the local alum and colloid concentrations. Movement of the alum through the liquid can be predicted by a mass-transfer model, and the reaction model and mass-transfer model together would predict the extent of or overall completeness of the reaction.

In most cases of interest in water and wastewater treatment, the mass-transfer model is quite simple. Uniform and instant mixing is commonly used in both coagulation and aeration processes, for example. In other cases (trickling filters and rotating biological contactors), we will find it quite useful to consider the combined mass-transfer and reaction problem.

A more general definition of homogeneous reaction is any reaction in which the reactants can be considered as being continuously distributed throughout the fluid. Thus any reaction in which one or more of the constituents is easily identified by location is nonhomogeneous. Trickling filters, for example, are clearly nonhomogeneous systems. Adsorption on surfaces and ion exchange often must be considered as nonhomogeneous systems. Ordinarily justification of an assumption of homogeneity results in a more easily analyzed model. For this reason, the form or structure of a reaction model should always be carefully considered, either in one's own work or in reviewing the work of others.

Batch-reaction Systems

Consideration of several reacting systems in the following paragraphs forces the introduction of the batch reactor. This type of reactor is nothing more than a tank or pot that is well stirred so that concentration gradients do not exist in the liquid phase. Perhaps the closest real representation of an ideal batch reactor is a food blender. Systems with far less violent stirring often provide very satisfactory approximations, however. In this reactor, reactants are added at time equal to zero and the reaction's progress is followed with time. A mass balance, using the general mass-balance statement [Eq. (2-7)], gives Eq. (2-8), where i is any reactant or product.

Rate of mass in + rate of mass produced = rate of mass out

$$+ \text{ mass accumulation rate} \qquad (2\text{-}7)$$

$$(QC_i)_{\text{in}} \pm V_1 r_i = (QC_i)_{\text{out}} \pm V \frac{dC}{dt} \qquad (2\text{-}8)$$

where Q = volumetric flow rate
C_i = mass concentration of component i
V = liquid volume in reactor
r_i = mass rate of formation of i

In the batch reactor the reactant materials are added before the reaction starts, and therefore there is no flow in or flow out ($Q = 0$). Equation (2-8) reduces to

$$r_i = \frac{dC_i}{dt} \qquad (2\text{-}9)$$

For the case of the batch reactor, *and only* the batch reactor, the rate of formation is equal to the accumulation rate. Thus batch reactors are inherently non-steady state; i.e., they change with time. While this may seem to be an over-emphasis of a rather trivial point, the literature of wastewater treatment rarely makes this distinction, with the result that in many cases steady-state terms are referred to as time-dependent differentials, and in some instances, two time differentials have appeared in the same expression.

Competing Homogeneous Reactions

Now consider a situation in which A may react to produce either B or C.

Competing reactions occur where two or more reactants are available for an active site (adsorption or ion exchange) or where products react with the original reactant (coagulation). Both reactions will be assumed to be first order with respect to A, and the following statements can be made for a batch reactor.

$$\mu_1 = k_1 C_A$$

$$\mu_2 = k_2 C_A$$

$$\frac{dC_A}{dt} = -(\mu_1 + \mu_2) = -(k_1 + k_2)C_A$$

$$\frac{dC_B}{dt} = \mu_1 = k_1 C_A$$

$$\frac{dC_C}{dt} = \mu_2 = k_2 C_A$$

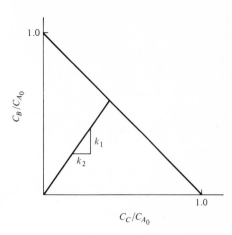

FIGURE 2-2
Attainable region for parallel first-order
reactions.

Integration of the three differential equations gives

$$C_A = C_{A_0} e^{-(k_1 + k_2)t}$$

$$C_B = \frac{k_1 C_{A_0}}{k_1 + k_2} \left(1 - e^{-(k_1 + k_2)t}\right)$$

$$C_i = \frac{k_2 C_{A_0}}{k_1 + k_2} \left(1 - e^{-(k_1 + k_2)t}\right)$$

From the integrated expressions, we can see that the relative amounts of B and C present at any time depend on the ratio k_1/k_2. A useful relationship is developed by plotting the relative amount of the original material A that has been converted to B, C_B/C_{A_0}, against the relative amount of the original material that has been converted to C, C_C/C_{A_0}. Such a relationship is shown in Fig. 2-2. The area enclosed by the triangle formed by the abscissa, ordinate, and the line extending from $(0, 1)$ to $(1, 0)$ is the *attainable region*. A reaction proceeds with time along a line of slope k_1/k_2 from the origin to the hypotenuse of the triangle. In the case of the first-order reactions described, infinite time would be needed to reach completion.

   The reaction-progress line will always be linear if the two competing reactions are of the same form (i.e., differ only in their rate constants). If one of the reactions were first order and the other were second order, the reaction-progress line and the attainable region would be nonlinear.

Consecutive Homogeneous Reactions

A reaction system of considerable interest in wastewater treatment is that of sequential reactions. Examples of this type of system include nitrification, denitrification, and acid fermentation–methane fermentation in anaerobic treatment.

Here the general concept will be developed. Further discussion will be given in the chapters dealing with the particular processes.

The simplest set of consecutive reactions is $A \xrightarrow{1} B \xrightarrow{2} C$, with both reactions being first order and irreversible. Using nomenclature similar to the previous example,

$$\mu_1 = k_1 C_A = -r_1$$

$$\mu_2 = k_2 C_B = -r_2$$

$$\frac{dC_A}{dt} = -\mu_1 = -k_1 C_A$$

$$\frac{dC_B}{dt} = \mu_1 - \mu_2 = k_1 C_A - k_2 C_B$$

$$\frac{dC_C}{dt} = \mu_2 = k_2 C_B$$

Integration of these equations using C_{A0} as the initial condition gives

$$C_A = C_{A0} e^{-k_1 t}$$

$$C_B = \frac{k_1 C_{A0}}{k_2 - k_1} (e^{-k_1 t} - e^{-k_2 t})$$

$$C_C = C_{A0} \left(1 + \frac{1}{k_2 - k_1}\right)(k_1 e^{-k_2 t} - k_2 e^{-k_1 t})$$

for all cases except $k_1 = k_2$, which is left as an exercise for the reader to evaluate.

Two sets of curves derived from these equations are of particular interest. Figure 2-3 indicates the concentration of reactants and products vs time, and Fig. 2-4 shows the ratio of the concentration of the final product to the initial reactant concentration vs the ratio concentration of the intermediate product to the initial reactant concentration. Note that the final ratio of C_C/C_{A0} is equal to 1. The example used here had very simple stoichiometry, which results in the total number of moles in the system remaining constant $(C_A + C_B + C_C = C_{A0})$ at any time t. This situation rarely occurs but is extremely useful in developing the concepts of reacting systems.

Figure 2-3 indicates that if the intermediate product B is desirable and the final product C is undesirable, careful control of the reaction process must be exercised. In batch processes, this entails operation at the maximum point on the attainable-region diagram. In wastewater treatment, intermediate products are generally undesirable because of their relatively less oxidized state. Thus the normal situation is to encourage reactions to go toward complete conversion (that is, $C_C/C_{A0} \rightarrow 1.0$). Equilibrium between reactants and products will eventually be reached, but, in general, satisfactory treatment processes have equilibrium concentrations that emphasize the product to the point of being irreversible.

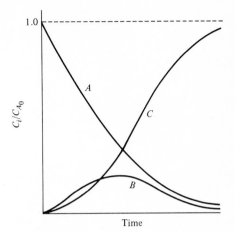

FIGURE 2-3
Reactant and product concentrations for consecutive reactions as a function of time.

The attainable region of the consecutive-reaction system does not include the entire triangle formed by joining points (0,1), (0,0), and (1,0). The moment B is formed, the second reaction immediately begins. Therefore there is no theoretical possibility of attaining $C_B/C_{A_0} = 1$. Characteristics of this curve defining the progress of the reaction are of interest, particularly the maximum value of C_B/C_{A_0} and the initial and final slopes. The slope at any point is defined by the ratio dC_C/dC_B.

$$\frac{dC_C}{dC_B} = \frac{k_2 C_B}{k_1 C_A - k_2 C_B}$$

At time equal to zero, $C_B = 0$, while $C_A = C_{A_0}$. Thus the initial slope of the reaction-progress line is equal to zero [i.e., the slope approaching (0,0)]. As the

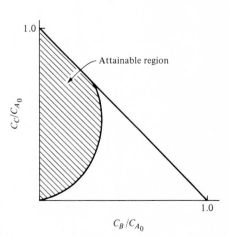

FIGURE 2-4
Attainable region for consecutive first-order reactions.

reaction progresses toward completion (irreversibility is assumed), both the numerator and the denominator approach zero, and the slope of the curve cannot be directly determined. Inverting the slope, $(dC_C/dC_B)^{-1}$, provides a more tractable form of the relationship

$$\frac{dC_C^{-1}}{dC_B} = \frac{k_1 C_A}{k_2 C_B} - 1$$

The ratio of C_A/C_B can be evaluated from the original expression.

$$\frac{C_A}{C_B} = \frac{k_1}{k_2 - k_1} \left(1 - e^{(k_1 - k_2)t}\right)$$

When $k_1 > k_2$, the value increases without bound as the reaction goes to completion ($t \to \infty$). Thus the inverse, C_B/C_A, approaches zero. When $k_1 < k_2$, the value approaches $k_1/(k_2 - k_1)$ as the reaction goes to completion. The value of the inverse, C_B/C_A, is then $k_2/(k_1 - 1)$. The two conditions result in two possible slopes as the reaction approaches completion. When the first reaction is the fastest ($k_1 > k_2$), the final slope is -1, and when the second reaction is fastest, the final slope is k_2/k_1.

Determination of the maximum intermediate product concentration is relatively straightforward. Differentiating the expression for C_B and solving for the time $t*$ associated with the maximum value of C_B/C_{A_0} gives

$$t* = \frac{\ln(k_2/k_1)}{k_2 - k_1}$$

The maximum value of C_B/C_{A_0} can now be found by substituting $t*$ for t.

$$\frac{C_{B, \max}}{C_{A_0}} = \frac{k_1}{k_2 - k_1} \left(e^{-k_1 t*} - e^{-k_2 t*}\right)$$

The Quasi-steady-state Hypothesis

Many catalyzed reactions can be treated as sequential reactions in which the intermediate product is a complex of the reactant and the catalyst. Enzymatic reactions are a major example. In this case, the decomposition of the complex may result in either the formation of the final product or the "reformation" of the original reactant. Three rate terms are of importance: the rate of formation of the catalyst-reactant complex AC, the rate of formation of the product B, and the rate of formation of the original reactant A. Each reaction has a separate rate constant as shown in Eq. (2-10).

$$A + C \underset{k_2}{\overset{k_1}{\rightleftharpoons}} AC \overset{k_3}{\to} B + C \qquad (2\text{-}10)$$

Assuming an excess of catalyst (C), the rates of formation of the materials can be written for first-order reactions.

$$\frac{dC_A}{dt} = k_2 C_{AC} - k_1 C_A$$

$$\frac{dC_{AC}}{dt} = k_1 C_A - (k_2 + k_3) C_{AC}$$

$$\frac{dC_B}{dt} = k_3 C_{AC}$$

The usefulness of the quasi-steady-state hypothesis occurs when the catalyst-reactant complex is extremely reactive, i.e., breaks down very quickly. When $k_2 + k_3 \gg k_1$, the rate of formation of the complex is approximately equal to zero ($dC_{AC}/dt = 0$), and an approximate or quasi steady state occurs. The concentration of the complex can be derived in terms of the rate coefficients and the reactant concentration.

$$C_{AC} = \frac{k_1}{k_2 + k_3} C_A$$

The rates of formation of A and B under these conditions are then

$$\frac{dC_A}{dt} = \frac{-k_1 k_3}{k_1 + k_2} C_A$$

and

$$\frac{dC_B}{dt} = \frac{k_1 k_3}{k_1 + k_2} C_A$$

In many situations, particularly in biological processes where enzymes act as catalysts, the catalyst concentration affects the rates of formation and results in second-order relationships.

$$\frac{dC_A}{dt} = k_2 C_{AC} - k_1 C_C C_A$$

$$\frac{dC_{AC}}{dt} = k_1 C_A C_C - (k_2 + k_3) C_{AC}$$

Applying the quasi-steady-state hypothesis ($dC_{AC}/dt = 0$) to the second-order models results in the following equation.

$$C_A C_C = \frac{k_2 + k_3}{k_1} C_{AC}$$

In a situation where the total quantity of catalyst present is nearly constant, we may state that the initial concentration C_{C_0} is always equal to the current

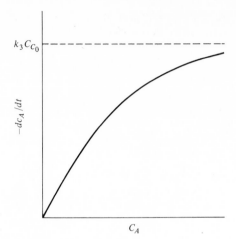

FIGURE 2-5
Rate of formation vs reactant concentration for sequential second-order reaction.

concentration C_C plus the complexed catalyst concentration C_{AC}, which gives the
following relation:

$$k_1 C_{C_0} C_A = (k_2 + k_3)C_{AC} + k_1 C_A C_{AC}$$

Dividing through by k_1 and solving for C_{AC} gives

$$C_{AC} = \frac{C_{C_0} C_A}{(k_2 + k_3)/k_1 + C_A}$$

The rates of formation of A and B can now be restated.

$$\frac{dC_A}{dt} = -\frac{dC_B}{dt} = \frac{-k_3 C_{C_0} C_A}{(k_2 + k_3)/k_1 + C_A}$$

Examination of this latter equation shows that a maximum rate of formation
exists and that this value is approached when $C_A \gg (k_2 + k_3)/k_1$. Plotting the
rate of formation against the reactant concentration gives the graphical relationship shown in Fig. 2-5. Another useful method of relating the rate of formation
to the reactant concentration can be developed by rearranging the rate equation
in reciprocal form.

$$-(k_3 C_{C_0})\left(\frac{dC_A}{dt}\right)^{-1} = \frac{k_2 + k_3}{k_1 C_A} + 1 \qquad (2\text{-}11)$$

This latter relationship provides a straight-line relationship (Fig. 2-6) and is used
to evaluate the coefficient groupings.

Equation (2-11) is used in describing simple enzymatic reactions and is
usually referred to as the *Michaelis-Menten equation* after the workers who first
utilized this relationship. Figure 2-6 is usually called a *Lineweaver-Burke plot*
in biological and biochemical texts.

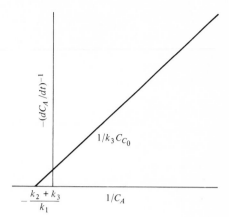

FIGURE 2-6
Inverse plot of rate of formation vs reactant concentration.

Monod[1] found that the "specific" growth rate of pure cultures of bacteria (mass or number increase of cells per mass or number present per unit time) could be described by an expression of the form of Eq. (2-11). Other workers have reported similar results, and the relationship (usually written as $\mu = \mu_{max} C/K + C$) is currently used extensively to describe bacterial growth rates in biological waste treatment systems. Mechanistic understanding of the growth rate relationship is not available, but two interesting concepts can be presented. One is that a single enzymatic reaction that conforms to the Michaelis-Menten model controls the overall reaction rate of the system. A second and seemingly more reasonable possibility is that nutrient adsorption on the surface of the bacterial cells controls the growth rate. This mechanism is suggested by the fact that the equation is of the same form as the Langmuir adsorption isotherm (see Chap. 3).

2-3 NONHOMOGENEOUS REACTIONS

As was stated above, two general types of reactions occur in wastewater treatment: homogeneous reactions, those distributed through the continuum, and heterogeneous reactions, those which occur at specific locations such as particulate surfaces, walls, or interfaces. Differentiation between the reaction types (i.e., assigning a type characteristic to a reactive system) is generally somewhat pragmatic in nature. Reactions occurring between solute molecules would clearly fall into the homogeneous category, while reactions occurring on fixed boundaries are heterogeneous in nature. A fluidized-bed ion-exchange process might be successfully treated as a homogeneous-reaction system even though the reactions occur on particulate surfaces because reaction sites are distributed throughout the fluid. This "pseudohomogeneous" type of model begins to break down when parameters other than reaction rate become rate-controlling variables. The BOD bottle provides an example of this type of situation. Rate models of BOD systems are traditionally assumed to be homogeneous because of the small size of bacteria, the uniform initial distribution of the organic material, and the assumption that all parts of the system are behaving identically. However, if molecular diffusion is the rate-controlling parameter, the homogeneous-reaction models become less satisfactory and a more "microscopic" viewpoint must be taken. Detection of this and other nontraditionally considered circumstances can usually be made by changing some rate-controlling parameters. Normally, we have three methods of accomplishing this: changing the concentration of one or more reactants, changing the physical environment (e.g., mixing conditions or temperature), or changing the chemical environment (for example, pH or oxidation-reduction potential). One of these effects will be demonstrated in the succeeding sections.

Another way of viewing the difference between homogeneous and hetero-geneous reactions is that in the homogeneous case the reaction term is included in the mass-balance equations, while in the heterogeneous case the reaction term serves as a boundary condition for the mass balance.[2] The differences can be more easily seen in an example problem.

EXAMPLE 2-2   A film of liquid covers a reactive surface (which could be a catalyst or a biological slime) to a depth δ as shown in Fig. 2-7. Oxygen from the air above the film reacts at the surface after being transported through the liquid. Assume that the liquid-phase concentration of oxygen varies with position in the z plane but is constant with time and equal to C_0 at the gas-liquid inter-face, i.e., steady state. The reaction rate is given by

$$r'_o = -k'C_\delta$$

where C_δ = oxygen concentration at solid-liquid interface
 r'_o = rate of formation of oxygen, m/l^2t

The rate coefficient, therefore, has units of length per time (l/t). Determine the oxygen-concentration profile in the liquid. To solve the problem, consider the flux of oxygen through a liquid layer of thickness Δz and encompassing the total cross section of the liquid.

A molar balance on the oxygen for the shell shown in Fig. 2-8 gives

$$N_z A + A \Delta z \, r_o = AN_{z+\Delta z} + A \Delta z \frac{\partial C}{\partial t}$$

where N_z = flux, mol/l^2t), of oxygen in z direction
 A = cross-sectional area

and r_o is the homogeneous molar reaction rate and is equal to zero in this

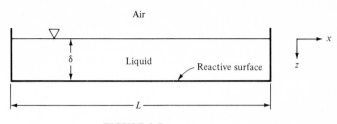

FIGURE 2-7
Heterogeneous-reaction-system schematic diagram.

problem, i.e., no reaction in liquid film. Collecting terms and allowing Δz to approach zero gives

$$\frac{\partial N_z}{\partial z} = -\frac{\partial C}{\partial t}$$

The flux N_z is related to the concentration by Fick's law[2]:

$$N_{A_z} = -\frac{\mathscr{D}_{AB}}{1 - x_A}\frac{\partial C_A}{\partial z} \qquad (2.12)$$

where A = solute in fluid B
$\quad x_A$ = mole fraction of A
$\quad \mathscr{D}_{AB}$ = diffusivity of A in B

For most wastewater constituents, x_A will be much less than unity and can be eliminated from the expression without introducing appreciable error. Because the system was assumed to be operating under steady-state conditions, $\partial C/\partial t$ can be eliminated, and the result is given by

$$\frac{\partial C}{\partial z} = \text{constant}$$

that is, the concentration gradient is constant. Because the flux of A at the solid-liquid interface must equal the rate at which A is consumed in the solid by the reaction

$$N_\delta = -k'C_\delta$$
$$= \mathscr{D}_{AB}\frac{\partial C}{\partial z}$$

FIGURE 2-8
Shell balance on liquid layer.

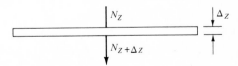

therefore
$$\frac{dC}{dz} = -\frac{k'}{\mathscr{D}_{AB}} C_\delta$$

Integrating between 0 and δ gives
$$C_\delta = \frac{C_0}{1 + (k'\delta)/\mathscr{D}_{AB}}$$

and
$$\frac{dC}{dz} = -\frac{k'}{\mathscr{D}_{AB}} \frac{C_0}{1 + (k'\delta)/\mathscr{D}_{AB}}$$

These results can be compared with those obtained for a similar physical system having *only* a homogeneous reaction ($r = -kC$) in the liquid film. In this case, the reaction rate enters into the molar-balance equation instead of appearing as a boundary condition for the steady-state situation, and we obtain

$$\frac{dN_z}{dz} - kC = 0$$

$$\frac{d^2C}{dz^2} = -\frac{k}{\mathscr{D}_{AB}} C$$

BC:
$$\text{At } z = \begin{cases} 0 & C = C_0 \\ \delta & N_z = 0 \quad \text{or} \quad \frac{dC}{dz} = 0 \end{cases}$$

The solutions to these equations are

$$\frac{C_z}{C_0} = \frac{\cosh\left[\sqrt{k\delta^2/\mathscr{D}_{AB}}(1 - z/\delta)\right]}{\cosh\sqrt{k\delta^2/\mathscr{D}_{AB}}}$$

$$\frac{C_\delta}{C_0} = \frac{1}{\cosh\sqrt{k\delta^2/\mathscr{D}_{AB}}}$$

The heterogeneous and homogeneous solutions indicated are difficult to compare quantitatively unless real values for k' and k are known. An item of interest can be seen by noting the effect of the ratio $k'\delta/\mathscr{D}_{AB}$. Assuming δ is a constant and is simply carried in the analysis, a plot of C_δ/C_0 vs $k'\delta/\mathscr{D}_{AB}$ and $k\delta^2/\mathscr{D}_{AB}$, respectively, can be used to illustrate the relative importance of reaction or diffusion rate control for treatment systems. As the ratio $k/\mathscr{D}_{AB}$ increases, the system becomes increasingly mass-transfer controlled; i.e., the rate at which reactant is transported through the liquid controls the conversion rate. If the diffusivity constant is very large (i.e., resistance to transport is very small), the concentration profile is less steep, C_δ/C_0 approaches 1, and the reaction rate becomes the controlling variable. Comparison of heterogeneous and homogeneous systems is shown in Fig. 2-9.

////

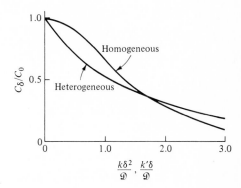

FIGURE 2-9
Comparison of mass-transfer effects in heterogeneous- and homogeneous-reaction systems.

2-4 Effectiveness-factor Model

Many processes used for wastewater treatment involve potential mass-transfer limitations somewhat different from those described in the example above. Biological slimes and flocs and activated-carbon particles are examples of reaction systems in which reactions take place within the material rather than just on the surface. The relative importance of mass transfer is measured by comparing the average reaction rate to that occurring on the particle surface. In wastewater treatment systems, this ratio is normally less than 1, but in typical industrial processes where reactions are strongly exothermic, the ratio is often much greater than 1 due to high internal temperature.

In describing the effectiveness-factor concept, an ideal spherical pellet will be used. The pellet (Fig. 2-10) may be thought of as having many very fine pores, like an activated-carbon particle, or being more gelatinous, like an activated-sludge floc.

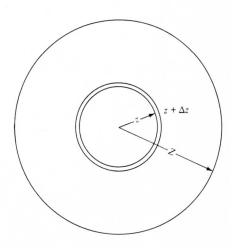

FIGURE 2-10
Ideal spherical pellet with concentric shells used to develop the effectiveness-factor model.

In describing this system, the reaction or adsorption process is assumed to occur homogeneously within the pellet, even though the activity is actually related to surface area. This assumption is justified by the extremely porous nature of activated carbon and the size and distribution of bacteria in flocs and slimes. Assuming a homogeneous model, the flux through a shell as shown in Fig. 2-10 can be described by the mass balance.

$$-\mathscr{D} \, \nabla C[4\pi(z + \Delta z)^2] + 4\pi z^2 \, \Delta z \, r = -\mathscr{D} \nabla C \, 4\pi z^2 + 4\pi z^2 \, \Delta z \, \frac{\partial C}{\partial t}$$

The mass balance assumes that there is no convective flux in the z direction, as was done in the previous example, and that the mole fraction of the reactant is very small. Adding the assumption of steady-state conditions $\partial C/\partial t = 0$ (which is not suitable for activated carbon), we obtain

$$\frac{\mathscr{D}}{z} \frac{d^2(zC)}{dz^2} = -r$$

For the purposes of this example, a simple relationship for the reaction rate term is in order, and $r = -kC$ will be used. The standard solution form of the resulting equation is

$$zC = A' \exp\left(\sqrt{\frac{k}{\mathscr{D}}} z\right) + B' \exp\left(-\sqrt{\frac{k}{\mathscr{D}}} z\right)$$

Because k and $\mathscr{D}$ are both positive, the root must be real, and the equation can be rewritten as

$$zC = A \sinh\left(\sqrt{\frac{k}{\mathscr{D}}} z\right) + B \cosh\left(\sqrt{\frac{k}{\mathscr{D}}} z\right)$$

When z equals zero, both sides of the equation must equal zero; therefore $B = 0$. Using the boundary condition $C = C_b$ at $z = Z$, the equation reduces to

$$C = \frac{ZC_b}{z} \frac{\sinh(\sqrt{k/\mathscr{D}} \, z)}{\sinh(\sqrt{k/\mathscr{D}} \, Z)}$$

The average rate of reaction is equal to the total rate of reactant mass entering the pellet divided by the pellet volume, thus

$$\bar{r} = \frac{1}{\frac{4}{3}\pi Z^3} \, \mathscr{D} 4\pi Z^2 \left. \frac{dC}{dz} \right|_{z=Z}$$

and $\quad \dfrac{dC}{dz} = -\dfrac{1}{z^2} \dfrac{ZC_b \sinh(\sqrt{k/\mathscr{D}} \, z)}{\sinh(\sqrt{k/\mathscr{D}} \, Z)} + \dfrac{ZC_b}{z} \sqrt{k/\mathscr{D}} \, (\tanh \sqrt{k/\mathscr{D}} \, z)^{-1}$

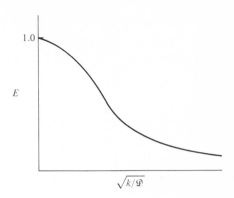

FIGURE 2-11
Catalyst effectiveness versus $\sqrt{k/\mathscr{D}}$.

Therefore

$$\bar{r} = \frac{-3\mathscr{D}}{Z} C_b \left[\sqrt{\frac{k}{\mathscr{D}}} \left(\tanh \sqrt{\frac{k}{\mathscr{D}}} Z \right)^{-1} - \frac{1}{Z} \right]$$

The ratio E of the average rate to the surface $\bar{r}/r_z$ is then

$$E = \frac{\bar{r}}{r_z} = \frac{\bar{r}}{-kC_b}$$

$$= \frac{3}{\sqrt{k/\mathscr{D}} \ Z} \coth \left(\sqrt{k/\mathscr{D}} \ Z \right) - \frac{3\mathscr{D}}{kZ}$$

Two conditions form a solution envelope for the effectiveness factor E. When $Z\sqrt{k/\mathscr{D}} \gg 1$, $\coth (\sqrt{k/\mathscr{D}} \ Z)$ approaches unity, the effectiveness approaches $3/(\sqrt{k/\mathscr{D}} \ Z)$, and the system is transport controlled. When $\sqrt{k/\mathscr{D}} \ Z \ll 1$, $\coth (\sqrt{k/\mathscr{D}} \ Z)$ can be evaluated by a series expansion, the effectiveness approaches unity, and the system is reaction controlled. Figure 2-11 provides a graphical description of this relation.

Based on the previous examples of the effects of microscopic scale limitations on reaction processes, one can readily recognize the importance of understanding what is actually happening within a reacting system. Many, if not most, real-world situations do not allow complete understanding of how a given reaction system operates, but a knowledge of the constraints on a system and the possible interactions as shown graphically in Figs. 2-9 and 2-11 are of great help to both design and operational engineers. A final example describing possible mass-transport limitations in the BOD bottle will summarize this section of Chap. 2.

Swilley, Bryant, and Busch[3] considered the possibility of reactant mass transport as the rate-limiting process in a BOD bottle. Their interest was aroused because the system was quiescent, and therefore molecular diffusion rather than convection would be the limiting mass-transport mechanism. While molecules such as oxygen and ammonia are small and diffuse rather rapidly, substrate

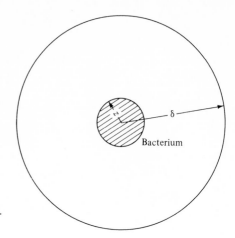

FIGURE 2-12
Idealized bacterium and sphere of influence.

molecules are often large, and the possibility of diffusion limitations seemed real. Their approach was to assume a spherical bacterial cell of radius Z with a sphere of influence of radius δ surrounding the cell. The cell was considered to draw all its nutrients from the sphere of influence (Fig. 2-12). A mass balance on a spherical shell gives

$$N_z = -\mathcal{D}_{sw} \frac{dC}{dz}$$

where N_z = flux of the substrate, mol/cm^2-s

For short periods of time, the rate of molar transport $(N_z 4\pi z^2)$ can be considered constant because bacterial growth is a relatively slow phenomenon compared with most chemical reactions. Based on this assumption, the differential equation can be integrated to yield

$$C_\delta - C_z = \frac{-M}{4\pi \mathcal{D}_{sw}} \left(\frac{1}{\delta} - \frac{1}{Z} \right)$$

where M is equal to the rate of molar transport in moles per second and is also equal to the rate of substrate conversion by the cell.

$$M = KC_z$$

Unfortunately, since bacterial cells are extremely small (on the order of 1 μm in radius), values of C_z cannot be directly measured. Assuming that the major fraction of the system has a concentration C_δ, the bulk concentration can be

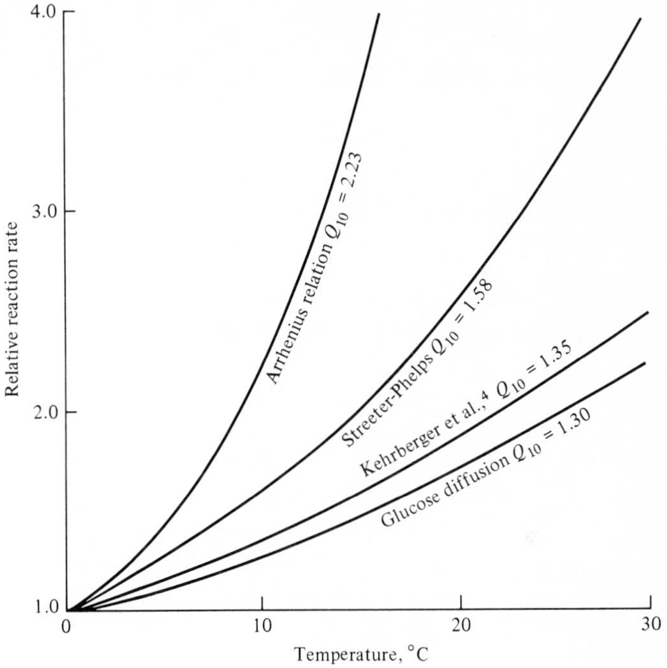

FIGURE 2-13
Effect of temperature on BOD exertion rate.

accurately measured, however, and a rate relationship may be based upon this bulk concentration.

$$M = K'C_\delta$$

Solving for K gives

$$K = \frac{K'}{1 + \dfrac{K'}{4\pi \mathscr{D}} \left(\dfrac{1}{\delta} - \dfrac{1}{Z} \right)}$$

If the rate of organic uptake is controlled by the mass transport rate, K will be greater than K'. Conversely, if K is equal to K', the overall uptake rate is dependent upon the rate of utilization at the surface rather than the mass transport rate to the surface. This would mean that the measured rate coefficient K' would be much smaller than $\left[\dfrac{1}{4\pi \mathscr{D}} \left(\dfrac{1}{\delta} - \dfrac{1}{Z} \right) \right]^{-1}$. The diffusivity of molecules such as glucose, a six-carbon sugar, is of the order of 10^{-5} cm/s. The sphere of influence most likely has a radius of the order of five bacterial radii. The term

$\dfrac{1}{4\pi\mathscr{D}}\left(\dfrac{1}{\delta}-\dfrac{1}{Z}\right)$ then has a value of the order 10^8 s/cm^3. Data taken during the maximal growth rate period in a BOD bottle indicate that the overall rate coefficient K' is of the order of 3×10^{-9} cm^3/s. Thus the coefficient K' is indicated to be of the same order of magnitude as $\left[\dfrac{1}{4\pi\mathscr{D}}\left(\dfrac{1}{\delta}\dfrac{1}{Z}\right)\right]^{-1}$, and diffusion is probably a rate-controlling process. This conclusion was strengthened considerably by Kehrberger et al.[4] in experiments with stirred and unstirred (quiescent) BOD systems. The significance of Swilley, Bryant, and Busch's work for us is that an example of actual rate limitation with a commonly employed wastewater evaluation system is provided.

Based on these observations, BOD systems would be expected to respond to temperature changes as a function of the diffusion coefficient rather than in the exponential manner of the Arrhenius expression, as commonly assumed. Kehrberger et al. demonstrated this by comparing rate coefficients for mixed and unmixed systems at various temperatures (Fig. 2-13). In practice, the question must be raised as to what the controlling rate is in the system of which the BOD is supposed to be a laboratory model. If BOD uptake rates are to be used for predicting oxygen sag values, this is of particular interest.

2-5 CONTINUOUS HOMOGENEOUS REACTORS

Because of the great volumes of water that must be treated, wastewaters are usually handled on a continuous basis. Continuous-flow reactors can be divided into three groups: ideal continuous-flow stirred-tank reactors (CFSTR), ideal plug flow, and backmixed reactors (the latter group accounts for the deviations from the ideal groups). A limited number of kinetic models can be treated here because of space limitations. Additional models and their consequences will be presented in later chapters in discussion of specific processes.

Continuous-flow Stirred-tank Reactors (CFSTR)

These reactors are similar to the batch reactors discussed earlier in that perfect mixing (i.e., no concentration gradients) is assumed throughout the tank. Unlike batch reactors, there are continuous inflow and outflow streams. The ideal mixing assumption imposes two important constraints on the system. One is that as material flows into the tank, concentrations are immediately diluted to the tank concentration. The second result is that all material concentrations in the outflow are the same as those in the tank. A mass balance on waste species A (where

A identifies a specific waste material or a grosser measure such as COD or BOD) is given in the following equation.

Mass rate of A into CFSTR

+ mass rate of generation of A in CFSTR

= mass rate of A leaving CFSTR

+ rate of accumulation of A in CFSTR

As in the case of batch reactors, the rate of formation function r_A must be known in order to proceed further. Most of the examples developed in this chapter will utilize first-order relationships (for example, $r_A = -kC_A$). Later chapters dealing with specific processes will make greater use of rate relationships that are known to apply to specific reactions or treatment systems. Thus, for the case of $r_A = -kC_A$, remembering that outflow concentration is the same as that in the tank,

$$QC_{A_i} + Vr_A = QC_A + V\frac{dC_A}{dt}$$

$$QC_{A_i} - VkC_A = QC_A + V\frac{dC_A}{dt} \qquad (2\text{-}12)$$

$$\frac{dC_A}{C_{A_i} - [1 + (V/Q)k]C_A} = \frac{Q}{V}\,dt$$

The term V/Q very often appears in reactor analysis and is usually referred to as the hydraulic residence time, or mean residence time, and is symbolized here by the Greek letter Θ. Substituting Θ for V/Q and integrating Eq. (2-12) between the limits of zero and t and zero and C_A, respectively, gives

$$\frac{t}{\Theta} = \frac{1}{1 + \Theta k}\ln\frac{C_{A_i} - (1 + \Theta k)C_A}{C_{A_i}} \qquad (2\text{-}13)$$

and

$$C_A = \frac{C_{A_i}}{1 + \Theta k}\left(1 - e^{-(t/\Theta)(1 + \Theta k)}\right) \qquad (2\text{-}14)$$

As time becomes very large, the exponential term of Eq. (2-14) becomes very small, and the steady-state solution is approached.

$$C_A = \frac{C_{A_i}}{1 + \Theta k} \qquad (2\text{-}15)$$

The time necessary to achieve a good approximation to steady-state conditions is dependent upon the rate coefficient, as shown in Fig. 2-14.

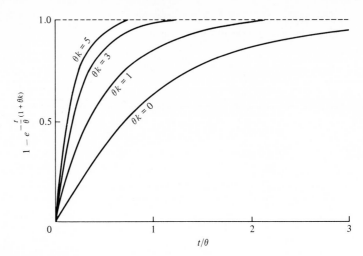

FIGURE 2-14
Time necessary to approach steady-state conditions in continuous-flow reactor.

Most treatment systems receive an inflow that varies in both flow rate and constituent makeup and concentration with time. Rate of change is relatively slow in most cases due to lags induced in transportation and the manner of input into the collection system. Treated effluent must consistently meet minimum quality standards, and thus processes must be designed for maximum loading rates. The relatively slow rate of change allows assumption of steady-state conditions for design purposes in most cases. The following developments for CFSTR processes are for steady-state conditions.

Previously we considered parallel and consecutive reactions under batch conditions. Considering the two cases again, we find for parallel first-order reactions with all stoichiometric coefficients being 1 in a CFSTR,

$$C_A = \frac{C_{A_i}}{1 + \Theta(k_1 + k_2)}$$

$$C_B = \frac{k_1 \Theta C_{A_i}}{1 + \Theta(k_1 + k_2)}$$

$$C_C = \frac{k_2 \Theta C_{A_i}}{1 + \Theta(k_1 + k_2)}$$

We note that as Θ, the residence time, becomes large, C_A decreases and C_B/C_{A_i}

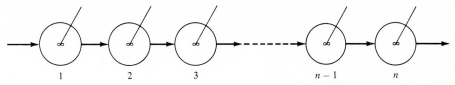

FIGURE 2-15
Schematic diagram of CFSTR cascade.

and C_C/C_{A_i} approach $k_1/(k_1 + k_2)$ and $k_2/(k_1 + k_2)$ as before. Performing similar operations for consecutive first-order reactions gives

$$C_A = \frac{C_{A_i}}{1 + \Theta k_1}$$

$$C_B = \frac{\Theta k_1 C_{A_i}}{(1 + \Theta k_1)(1 + \Theta k_2)}$$

$$C_C = \frac{\Theta^2 k_1 k_2 C_{A_i}}{(1 + \Theta k_1)(1 + \Theta k_2)}$$

Again as C becomes large, both A and B disappear and C_C approaches C_{A_i}.

CFSTR Cascades

Sequences of CFSTRs (Fig. 2-15), called *cascades*, are useful both from a conceptual and design viewpoint. A steady-state balance on the nth reactor (all of which are assumed identical here) gives

$$QC_{n-1} + Vr = QC_n$$

where C is the concentration of the reactant. Assuming a first-order reaction $(r = -kC)$, we find

$$C_n = \frac{C_{n-1}}{1 + \Theta k}$$

The concentration of C leaving any reactor in terms of the influent concentration is

$$C_n = \frac{C_i}{(1 + \Theta k)^n} \qquad (2\text{-}16)$$

Now remembering that Θ is the residence time for one of the n reactors and solving for the total residence time necessary to produce an effluent concentration C_n,

$$n\Theta = \frac{n}{k} \left[\left(\frac{C_i}{C_n} \right)^{1/n} - 1 \right] \qquad (2\text{-}17)$$

or

$$V = \frac{Q}{k} \left[\left(\frac{C_i}{C_n} \right)^{1/n} - 1 \right]$$

Thus, for a given value of C_n, the greater n is the smaller the total reactor volume needed will be. The reason that the total residence time and, therefore, the total tank volume decreases is that the average concentration of reactant C in each tank is higher than that for a single CFSTR producing an effluent of concentration C_n. Thus the average reaction rate r is higher, and the reaction time necessary is less.

We will also find the cascade of CFSTRs useful in describing nonideal reactor hydraulics. If a tracer such as salt or a dye is fed into a cascade, we find for the nth reactor,

$$Q(C_t)_{n-1} + Vr_t = Q(C_t)_n + V \left(\frac{dC_t}{dt} \right)_n$$

where C_{ti} signifies the tracer concentration and r_t is assumed to equal 0 (i.e., the tracer is conservative). If a slug of tracer is injected into the first reactor such that the concentration goes from zero to C instantaneously, the outflow concentration from the second through nth reactors is

$$C_{tn} = \frac{C_{ti}}{(n-1)!} \left(\frac{t}{\Theta} \right)^{n-1} e^{-t/\Theta} \qquad (2\text{-}18)$$

Equation (2-18) has a maximum value for any n equal to

$$C_{n,\,max} = \frac{C_i}{(n-1)!} (n-1)^{n-1} e^{-(n-1)}$$

which occurs when $t/\Theta = n - 1$. The predicted effluent tracer concentration is shown by the solid lines in Fig. 2-16. Experimental data (in Fig. 2-16) are fitted to the curves, and if experimental data result in a reasonable fit, the real system may be assumed to behave as n identical, ideal reactors in series. If tracer data from the real system do not fit the model, two courses can be followed. The first is to assume some combination of reactors of different sizes that will fit the data, and the second is to attempt to fit a plug-flow (with dispersion) model. This latter model will be discussed later in the chapter.

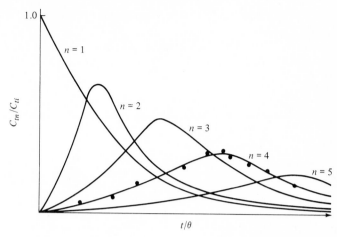

FIGURE 2-16
Response of CFSTR cascade to a pulse input of tracer.

Plug-flow Reactors

The basic premise of ideal plug-flow-reactor models is that mixing is ideal in the lateral plane and absent in the longitudinal direction. A tracer input to an ideal plug-flow reactor has exactly the same shape leaving as entering and is only translated in time (Fig. 2-17).

Because concentrations of reactive (nonconservative) materials decrease in the direction of flow, overall mass balances on a differential volume are required as indicated in Fig. 2-18 and Eq. (2-19).

$$QC_z + A\,\Delta z\,r_z = QC_{z+\Delta z} + A\,\Delta z\,\frac{\partial C_z}{\partial t} \qquad (2\text{-}19)$$

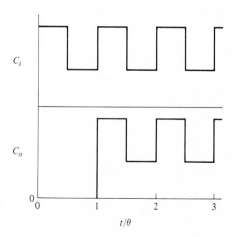

FIGURE 2-17
Input and output from ideal plug-flow
reactor for a conservative tracer.

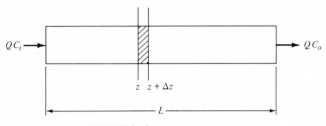

FIGURE 2-18
Plug-flow-reactor schematic with differential section.

Collecting terms and allowing the differential term to approach zero gives

$$\frac{-Q}{A}\frac{\partial C_z}{\partial z} + r_z = \frac{\partial C_z}{\partial t}$$

Both the flow rate Q and the cross-sectional area A are assumed constant. Therefore they can be brought inside the differential:

$$\frac{-\partial C}{\partial \Theta} + r_z = \frac{\partial C}{\partial t} \qquad (2\text{-}20)$$

where Θ is a form of the residence time. Under steady-state conditions, Eq. (2-20) reduces to a form identical to the expression for a batch reactor, except that the time term Θ represents a position in the reactor not real time.

$$r_z = \frac{dC}{d\Theta}$$

Because plug-flow reactors operating at steady state and batch reactors (which are always nonsteady state) have identical mass balances, there is no need to repeat the examples for each specific type of reaction. The reader should review the batch-reaction section and note that when Θ is substituted for t, a plot of C/C_i vs Θ becomes essentially a plot of concentration vs position.

Plug Flow with Dispersion

As stated previously, most real-world reactors do not behave as the ideal CFSTRs and plug-flow reactors described. In practice, most reactors have a certain amount of dead volume that effectively does not enter into the process operation, and channeling or short circuiting is also often a problem (in one sense the presence of dead volume means short circuiting exists). Finally, there is the process of dispersion in which material is transported from regions of high concentration to regions of low concentration by turbulent eddies. The process is analogous to molecular diffusion because the eddy motion in any direction is random, and the net transport takes place because of the difference in concentration at the eddy sources. Because of the analogy, dispersion can be described by an expression similar to Fick's law:

$$\frac{\partial C}{\partial t} = D \frac{\partial^2 C}{\partial Z^2} - U \frac{\partial C}{\partial Z}$$

where U = mean or bulk velocity in z direction
D = dispersion coefficient, cm^2/s

Dividing this expression by U/L (where L is the reactor length) and collecting terms as $t^* = tU/L$ and $Z = z/L$ gives

$$\frac{\partial C}{\partial t^*} = \frac{D}{UL} \frac{\partial^2 C}{\partial Z^2} - \frac{\partial C}{\partial Z} \qquad (2\text{-}21)$$

When the dimensionless term D/UL is small, the system approaches ideal plug flow, and when D/UL is large, ideal mixed flow is approached. Only one-dimensional dispersion has been discussed here; thus an assumption of homogeneity in the lateral direction has been made. Clearly, this assumption does not hold true everywhere. It does work reasonably well for an unexpectedly wide variety of systems. The dispersion model has been successfully used in a number of studies,[5-7] and the reader is encouraged to explore them.

For small values of D/UL, the tracer curve shape does not change significantly as it moves through the reactor, and the symmetrical solution to Eq. (2-21) can be satisfactorily represented by

$$C_{t^*} = \frac{1}{2\sqrt{\pi(D/UL)}} \exp\left[-\frac{(1-t^*)^2}{4D/(UL)}\right] \qquad (2\text{-}22)$$

Levenspiel[8] states that this solution gives an error of less than 5 percent when D/UL is less than 0.01 and of less than 0.5 percent when D/UL is less than 0.001. As the dispersion number increases, factors such as boundary conditions become more important. Levenspiel and Smith[9] give the following solution.

$$C_{t^*} = \frac{1}{2\sqrt{\pi t^* D/(UL)}} \exp\left[-\frac{(1-t^*)^2}{4t^* D/(UL)}\right]$$

with the mean occurring at

$$t^* = 1 + 2\frac{D}{UL}$$

and a variance of

$$\sigma_{t^*}^2 = 2\frac{D}{UL} + 8\left(\frac{D}{UL}\right)^2$$

When chemical reactions occur in a plug-flow reactor with dispersion, Eq. (2-21) must be modified to include the reaction term. A mass balance for

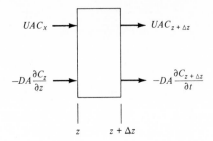

FIGURE 2-19
Convective and dispersive transport into
and out of a differential section of plug-
flow reactor.

the differential element of Fig. 2-19 results in Eq. (2-23).

$$D \frac{\partial^2 C}{\partial z^2} - U \frac{\partial C}{\partial z} + r = \frac{\partial C}{\partial t} \qquad (2\text{-}23)$$

Wehner and Wilhelm[10] solved Eq. (2-23) analytically for steady-state conditions
$(\partial C/\partial t = 0)$ with a first-order reaction $(r = -kC)$ to yield

$$\frac{C}{C_i} = \frac{4\gamma \left(\frac{1}{2} \frac{UL}{D} \right)}{(1 + \gamma)^2 \exp \left(\frac{\gamma}{2} \frac{UL}{D} \right) - (1 - \gamma)^2 \exp \left(-\frac{\gamma}{2} \frac{UL}{D} \right)} \qquad (2\text{-}24)$$

where

$$\gamma = \left(1 + 4k\Theta \frac{D}{UL} \right)^{1/2}$$

2-6 CONVERSION

A useful method of comparing reactor characteristics and the extent of reaction
is to write the concentrations in terms of conversion, which is defined as

$$\eta(t) = \frac{C_i - C(t)}{C_i} \qquad (2\text{-}25)$$

We can see the usefulness of the conversion term by comparing ideal batch,
plug-flow, and CFSTR systems. A batch process will convert a certain amount
of reactant in a given time t. An additional time t_r must be allowed to empty
and refill the reactor, however. The amount of reactant produced (or removed)
per unit time is, therefore,

$$Pr_b = \frac{V_b C_i \eta}{t + t_r}$$

where Pr_b is the batch production rate, m/t. Substituting E for t_r/t, we can rewrite the expression as

$$Pr_b = \frac{V_b C_i \eta}{t} \frac{1}{1 + E} \qquad (2\text{-}26)$$

For the purposes of our explanation, we will assume a first-order reaction. The reaction time t can then be written in terms of η and k as

$$\frac{d\eta}{dt} = k(1 - \eta)$$

which on integration becomes

$$t = \frac{1}{k} \ln \frac{1}{1 - \eta}$$

Substituting this last expression into (2-26) and rearranging to make later comparison simple gives

$$kV_b C_i \eta (1 + E) = \ln \left(\frac{1}{1 - \eta} \right) \qquad (2\text{-}27)$$

for batch systems. Clearly, an ideal plug-flow reactor with first-order reactor will give the same relationship except for the term $1/(1 + E)$.

$$\frac{kV_p C_i \eta}{Pr_p} = \ln \frac{1}{1 - \eta} \qquad (2\text{-}28)$$

Therefore, the ratio of volumes of batch and plug-flow reactors is equal to $1 + E$, which is always greater than 1.

Performing similar operations on a CFSTR with first-order reaction gives the following expressions:

$$Pr_{\text{CFSTR}} = \frac{V C_i \eta}{\Theta}$$

$$\frac{1}{1 + \Theta k} = 1 - \eta$$

$$\Theta k = \frac{1}{1 - \eta}$$

$$\frac{kV_{\text{CFSTR}} C_i \eta}{Pr_{\text{CFSTR}}} = \frac{\eta}{1 - \eta}$$

Comparing CFSTR and plug-flow-reactor volumes for a given value of conversion,

$$\frac{V_{\text{CFSTR}}}{V_p} = \frac{\eta}{(\eta - 1) \ln (1 - \eta)} \qquad (2\text{-}29)$$

Remembering that by definition the conversion is always greater than or equal to 0 and less than or equal to 1, we can see that for first-order reactions V_{CFSTR}/V_p is always greater than 1. In water and wastewater treatment, conversions of the order of 0.9 or greater are normally required, and for these conditions the minimum required volume ratio is 3.9. For a conversion of 0.99 the ratio is 21.5. These ratios mean that an ideal CFSTR with a first-order reaction used for treatment processes will always be considerably larger than an ideal plug-flow reactor used for the same purpose. As will be discussed in a later chapter, this is of particular interest because activated-sludge-process designers have been moving from nominal plug-flow processes to nominal stirred-tank systems in recent years without changing detention-time requirements. Clearly, either the long narrow tanks traditionally designed were actually CFSTRs and/or the first-order reaction model does not come close to applying and/or the systems are very overdesigned from a reaction point of view. Probably all these factors are involved, but, as will be explained later, overdesign with respect to reaction rate is most likely the principal factor. However, the reader should remember that quite often a number of constraints must be considered in a design with the result that several parts or features are overdesigned in order to satisfy a controlling constraint.

Now consider other reaction rates: zero order and second order. The respective volume ratios are given below, and the algebra is left to the reader as an exericse.

$$\frac{V_{CFSTR}}{V_p} = \begin{cases} 1 & \text{zero-order reaction} \\ \dfrac{1}{1-\eta} & \text{second-order reaction} \end{cases}$$

Earlier in this chapter a brief discussion of the Michaelis-Menten type of reaction model was given. In treatment-process-design practice, the model is often more complex than presented, being multiplied by the cell mass concentration. As presented the reaction rate falls between zero- and first-order reaction models, however. When reactant concentration is high, the rate approaches a constant value (zero order), and when reactant concentration is low ($C \ll K$) the reaction is approximately first order.

The reader should also note that the family of curves created for volume ratio vs conversion as a function of reaction order n ($r = -kc^n$) increases in slope with n. Thus, the larger the reaction order, the greater the volume ratio for a given conversion.

The value of using cascades of stirred-tank reactors can again be illustrated using the conversion term and the volume ratio. Rewriting Eq. (2-16) in terms of conversion for a first-order reaction,

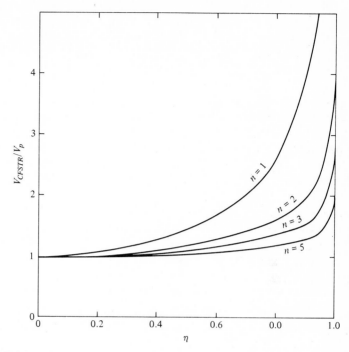

FIGURE 2-20

Comparison of CFSTR and plug-flow-reactor volumes for various cascades and conversions.

Solving for the residence time Θ and performing the same steps as before,

$$\frac{V_{cascade}}{V_p} = \frac{n\left[1 - \left(\dfrac{1}{1-\eta}\right)^{1/n}\right]}{\ln(1-\eta)}$$

where $V_{cascade}$ is the total volume of all the reactors. As before, the ratio is always greater than 1. When the number n of tanks becomes large, this ratio approaches 1 for any value of η. Proof of this is left to the reader, but the result is demonstrated in Fig. 2-20.

2-7 NONHOMOGENEOUS REACTORS

A number of nonhomogeneous reactors are used for water and wastewater treatment. Ion exchange and carbon adsorption have been discussed to some extent already. Coagulation-flocculation, the process through which solid particles grow by colliding and forming larger particles, can be treated as a heterogeneous-reaction process in many cases. Gas-exchange processes such as stream aeration

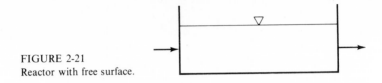

FIGURE 2-21
Reactor with free surface.

or ammonia stripping can also be treated as heterogeneous-reaction systems. Perhaps the most obvious example is the trickling filter, where the liquid film flows over the "reactive" surface in a manner very analogous to flow over solid catalysts.

Development of generalized models for heterogeneous-reaction systems will not be discussed in depth here because each system of interest has its own peculiarities and constraints and because a fairly generalized development was presented earlier in this chapter. Instead the following summation is made to tie the conceptual processes together. Two scales of heterogeneous-reaction systems exist. In one the reactions are distributed over the liquid phase much as in a homogeneous system. Mass transport is of concern, however, and the mathematical model is developed as for a heterogeneous system. The porous catalyst pellet falls into this category. Clearly, once the relation between total conversion rate and concentration is developed, an overall rate parameter incorporating diffusion rate, reaction rate, and particle size can be used, and the model becomes "pseudohomogeneous" for practical purposes. A second type of model develops where the reaction sites are not distributed through the liquid phase. In this case the hydraulic and reaction models are not easily coupled, and the descriptive expressions are generally not as adaptable to the development of a pseudohomogeneous type of model.

An example of the differences between the two systems described is provided by considering the effect of increasing the flow rate through two geometrically similar reaction vessels with free surfaces (Fig. 2-21). If the reaction is carried out with the aid of large porous catalyst particles evenly distributed through the liquid phase, the effect of increasing flow rate should be quite different than with a system in which the catalyst is bound in a layer on the bottom of the tank.

PROBLEMS

2-1 Plot the conversion as a function of time in a batch-reaction system for the reactions given below. Note any asymptotes of interest.

$$A \rightarrow B$$
$$r = kC_A^{\alpha}$$

(a) $0 < \alpha < 1$
(b) $\alpha = 1$

(c) $1 < \alpha < 2$

(d) $2 < \alpha$

(e) $\alpha = 0$

2-2 A particular batch-fermentation system can be described by the overall expression

$$2A \rightarrow C + CO_2$$

where A is the organic nutrient, and C is cells produced. The reaction rate is given by the expression

$$R = kC_A C_C$$

Given that at time $t = 0$, $C_A = C_{AO}$ and $C_C = C_{CO}$, and that $k = 1 \ \text{l}^3/\text{mg-h}$, draw a graph of C_A/C_{AO} and C_C/C_{CO} as functions of time.

2-3 The data presented in the table below were obtained from a batch-reaction experiment with an initial reaction concentration of 100 mg/l. Determine the order of reaction, the value of the activation energy for Eq. (2-6), and the rate coefficients for each temperature.

Time, d	Reactant concentration, mg/l				
	10°C	20°C	25°C	30°C	40°C
0.1	94	90	88	85	77
0.5	74	61	53	44	27
1.0	54	37	28	20	7
1.5	40	22	15	9	2
2.0	30	14	8	4	
3.0	16	5	2	1	
4.0	9	2			

2-4 Quite often temperature-rate relationships are given as functions of a constant to the $T - T_r$ power, where T_r is a reference temperature (usually 20°C). For example, a common expression is

$$k_T = k_{20} \, \Psi^{T-20}$$

with T in degrees Celsius. Using the data of Prob. 2-3, determine the value of Ψ.

2-5 A common name for the saturation coefficient K in the Michaelis-Menten rate expansion is the *half velocity constant* because the concentration of the reactant, C, is equal to K when $\mu = \mu_{max}/2$. Derive this relationship.

2-6 A chemical reaction $A + O_2 \rightarrow 2AO$ occurs at the liquid-solid interface of the system shown in Fig. 2-22. The reactant A is provided in excess by the inflow, and the reaction rate is given by

$$\mu' = k'C_{O_2, \delta}$$

where μ' has units of mg/cm²-s and k' is known to have a value of 0.0001 cm/s. The influent concentration of A is 500 mg/l, and the gas-liquid interface oxygen concentration can be assumed to be 9 mg/l. For a flow rate of 1 l/s, determine the effluent product concentration C_{AO_2} per unit reactor width. The liquid depth is 3 cm and the diffusion coefficient is 1×10^{-5} cm²/s. Assume that there is no longitudinal concentration gradient of oxygen.

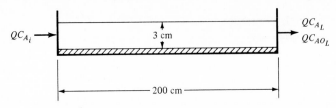

FIGURE 2-22
Heterogeneous-reaction system of Prob. 2-6.

2-7 The system described in Prob. 2-6 can also be run in a pseudohomogeneous manner by using a fine suspension of catalyst. The catalyst is separated from the process effluent and reused. The reaction rate for this system is $\mu = kC$, where $k = 0.001$ s^{-1}. Determine the effluent concentration of AO and the effectiveness factor for the system.

2-8 A plug-flow reactor and an ideal stirred-tank reactor are to be operated in series. For a reaction with the rate relationship

$$\mu = kC^{\alpha}$$

compare the conversion obtained for a plug-flow reactor followed by a CFSTR with a CFSTR followed by a plug-flow reactor, where

(a) $\alpha = 0$
(b) $0 < \alpha < 1$
(c) $\alpha = 1$
(d) $\alpha > 1$

2-9 The tracer data below were obtained by increasing the influent tracer concentration from 0 to 100 mg/l. Use the data to predict the effluent reactant concentration from the reactor if the influent reactant concentration C_i is 100 mg/l and the reaction rate is given by

$$r = -0.1C \text{ mg/l-h}$$

Sampling time, min	Effluent tracer concentration, mg/l
0	0
2	0
4	0
6	0
8	0
10	0.1
11	3.3
12	6.4
14	12.5
16	18.1
20	28.3
25	39.3
40	63.2
60	81.1

2-10 A reaction $A \to B$ is known to proceed according to the rate expression

$$r_A = - \frac{kC_A}{K + C_A}$$

Use the data presented in the following table to determine the rate coefficients, and then determine the CFSTR volume needed to produce an effluent concentration C_A of 0.1 mg/l from an influent stream with a flow rate of 10 l/s and a concentration C_{Ai} of 1000 mg/l.

PILOT PLANT DATA (STEADY STATE)
Tank volume = 100 l; C_i = 100 mg/l

Q, l/h	C, mg/l
0.39	5.8
0.78	17.9
1.56	46.0
3.13	71.9
6.25	85.7
12.50	92.8
25.00	96.4
50.00	98.1
100.00	99.1

REFERENCES

1 MONOD, J.: The Growth of Bacterial Cultures, *Annu. Rev. Microbiol.*, vol. 3, 1949.
2 BIRD, R. B., T. STEWART, and E. LIGHTFOOT: "Transport Phenomena," John Wiley & Sons, Inc., New York, 1960.
3 SWILLEY, E. L., J. O. BRYANT, and A. W. BUSCH: A Significance of Transport Phenomena in Wastewater Treatment Processes, *Proc. 19th Ind. Waste Conf.*, 1964.
4 KEHRBERGER, G. J., J. D. NORMAN, E. D. SCHROEDER, and A. W. BUSCH: BOD Progression in Soluble Substrates VIII, Temperature Effects, *Proc. 19th Ind. Waste Conf.*, 1964.
5 MURPHY, K. L., and P. L. TIMPANY: Design and Analysis of Mixing for an Aeration Tank, *J. Sanit. Eng. Div.*, ASCE, vol. 93, SA5, p. 1, 1967.
6 O'CONNOR, D. J.: Estuarine Distribution of Nonconservative Substances, *J. Sanit. Eng. Div.*, ASCE, vol. 91, SA1, p. 23, 1965.
7 THOMAN, R. V.: "Systems Analysis and Water Quality Management," Environmental Sciences Service Corporation, Stamford, Conn., 1972.

8   LEVENSPIEL, O.: "Chemical Reaction Engineering," 2d ed., John Wiley & Sons, Inc., New York, 1972.

9   LEVENSPIEL, O., and W. K. SMITH: Notes on the Diffusion Type Model for Longitudinal Mixing of Fluids in Flow, *Chem. Eng. Sci.*, vol. 6, p. 227, 1957.

10 WEHNER, J. F., and R. F. WILHELM: Boundary Conditions of Flow Reactor, *Chem. Eng. Sci.*, vol. 6, p. 89, 1958.

3

PHYSICAL-CHEMICAL REMOVAL OF DISSOLVED MATERIALS

Two general methods of removal of dissolved materials are in use. These can be most easily grouped as biological conversion and nonbiological or physical-chemical removal processes. Among the physical-chemical processes in use are adsorption, ion exchange, reverse osmosis, chemical oxidation, precipitation, and gas stripping. Because of overlap with other sections, the latter subjects will be discussed in a later chapter. The type of process considered for use in any particular case must take into account the characteristics of the waste materials. For example, nonpolar materials such as most organic compounds can be readily adsorbed onto activated carbon (a nonpolar sorbent), while polar materials require a polar sorbent such as silica gel. Each process has definite advantages and disadvantages, particularly in the area of types or ranges of materials for which it can be applied.

3-1 ADSORPTION

In practice adsorption is nearly limited to the use of activated carbon for the removal of nonpolar materials. The reason for this situation is economic. On a relative scale activated carbon is an inexpensive, nonselective adsorbent. Because of the large volumes of water that must be treated and the wide range of compounds found in most waters, both factors are important. In addition, a successful sorbent should not attract water (as silica gel does), and thus adsorption from water and wastewater is limited to nonpolar materials.

Strictly speaking, adsorption is the accumulation of materials at an interface. The interface may be liquid-liquid, liquid-solid, gas-liquid, or gas-solid. Of these types of adsorption, only liquid-solid adsorption is widely used in water and wastewater treatment. Dissolved materials tend to either accumulate at interfaces or to disperse away from interfaces depending on their relative strength of attraction for themselves or for the solvent. Thus nonpolar molecules tend to migrate toward interfaces. Two significant results of this accumulation are the lowering of the surface tension of the solvent, which allows wetting of the solid (sorbent) surface, and the concentration of the dissolved material (sorbate) in close proximity to the sorbent surface. The latter result is of particular importance because the rate of adsorption is proportional to the concentration of the sorbate.

Many organic compounds are both hydrophobic (water hating) and hydrophilic (water loving). Detergents are the best known and most common example of these compounds. These "amphoteric" compounds tend to become oriented at the surface with the hydrophobic portion at the interface and the hydrophylic portion remaining in solution. Their function is to produce wetted surfaces without being removed themselves.

Equilibrium Adsorption Models

Adsorption processes are usually described by the equilibrium isotherm they follow. These isotherms are simply relationships between the moles of sorbate adsorbed per unit mass of sorbent and the concentration of sorbate remaining in solution at equilibrium at a constant (isothermal) temperature. Experimental determination of an isotherm is usually accomplished by mixing a known amount of adsorbent with a given volume of liquid of known initial sorbate concentration. The system is allowed to come to equilibrium at a selected temperature, and the final liquid-phase sorbate concentration is measured. The concentration change is then used to calculate the moles of sorbate adsorbed.

$$x = \text{moles of substrate adsorbed}$$
$$= (C_i - C_e)V \qquad (3\text{-}1)$$

where C_i, C_e = initial and equilibrium molar concentrations of sorbate, respectively

V = liquid volume

Moles adsorbed are then divided by the mass of sorbent M, and the result is plotted against the equilibrium concentration C_e.

Most equilibrium data follow, or can be made to follow, one of three commonly used models: the Brunauer, Emmett, Teller (BET) isotherm; the

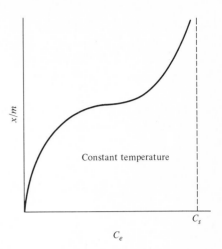

FIGURE 3-1
Form of BET adsorption isotherm.

Langmuir isotherm; or the Freundlich isotherm. Both BET and Langmuir isotherms are based on theoretical developments, while the Freundlich isotherm is an empirical relationship. The BET isotherm is based on the concept of multiple-layer adsorption, i.e., multiple layers of materials being adsorbed on the surface, while the Langmuir model assumes that only a single (mono) layer can be adsorbed.

The BET Adsorption Model

The BET model assumes that layers of molecules are adsorbed on top of previously adsorbed molecules. Each layer adsorbs according to the same Langmuir-type model.

$$\frac{x}{m} = \frac{BC_e(x/m)^0}{(C_s - C_e)(1 + B - 1)(C_e/C_s)} \qquad (3\text{-}2)$$

where $(x/m)^0$ = value of x/m when mono layer has been completed

C_s = saturation concentration in the liquid

B = constant related to energy of interaction between sorbent and sorbate

A plot of x/m versus C_e results in a curve of the form shown in Fig. 3-1.

Theoretically, as C_e approaches the saturation value, the moles adsorbed become very large because the model does not constrain the number of layers adsorbed. A linear representation of the BET model can be obtained by plotting $C_e/[(C_s - C_e)x/m]$ versus C_e/C_s as shown in Fig. 3-2.

In practice the saturation concentration C_s can only be estimated, thus making the development of a curve such as Fig. 3-2 an iterative process.

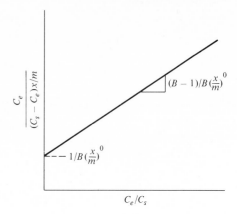

FIGURE 3-2
Linear form of BET adsorption isotherm.

Initially, data would be plotted in the form of Fig. 3-1, and a value of C_s would be approximated. The linear form would then be plotted. If the data did not fit a straight line satisfactorily, an improved estimate of C_s would be made and a new curve plotted. Estimated values of C_s that are greater than the true value produced a curve which bends downward as C_e/C_s increases, while low estimates of C_s result in an upward-bending (concave) curve as shown in Fig. 3-3.

Langmuir Adsorption Isotherms

The Langmuir isotherm results from assuming that adsorption is reversible and occurs only for a monolayer on the sorbent surface.

$$\frac{x}{m} = \frac{b(x/m)^0 C_e}{1 + bC_e} \qquad (3\text{-}3)$$

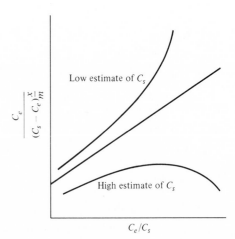

FIGURE 3-3
Effect of error in estimation of C_s on linear BET plot.

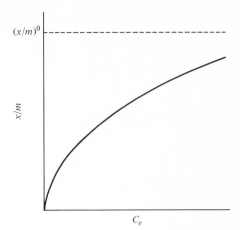

FIGURE 3-4
Form of Langmuir adsorption isotherm.

where b is an adsorption coefficient. Data are plotted either as shown in Fig. 3-4 or in one of the two alternate linear forms shown in Fig. 3-5. The second linear form gives extra weight to higher values of C_e and is useful because quite often these are far more reliable due to poor analytical sensitivity at low concentrations.

Approximation of the linear form is not necessary for the Langmuir isotherm, and the first plot of the data (assuming the data are of satisfactory quality) will demonstrate whether or not the model is applicable and also will allow determination of the coefficients. In many cases a single model will not be

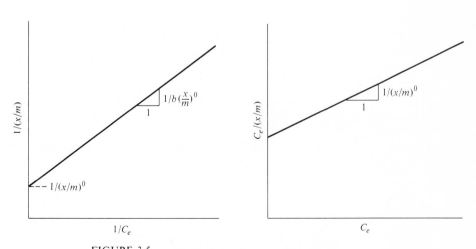

FIGURE 3-5
(a) Conventional linear form of Langmuir adsorption isotherm. (b) Modified form of Langmuir adsorption isotherm emphasizing higher concentration data.

satisfactory for a wide range of concentrations but will serve in narrow regions. An example of this can be seen by noting that at low concentrations ($C_e/C_s \ll 1$) the BET model reduces to a Langmuir isotherm-type expression.

Freundlich Adsorption Isotherms

The Freundlich isotherm, an exponential model, is the third commonly used adsorption expression.

$$\frac{x}{m} = K_f C_e^{1/n} \qquad (3\text{-}4)$$

where K_f is the Freundlich adsorption coefficient. As noted earlier, Eq. (3-4) is empirical but has been found to fit experimental data quite often. Usually the Freundlich isotherm is plotted on log-log paper to facilitate determination of the model's validity and the values of the coefficients K_f and n.

Rate of Adsorption

Because most models used in wastewater treatment are reaction models in which a nonequilibrium situation is inherent, there is often some difficulty in interpretation of the adsorption isotherms. An effort must be made to remember that the concentration values C_e and the adsorption values x/m are for equilibrium conditions and that in real, continuous-flow situations these conditions are probably only approximated. A second conceptual problem is the development of a rate model to describe the adsorption process. This model is, in reality, a reaction rate expression and occupies the same place in a reactor mass balance as the reaction rate term discussed in Chap. 2. In most cases, a satisfactory assumption is that the sorbent comes to equilibrium with the liquid adjacent to it very quickly, and thus the interface concentration is always assumed to be the equilibrium concentration C_e.

EXAMPLE 3-1   A CFSTR has a wastewater feed rate of 100 l/s (0.23 mgd) and a tank volume of 6000 l (1585 gal). The wastewater contains 20 mg/l of organic material that has been found to adsorb on activated carbon, according to the Langmuir isotherm with coefficient of $b = 0.13$ l/mg and $(x/m)^0 = 0.345$, where all measurements were made as mass rather than moles. Twenty grams per liter of clean granular activated carbon is added to the CFSTR at time equal to 0. Carbon leaving the reactor is separated from the effluent and returned as shown in the schematic diagram (Fig. 3-6).

(a) Determine the length of time the system can run before the effluent concentration exceeds the standard value of 1 mg/l.

(b) Determine the rate of carbon addition needed to meet the standard.

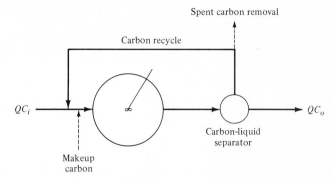

FIGURE 3-6
CFSTR with recycle carbon-adsorption system.

Because the carbon is always assumed to be at equilibrium with the adjacent liquid and because there are no concentration gradients in an ideal CFSTR, the effluent concentration is equal to the equilibrium concentration at time t. The fast reaction implies that the adsorbed organic will increase proportionally with the amount entering.

$QC_i t^*$ = mass of organic carbon entering from $t = 0$ to t^*

$$\left(\frac{x}{m}\right)_{t^*} = \frac{0.345(1)}{\frac{1}{0.13} + 1} = 0.0397$$

$$x_{t^*} = 0.0397(20)(6000) = 4763 \text{ g}$$

This value does not account for all the organic entering because at any time t some organic is leaving in the effluent. An approximation, t' to t^*, is made.

$$Qt'\left(C_i - \frac{C_e}{2}\right) = x_{t^*}$$

$$t' = \frac{4763}{100(0.0195)}$$

$$= 2442 \text{ s} = 40.7 \text{ min}$$

The rate of carbon addition necessary is simply enough to match the rate or organic input.

$$\frac{x}{m} = 0.0397 \text{ g organic/g carbon}$$

$$QC_i\left(\frac{x}{m}\right)^{-1} = 100(20)(0.0397)^{-1}$$

$$= 50,377 \text{ mg/s}$$

$$= 50.38 \text{ g/s}$$

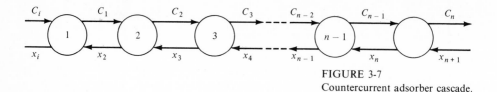

FIGURE 3-7
Countercurrent adsorber cascade.

Consider now an alternative configuration in which the carbon is introduced countercurrently to the wastewater. Assume that the carbon is clean ($x_i = 0$) on entering and in equilibrium with the incoming wastewater when it leaves. A mass balance can now be written for this nominally plug-flow situation.

$$Q(C_i - C_o) = M\left[\left(\frac{x}{m}\right)_o - \left(\frac{x}{m}\right)_i\right]$$

where M is the mass flow rate of the carbon and the subscripts o and i relate to out and in, respectively. The assumption that the incoming carbon is clean means that $(x/m)_i = 0$. The value of $(x/m)_o$ is known from the isotherm data.

$$\left(\frac{x}{m}\right)_o = \frac{0.13(0.345)(20)}{1 + 0.13(20)} = 0.249$$

In order to meet the effluent standard of 1 mg/l, the carbon input rate must be at least

$$M = \frac{100(20 - 1)}{0.249} = 7630 \text{ mg/s} = 7.63 \text{ g/s}$$

Clearly, the concurrent CFSTR system is much less satisfactory than the countercurrent process—at least theoretically. ////

In actual fact, the assumption of equilibrium between the influent wastewater stream and the effluent carbon stream is incorrect. Physically, the process must be operated in the same manner as the CFSTR because the two streams are intermixed. For this reason, a countercurrent flow sheet must be a series or cascade of reactors as shown in Fig. 3-7. An assumption of equilibrium is made for each stage. For example, C_i and x_i are equilibrium values. A mass balance on this system gives

$$Q(C_i - C_n) = M\left[\left(\frac{x}{m}\right)_1 - \left(\frac{x}{m}\right)_{n+1}\right]$$

Again assuming that the incoming carbon is clean, the sorbent flow rate can be determined.

$$\frac{M}{Q} = \frac{C_i - C_n}{\left(\dfrac{x}{m}\right)_1}$$

where $(x/m)_1$ is defined by the appropriate isotherm. While C_i and C_n are normally given values in a problem, the value of $(x/m)_1$ must be determined in each case. Where an isotherm can be expressed analytically (e.g., Langmuir or Freundlich), determination of $(x/m)_1$ is an algebra problem. Making a mass balance on the last (nth) stage,

$$Q(C_{n-1} - C_n) = M\left[\left(\frac{x}{m}\right)_n - \left(\frac{x}{m}\right)_{n+1}\right]$$

where

$$\left(\frac{x}{m}\right)_{n+1} = 0$$

and $(x/m)_n$ is defined by the isotherm. Thus successive values of concentration may be determined back through the first stage. A simple but very practical example is the two-stage adsorber.

$$Q(C_1 - C_2) = M\left(\frac{x}{m}\right)_2$$

The quantity $(x/m)_2$ is known because the isotherm is known, and C_2 is specified. Therefore C_1 can be determined.

$$C_1 = \frac{M}{Q}\left(\frac{x}{m}\right)_2 + C_2$$

Thus $(x/m)_1$ can be determined, and the problem is solved.

The two-stage adsorption system is practical because separating processes into stages requires expensive separation units between stages. Cost evaluation usually limits the number of "real" stages to two or three.

Using a two-stage countercurrent adsorber in Example 3-1 would result in the following:

$$\left(\frac{x}{m}\right)_2 = \frac{0.13(0.345)(1)}{1 + 0.137(1)}$$

$$= 0.040$$

$$C_1 = \frac{M}{Q}(0.040) + 1$$

$$\left(\frac{x}{m}\right)_1 = \frac{0.13(0.345)[(M/Q)(0.040) + 1]}{1 + 0.13[(M/Q)(0.040) + 1]}$$

$$Q(C_1 - C_2) = M\left(\frac{x}{m}\right)_1$$

$$19 = \frac{M}{Q} \frac{0.13(0.345)[(M/Q)(0.040) + 1]}{1 + 0.13[(M/Q)(0.040) + 1]}$$

$$\frac{M}{Q} = 125 \text{ mg carbon/l wastewater}$$

$$M = 12,500 \text{ mg/s} = 12.5 \text{ g/s}$$

Thus even a two-stage countercurrent system is far superior to a single-stage unit. If more stages were feasible, additional savings down to the theoretical value of 7.63 g/s could be obtained.

A graphical technique for solving the multistage-adsorption problem is often used. This method is described in Coulson and Richardson[4] and other standard chemical engineering design texts.

If the granules in Example 3-1 are large, there is a possibility that the mass transport rate inside the granules would control the sorption rate, and a problem similar to the porous catalyst problem of Chap. 2 develops. In this case, the situation is inherently nonsteady state because a carbon granule has a finite sorption capacity. Thus the concentration of organic material at any point in the particle would change with time. A mass balance on even a system with extremely simple geometry results in a nonlinear partial differential equation whose solution is beyond the scope of this text. The effective result will be a slower overall adsorption process and a higher than expected bulk-liquid organic concentration. The run time t^* will be shorter than predicted in Example 3-1, but the carbon will have sorptive capacity remaining when the liquid concentration reaches 1 mg/l.

Mass transport through the bulk liquid could also be limiting in some cases. Such a situation might occur if the granules were small, if a great deal of excess carbon existed, or if the liquid phase were quiescent, i.e., no convective transport. These conditions often are approximated in water treatment but would be extremely rare in wastewater treatment. The resulting model could be approximated by a steady-state expression in most cases and would be similar to the heterogeneous case described in Example 2-2.

Effect of Environmental Conditions

Both the rate and extent of adsorption are affected by environmental conditions. A primary factor is pH. In general, sorption capacity is increased by lowering liquid pH. Temperature also affects adsorption but is not of great importance in water and wastewater treatment. Adsorption equilibrium moves toward the sorptive phase with decreasing temperatures, and therefore capacity can be expected to increase at lower temperatures. Temperature variation in wastewater treatment is not great enough to make this a significant parameter. Temperature of carbon activation affects the sorptive qualities of the material. Weber[1] notes that carbon activated near 400°C is a good alkali adsorber, while carbon activated near 1000°C is a good acid adsorber.

Contaminants in water and wastewater are nearly always mixtures of organic materials, and, not surprisingly, the nature of the mixture has an effect on the sorptive characteristics. Some mixtures result in enhanced adsorption, others in decreased adsorption. With the array of compounds normally present, there seems little chance of predicting interactions solely on theoretical grounds, and we will generally have to deal with whatever comes down the pipe. Most waters and wastewaters are reasonably consistent in their makeup, and thus experimental determination of the sorptive characteristics should prove satisfactory in most cases.

Adsorption Process Design

Three types of activated-carbon adsorption process are in common use: fixed bed with gravity (down) flow, fixed bed with pressure (upflow), and fluidized beds (upflow). All three systems are nominal plug-flow processes, and all three can be designed in the same fashion. This latter fact is only true because carbon density increases significantly as adsorption progresses, causing the more saturated granules to migrate to the bottom of the bed. Thus in all three cases incoming wastewater is contacted with the carbon that has adsorbed the greatest amount of organic material (i.e., processes are countercurrent).

The three systems do have different operating characteristics, particularly with respect to plugging. Downflow gravity processes are the most susceptible to plugging and, consequently, should be used only for waters of low turbidity. Pressurized systems plug less because of the increased force applied to particles, and upflow systems also tend to plug less because of the direction of flow. Plugging is not a problem with fluidized beds, but problems associated with abrasion and particle breakup can be significant.

Breakthrough Curves

Processes are most easily designed through the use of breakthrough curves obtained in pilot plant studies. These curves (Fig. 3-8) relate the effluent organic concentration to the volume of water treated and can be used to develop column characteristics. In the downflow situation described by Fig. 3-8, the saturated zone is shown to move through the bed preceded by a zone of active adsorption with depth δ. An assumption is made for the purposes of the analysis that all the adsorption takes place in this latter zone and that the region behind the active zone is completely saturated. When the front of the active zone reaches the bottom of the bed, breakthrough occurs, and the effluent concentration begins to rise rapidly. Eventually, the carbon will be completely saturated and the effluent concentration will approach C_i, the influent

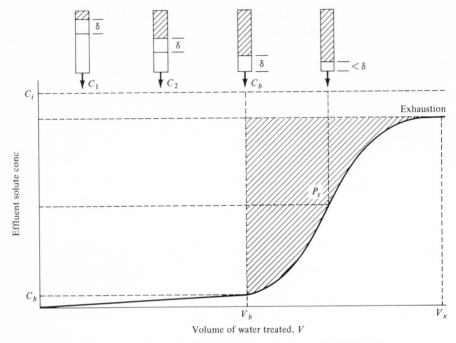

FIGURE 3-8
Carbon breakthrough curve.

concentration. In practice, the determination of breakthrough is somewhat arbitrary, and usually either the effluent requirement or $0.05C_i$ is used as a value. Exhaustion is usually defined as $C_{eff} = 0.95C_i$.

When breakthrough occurs, the carbon in the adsorption zone is partially spent. A major part of breakthrough-curve analysis is concerned with the determination of the fractional capacity f and the depth δ of the adsorption zone. This information can then be used to predict the effective column capacity. These parameters will change with changes in flow rate because dispersion, diffusion, and channeling are related to flow rate.

We begin our adsorption-zone computations by defining the time to exhaustion t_x and the time of passage through the adsorption zone t_δ.

$$t_x = \frac{V_x}{Q} = \frac{V_x \rho_l}{F_m A} \tag{3-5}$$

$$t_\delta = \frac{V_x - V_b}{Q} = \frac{(V_x - V_b)\rho_l}{F_m A} \tag{3-6}$$

where V_x, V_b = volumes of liquid passed through column at exhaustion and break-
       through, respectively
    Q = volumetric flow rate
    ρ_l = liquid density
    F_m = mass liquid flux
    A = cross-sectional area of column

    The velocity at which the adsorption zone moves through the column is constant except during the period it is being formed. This velocity defines the adsorption-zone height δ.

$$\delta = u_\delta\, t_\delta$$

$$= \frac{L}{t_x - t_f}\, t_\delta \qquad (3\text{-}7)$$

where L = column length
    t_f = zone formation time
    u_δ = adsorption-zone velocity

    The formation time can be satisfactorily estimated as

$$t_f = (1 - f)t_\delta$$

where f is the fractional capacity of the adsorption zone. Fractional capacity is the ratio of the quantity of material adsorbed in the adsorption zone from breakthrough to exhaustion P_s, to the total capacity of the carbon in the adsorption zone P_{tc}.

$$P_s = \int_{V_b}^{V_x} (C_i - C)\, dV \qquad (3\text{-}8)$$

$$P_{tc} = (V_x - V_b)C_i \qquad (3\text{-}9)$$

$$f = \int_{V_b}^{V_x} \frac{(C_i - C)\, dV}{(V_x - V_b)C_i}$$

$$= \int_0^1 \left(1 - \frac{C}{C_i}\right) d\,\frac{V - V_b}{V_x - V_b} \qquad (3\text{-}10)$$

We can see that the integral in Eq. (3-8) is equal to the cross-hatched area in Fig. 3-8. In addition, we note that as the fractional capacity approaches 1, t_f approaches zero and ideal plug-flow conditions are approximated. When channeling or mass-transfer limitations prevail, the curve rises very quickly following breakthrough until it nears C_x, the exhaustion concentration. Under these conditions, P_s is small, and the fractional capacity approaches zero.
    The depth of the adsorption zone can now be rewritten in terms of fractional capacity.

$$\delta = \frac{L(V_x - V_b)}{V_b + f(V_x - V_b)} \qquad (3\text{-}11)$$

When breakthrough occurs, the only section of the bed which has not been saturated (i.e., in equilibrium with the input concentration C_i) is the final portion of depth δ. Total sorptive capacity to breakthrough can then be defined as

$$S_b = \left(\frac{x}{m}\right)_{C_i} \rho_p[(L - \delta) + f\delta] \qquad (3\text{-}12)$$

where S_b = mass of organic adsorbed at equilibrium per unit of cross-sectional area, m/l^2

$\left(\dfrac{x}{m}\right)_{C_i}$ = mass of organic adsorbed per unit mass of carbon at the equilibrium concentration C_i

ρ_p = *apparent* packed density of the adsorbent, m/l^3
L = bed depth, l

The effective sorption capacity of a column is clearly strongly dependent upon the fractional capacity f. As f decreases, δ increases in size and S_b decreases in magnitude. Thus maximizing the fractional capacity of the adsorption zone is very beneficial. Earlier the relationship between fractional capacity and mass transfer rate into the carbon particles was discussed briefly. We will now lump all the mass-transfer effects (and quite possibly channeling also) into a simple mass transfer rate term.

$$F_m \frac{dC}{dy} = ka(C - C_e) \qquad (3\text{-}13)$$

where dy = differential column length
k = mass transfer rate coefficient, m/l^2t
a = external area of the adsorbent per unit volume, l^{-1}
C_e = equilibrium concentration for the amount of organic adsorbed

Rearranging Eq. (3-13) into a form suitable for integration gives

$$\int_0^y \frac{ka}{F_m} \, dy = \int_{C_b}^C \frac{dC}{C - C_e}$$

Dividing by the integral taken over the adsorption zone gives

$$\frac{y}{\delta} = \frac{V - V_b}{V_x - V_b} = \frac{\displaystyle\int_{C_b}^C \frac{dC}{C - C_e}}{\displaystyle\int_{C_b}^{C_x} \frac{dC}{C - C_e}} \qquad (3\text{-}14)$$

We now have two expressions that must be graphically (or numerically) integrated. Because data are inevitably in short supply, the integrations will be somewhat crude but justified when considering actual process-operation capability. The earlier applied mass-transfer model [Eq. (3-13)] can also be justified on this basis.

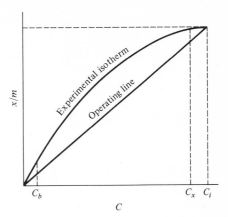

FIGURE 3-9
Operating line and equilibrium isotherm
for adsorption system.

Over the entire column the rate of adsorption can be defined as

$$F_m C_i = p_s \left(\frac{x}{m} \right)_{C_i} \qquad (3\text{-}15)$$

where p_s = superficial or apparent rate of saturation, $m^2/l^5 t$

Equation (3-15) gives one point on the operating curve of the adsorber. We would expect the curve to be linear for x/m versus C, and in most cases, the curve passes very closely to the origin. Thus we can write

$$F_m C = p_s \frac{x}{m} \qquad (3\text{-}16)$$

A plot of the operating curve on the same graph as that of the equilibrium curve is shown in Fig. 3-9.

A horizontal line drawn between the isotherm and the operating line gives the difference between the actual and the equilibrium value of any point. By definition then $C - C_e$ is always greater than zero, and this difference is the driving force for adsorption. If we assume that the operating curve goes through the origin, the curve is defined by the choice of C_i and the characteristic isotherm. This information can then be used for graphical integration to determine the column characteristics as shown in Example 3-2.

EXAMPLE 3-2   An adsorption column 10 m in depth is to be designed for an organic-containing wastewater. Equilibrium data are given in Table 3-1 for the organic adsorption on activated carbon. Wastewater flow is to be 50 l/min-m² (50 kg/min-m²). Organic concentration of the wastewater is expected to be 3000 mg/l ($\pm$ 100 mg/l). Exhaustion is considered to occur when $C = 2800$ mg/l. Separate experiments have shown the mass-transfer coefficient ka to be equal to

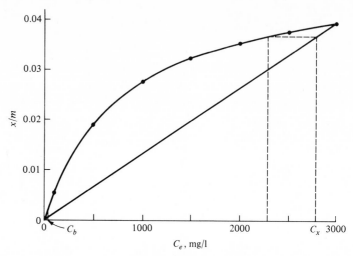

FIGURE 3-10
Equilibrium and operation curve for Example 3-2.

7000 kg/min-m^3 at 20°C, the operating temperature. Dry packed density of the carbon is 300 kg/m^3.

The latter two columns of Table 3-1 are not necessary but can be used to calculate isotherm coefficients. We also see that the experiment was run by adding solute until a selected equilibrium concentration C_e was reached. This is relatively reasonable as gravimetric determinations are, in general, more accurate than wet chemical analysis. Mass determinations are probably only accurate to ± 0.02 mg, however.

Table 3-1 ADSORPTION DATA TAKEN IN 1-LITER BEAKERS USING 1 GRAM OF ACTIVATED CARBON AND VARYING AMOUNTS OF SOLUTE

Mass organic added, mg	C_e, mg/l	x, mg	$(x/m)^{-1}10^{-3}$	C_e^{-1}, (mg/l)$^{-1}$
10.062	10	0.062	16.13	0.100
50.290	50	0.292	3.45	0.020
105.56	100	5.560	0.180	0.010
520.00	500	19.230	0.0520	0.002
1027.78	1000	27.780	0.0360	0.001
1532.61	1500	32.610	0.0307	0.00067
2035.71	2000	35.710	0.0280	0.00050
2537.88	2500	37.880	0.0264	0.00040
3039.47	3000	39.470	0.0253	0.00033

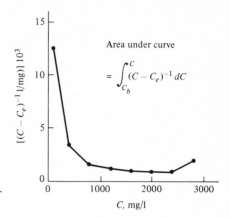

Area under curve

$$= \int_{C_b}^{C} (C - C_e)^{-1}\, dC$$

FIGURE 3-11
Curve for graphical intergration to determine adsorber fractional capacity f.

A plot of x/m versus C_e can now be made, and the operating curve is drawn as shown in Fig. 3-10. These two curves are used to develop the first three columns in Table 3-2 and Fig. 3-11. Values in column 4 are taken from Fig. 3-11, and values in column 5 are those in column 4 divided by 5.338 (the integral value at exhaustion).

The area under the curve in Fig. 3-11 is the integral that can be used to calculate the adsorption zone depth δ.

$$\delta = \frac{F_m}{ka} \int_{C_b}^{C} \frac{dC}{C - C_e} = \frac{50}{7000}\, 5.34 = 0.0381 \text{ m}$$

From Eq. (3-10) we recognize that a plot of $(V - V_b)/(V_x - V_b)$ versus C/C_i will allow the calculation of the fractional capacity f. The cross-hatched area in Fig. 3-12 is the integral of Eq. (3-10), that is, the fractional capacity, and the curve is a modified form of the breakthrough curve.

Table 3-2   **VALUES FOR NUMERICAL INTEGRATION TO DETER-
MINE FRACTIONAL CAPACITY OF ADSORPTION UNIT**

C	C_e	$(C - C_e)^{-1}$	$\int_{C_b}^{C} (C - C_e)^{-1}\, dC$	$\dfrac{V - V_b}{V_x - V_b}$
100 (C_b)	20	0.0125	0	0
400	100	0.00333	2.016	0.378
800	250	0.00182	2.936	0.550
1200	370	0.00120	3.536	0.662
1600	600	0.00100	3.736	0.737
2000	900	0.00091	4.316	0.809
2400	1350	0.00095	4.716	0.883
2800 (C_x)	2300	0.00200	5.338	1.000

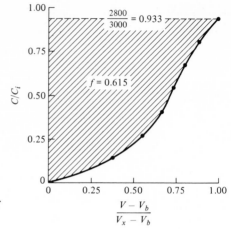

FIGURE 3-12
Determination of fractional capacity of
adsorption column.

Percent saturation at breakthrough is given by Eq. (3-17).

$$\text{Percent sat.} = \frac{L + \delta(f - 1)}{L} 100 \qquad (3\text{-}17)$$

For the example problem, the percent saturation is

$$\text{Percent sat.} = \frac{10 - 0.0381(0.385)}{10} 100$$

$$\approx 100 \text{ percent}$$

The column contains 10 m^3 carbon per square meter of cross section. Mass of the carbon per square meter is 3000 mg. At exhaustion this amount of carbon will have adsorbed $3000(39.47 \times 10^{-3}) = 118$ kg/m^2 of organic material, and we can now calculate the run time before regeneration is necessary and the volume of wastewater passed through to that point.

$$\frac{F_m C_i}{\rho_1} = 50(3.0) = 150 \text{ g/m}^2\text{-min}$$

$$t_b = \frac{118}{150 \times 10^{-3}}$$

$$= 787 \text{ min} = 13.1 \text{ h}$$

$$V_b = 787(50)$$

$$= 39,350 \text{ l/m}^2 \qquad ////$$

Example 3-2 was developed entirely from an isotherm based on batch data. No consideration was given to the possibility of diffusion, dispersion, or channeling.

In large processes, effects of these variables will always be present. If the determination of the mass-transfer coefficients was done on a continuous-flow system, some effects of flow-rate variables would be introduced. For this situation, we would expect that the mass-transfer coefficients would become pseudo-mass-transfer coefficients and hold only over certain flow rate ranges.

Carbon Regeneration

Once exhausted, carbon columns can be regenerated in several fashions. Solvent, acid, or caustic washing is possible and may be feasible in plants where the secondary disposal problem would not be an impediment. The purpose of the wash is to produce a concentrated solution of the sorbate by introducing a liquid that has a higher affinity than carbon for the sorbate. Steam regeneration is also possible. In this case, the increased temperature changes the equilibrium conditions, and the sorbate moves into the gas-liquid phase from the solid phase.

Virtually all current wastewater treatment activated-carbon processes use thermal regeneration, which is accomplished by passing the exhausted carbon through a continuous furnace. Drying takes place in the initial phase followed by baking and/or pyrolysis. Finally the sorbate fractions are oxidized, either becoming part of the activated carbon or being driven off as oxidized gaseous end products.

Carbon losses are generally fairly substantial in a thermal-regeneration process and often are greater than 10 percent per cycle. Culp and Culp[2] report that losses at South Lake Tahoe have averaged about 5 percent and have been as low as 2 percent. They note that in the same operation carbon costs (1970) were about $0.35/lb for new carbon and $0.032/lb for regenerated carbon. Three cents per pound is still an expensive material in a commercial operation, and therefore carbon costs make adsorption an expensive proposition in most cases. Saturation values of x/m are of the order of 2.1 g COD/g carbon, and thus carbon costs (including makeup carbon) will be of the order of $15/million l ($57/mg) treated for a wastewater similar to settled domestic sewage and $1.5/million l ($6/mg) for a good quality secondary effluent. In addition to carbon costs, tankage, power, and pretreatment costs (e.g., filtration for downflow columns) must be considered.[1]

Process Stability

Activated-carbon adsorption has been noted for its ability to undergo "shock" loads with little change in effluent quality. The reason for this is that adsorption, like filtration and ion exchange, is basically an "integrative" process. By integrative it is meant that they tend to sum up the total input over a period of time and thus are not as directly affected by the local time input. An activated-carbon column, ion-exchange unit, or filter has a finite capacity and must be taken out of operation and regenerated when that capacity has been reached. Up to that point the input rate has little effect *within a fairly wide range* of values. Water treatment plant operators have known for many years that they could increase the rate of filtration in their plants by sacrificing run time but without sacrificing product quality. There is a hydraulic limit at which material begins to wash through, of course, but it is usually at least double the conventional design value.

This brings to attention the choice of flow rate for design processes. Quite clearly, an *integrative process* such as a sorption unit, ion-exchange unit, or filter should be designed for average flow. A differential unit such as a settling tank should be designed for peak flow because the settling particle is affected by local velocities.

3-2 ION EXCHANGE

Ion exchange is a process primarily used for selective ion removal and as such is most often used for water treatment rather than wastewater treatment. The process can be used to remove specific toxic ions or nutrients such as NH_4^+ if the environmental hazards are great, for general demineralization of water supplies or for concentration and recovery of costly materials. Mechanisms and fundamental theories associated with ion exchange are discussed by Weber[1] and others[3-5] and will not be discussed in detail here. Basically, the process is similar to an adsorption system in which a solid, usually porous, particle with reactive sites on its surface comes to equilibrium with ions in solution. The sites have an exchangeable ion such as H^+, OH^-, Na^+, or Cl^- attached. When placed in a liquid with ions with a stronger affinity for the reactive (adsorption) site than the attached group, an exchange process takes place that results in an equilibrium with the stronger affinity groups predominating on the adsorption sites. Remember that this is an equilibrium process and is reversible. Therefore, both adsorbable groups will appear in both phases, and the proportion will depend on the equilibrium coefficient. Two important features of this situation are readily apparent. First, we can model ion exchange in much the same manner as for carbon adsorption, and second, we can regenerate ion-exchange materials by altering the equilibrium conditions, e.g., passing a brine of the originally adsorbed group through the exchange bed. This is exactly what is done in home water softeners. A tank of ordinary salt (NaCl) is the brine source. The brine is passed through the exchange bed, and the equilibrium between the attached Ca^{2+}, Mg^{2+}, and Na^+ ions changes to favor Ca^{2+} and Mg^{2+} in solution. The Ca^{2+}- and Mg^{2+}-rich brine is then wasted to the sewer, resulting in a wastewater treatment problem.

Many naturally occurring materials such as clays and aluminosilicates act as ion exchangers. Krone[6] has noted that clays washed down from the Sierra

Nevada and Siskiyou Mountains in California act as an ion-exchange bed for removing toxic materials from San Francisco Bay. Aluminosilicates have been used commercially for water softening,[7] although synthetic materials have largely replaced them today. Synthetic ion-exchange materials became predominant as the result of the work of two English chemists', Adams and Holmes, discovery in 1935 that crushed phenolic phonograph records acted as ion exchangers. This may be a good argument for parents of teenagers being research scientists but more likely was the result of sound scientific reasoning. Since that time, a large number of highly ion-selective synthetic resins have been developed.

Nearly all the commercially available resins are long-chain polymers with a great deal of branching and intertwining to form a matrix that is porous to molecules and ions of certain sizes as well as carrier fluid. Thus the ion selectivity of ion-exchange resins is partially due to a sieving action. A second mode of selectivity is the exchange equilibrium discussed above. In this case, we see a selectivity range; i.e., a resin is highly selective for one molecule, less selective for another, and prefers the originally attached group to a third solute ion. An important aspect of equilibrium selectivity is the magnitude of the solute-ion charge. In general, molecules with higher charge have a greater affinity for a resin. For example, PO_4^{3-} has a higher affinity than SO_4^{2-}, which in turn has a greater affinity than Cl^-. This situation is complicated by the tendency of ions to attract water molecules around them. The hydrated radius of an ion is the effective radius with respect to ion exchange, and therefore ions with smaller hydrated radii are preferentially adsorbed. This is of particular interest in a group of ions of the same charge. Thus for monovalent cations the generally followed adsorption series is

$$Cs^+ > Rb^+ > K^+ > Na^+ > Li^+$$

and for divalent cations

$$Ba^{2+} > Sr^{2+} > Ca^{2+} > Mg^{2+} > Be^{2+}$$

A similar list for monovalent anions is

$$CNS^- > ClO_4^- > I^- > NO_3^- > Br^- > CN^- > HSO_4^- >$$
$$NO_2^- > Cl^- > HCO_3^- > CH_3COO^- > OH^- > F^-$$

Finally, there are considerable effects of chemical interaction between the functional groups and the ions in solution. Affinities are related to the type of functional group and the solute. Thus, when a weak acid is the solute, a sulfonic resin will prefer other monovalent cations to the hydronium ion. A second example is that if the adsorbent-site group is similar to a precipitating ion for a solute ion (e.g., a phosphonic resin and solute Ca^{2+}), there is a marked increase in affinity.

Types of Ion-exchange Resins

We will classify three basic types of resin: cationic, anionic, and mixed. The cationic group includes

$R-SO_3H$	sulfuric
$R-OH$	phenolic
$R-COOH$	carboxylic
$R-PO_3H_2$	phosphonic

The anionic resins include

$R-NH_2$	primary amine
$R-R^1NH$	secondary amine
$R-R_2^1N$	tertiary amine
$R-R_3^1N^+OH^-$	quaternary amine

where in each case R indicates a hydrocarbon polymer and R^1 indicates a specific group, for example, CH_2. The mixed-bed resins are combinations of these two types. Note also that the salts of these resins can serve as the exchange medium and actually more commonly do. A particular example is the use of the sodium salt in water softening because of the feature of simple recharging. A note should be made that this results in a product water with high sodium concentration, which is not recommended for cardiac patients.

Exchange reactions can be classified as strongly acidic, weakly acidic, strongly basic, or weakly basic. Examples are below.

Strongly acidic

$$2R-SO_3H + Ca^{2+} \rightleftharpoons (R-SO_3)_2Ca + 2H^+$$
$$2R-SO_3Na + Ca^{2+} \rightleftharpoons (R-SO_3)_2Ca + 2Na^+$$

Weakly acidic

$$2R-COOH + Ca^{2+} \rightleftharpoons (R-COO)_2Ca + 2H^+$$
$$2R-COONa + Ca^{2+} \rightleftharpoons (R-COO)_2Ca + 2Na^+$$

Strongly basic

$$2R-R_3^1NOH + SO_4^{2-} \rightleftharpoons (R-R_3^1N)_2SO_4 + 2OH^-$$
$$2R-R_3^1NCl + SO_4^{2-} \rightleftharpoons (R-R_3^1N)_2SO_4 + 2Cl^-$$

Weakly basic

$$2R-NH_3OH + SO_2^{2-} \rightleftharpoons (R-NH_3)_2SO_4 + 2OH^-$$
$$2R-NH_3Cl + SO_4^{2-} \rightleftharpoons (R-NH_3)_2SO_4 + 2Cl^-$$

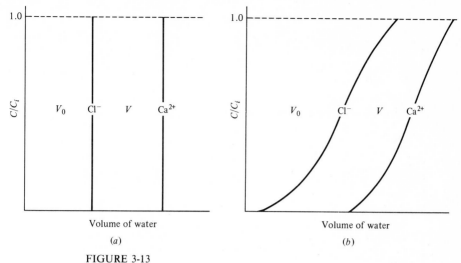

FIGURE 3-13
Ion-exchange-column breakthrough curves. (*a*) Ideal response, (*b*) typical response.

Exchange Capacity

Exchange capacity of resin can be measured as was done with activated carbon. This will result in an isotherm of the type shown in Fig. 3-5. Another method suggested by Tchobanoglous[8] provides the total capacity more directly. An example for a strongly acidic resin is as follows.

EXAMPLE 3-3

(*a*) Put resin in R–H form by washing with HCl.
(*b*) Wash off excess acid with distilled water.
(*c*) Measure resin volume.
(*d*) Add distilled water and titrate with base.

$$\text{Milliequivalents} = (\text{normality of base})(\text{ml of base used}) \qquad (3\text{-}18)$$

$$\text{Exchange capacity} = \text{meq/resin volume} \qquad (3\text{-}19)$$

////

A continuous-flow breakthrough curve is also of considerable interest as discussed in the section on sorption. A typical method of developing the breakthrough curve is to wash the resin in a column with a salt such as NaCl for the acid base, placing the resin in a known form. Excess chloride would then be washed out with water. A known concentration C_i of calcium chloride is then added and the breakthrough curve plotted. Ideally, the curves would be vertical as shown in Fig. 3-13*a*, but because of mass transfer rate limitation, channeling, and dispersion, results similar to Fig. 3-13*b* are usually obtained. In either case, the volume V_0 is the volume of liquid necessary to fill the pores, and the total capacity is given by VC_i. In practice the curves are normally considered to be symmetrical about their midpoints, allowing the volumes to be calculated from those values.

As with carbon adsorption, the properties of the breakthrough curve are of interest and are determined in a similar manner.

3-3 MEMBRANE PROCESSES

Membranes are materials that selectively stop or slow the passage of particular types of molecules (for example, stopping NaCl but not H_2O). They may consist of dry solids, solvent swollen gels, or immobilized liquids and for practical purposes can be considered to be constructed of polymers. In general, membranes of interest have a highly porous structure, although the pores may be as small as 10 Å in size. Pore shape is generally irregular, although certain gel membranes have both a highly regular and highly uniform diameter.[9] Basic mechanisms of selectivity by uncharged membranes are based on sieving, diffusion, and solubility effects. Charged membranes also select by ion charge and therefore can be used as ion-exchange systems.

The classical membrane process is osmosis. Here the potential of two solutions separated by a membrane is different because of a difference in salt concentrations. Water diffuses through the membrane until the potential is equal on both sides (Fig. 3-14*a*). There is an interaction between the pressure potential and the chemical potential caused by the concentration gradient, which results in balance being made by a combination of the two. Osmotic pressure data for several solutions are presented in Table 3-3. The osmotic process can be reversed [i.e., reverse osmosis (RO)] by increasing the pressure on the saline side of the membrane. A similar result can be achieved by introducing an electrical potential across the membrane (electrodialysis).

Selectivity of a given membrane is strongly affected by the method of manufacture. For example, cellulose acetate membranes are normally about 100 μm thick with a 0.2-μm thick surface skin that is much denser than the core, i.e., less porous or with smaller pores. This skin is formed during casting by exposing the surface to air. At this stage of manufacturing, the skin allows permeation of sodium chloride. Further treatment of the membrane by an annealing process increases the density of the skin and greatly decreases the permeability to sodium chloride. Because selectivity is basically a property of manufacturing methods, the subject will not be discussed further here. The manufacturers' literature is the best source of information on selectivity of specific membranes.

Flux of water through a membrane is proportional to the pressure gradient

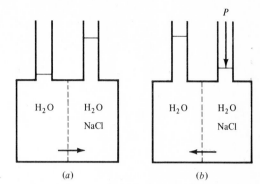

FIGURE 3-14
(a) Osmosis; (b) reverse osmosis (RO).

(Hagen-Poiseville type of flow) and can generally be described by Eq. (3-20).

$$F_m = W(\Delta P - \Delta P_o) \qquad (3\text{-}20)$$

where F_m = flux of water, g/cm^2-min
 ΔP = imposed pressure differential
 ΔP_o = osmotic pressure differential
 W = coefficient involving temperature, membrane thickness, and concentration of water, g/dyn-min

Some solute passes through the membrane by molecular diffusion in most cases, and the flux is proportional to the concentration gradient.

$$N_i = K_i \Delta C_i \qquad (3\text{-}21)$$

Table 3-3[†] OSMOTIC PRESSURES OF SELECTED SOLUTIONS AT 25°C, MPa[‡]

Solute mole fraction × 10³	Sucrose-water	Glycerol-water	NaCl–water	MgSO₄–water
1.798	0.248	0.248	0.462	0.296
3.590	0.503	0.496	0.917	0.558
5.375	0.758	0.745	1.372	0.800
7.154	1.020	0.993	1.820	1.048
8.927	1.282	1.248	2.282	1.289
10.693	1.551	1.496	2.744	1.538
12.453	1.827	1.744	3.213	1.793
14.207	2.103	1.999	3.682	2.055
15.955	2.379	2.248	4.158	2.317
17.696	2.668	2.503	4.640	2.606

† Data are from Ref. 10.
‡ One megapascal (MPa) = 145 psi.

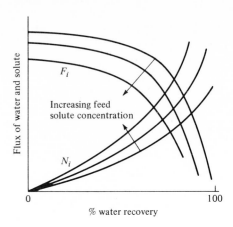

FIGURE 3-15
Solute- and water-flux relationships.

where N_i = flux of the solute i, mol/cm²-min
$\quad C_i$ = molar concentration of the solute i, mol/cm³
$\quad K_i$ = overall mass-transfer coefficient that includes effects of membrane thickness, cm/min

This latter coefficient is often referred to as the *coefficient of permeability*.

Because water flux is related to the osmotic pressure, flux decreases as water recovery (the fraction of feedwater in the product stream) increases. Thus, for constant applied pressure, we find that a practical limit on recovery will exist as shown in Fig. 3-15. Solute flux increases with product recovery for the same reason—increasing salt concentration in the wastewater side of the membrane, thus increasing the gradient across the membrane.

A measure of process efficiency often used is solute rejection ξ_i, the fraction of solute i remaining in the feed stream.

$$\xi_i = \frac{C_{if} - C_{ip}}{C_{if}} \qquad (3\text{-}22)$$

Solute rejection has the unfortunate property of not being based on a mass balance, and as a consequence, there is no information in the term on the mass input and output rates. A better method of defining an efficiency term would be Eq. (3-23) based upon Fig. 3-16.

$$E = \frac{C_{ib}}{C_{if}}\left(1 - \frac{Q_p}{Q}\right) \qquad (3\text{-}23)$$

where C_{ib} = concentration of solute i in brine
$\quad C_{ip}$ = concentration of solute i in product stream

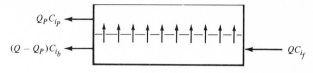

FIGURE 3-16
Simplified RO process schematic diagram.

In terms of the flux rates defined by Eqs. (3-20) and (3-21), this efficiency term becomes

$$E = \left[\left(\frac{2W(\Delta P - \Delta P_o)}{K} + 1\right) \frac{C_i}{C_{if}} - 1\right]\left[1 - \frac{AW(\Delta P - \Delta P_o)}{Q}\right] \qquad (3\text{-}24)$$

where A = membrane area
K = coefficient of permeability

and the solute flux is assumed to be related to the average concentrations $[(C_{if} + C_{ib})/2$ and $C_{ip}]$ on each side of the membrane.

Equations (3-23) and (3-24) provide direct information on the actual water production efficiency. When E approaches 1, there is *both* a high rejection as defined by Eq. (3-22) and a high fraction of input water going to product water.

A number of factors affect rejection or conversely rate of permeation for a specific solute. As mentioned previously, the membrane structure is the basic selection and control mechanism. Other characteristics of the system, particularly interaction between solutes or solute co-ions, are also very important. Multivalent ions are rejected better than monovalent ions, and undissociated or partially dissociated materials are, in general, poorly rejected except by sieving. Acids and bases are not as well rejected as might be expected from the discussion above and the preceding discussion of ion exchange.

In addition to these factors, there is the physical problem of concentration gradients resulting from the separation occurring at the membrane-liquid interface. Thus a higher-than-bulk-liquid concentration exists at the interface, and it is this concentration that controls the solute flux across the membrane. A second negative feature of this situation is that the osmotic pressure is also increased by the concentration at the surface. As a result, water flux decreases because $\Delta P - \Delta P_o$ decreases, and solute flux increases since ΔC is greater.

Other operational problems with reverse osmosis (RO) units include membrane binding due to turbidity in the feed stream, bacterial slime growth, plugging by oil or grease, precipitation of materials such as calcium carbonate and/or sulfates, or negative effects of low or high pH on acetate membrane hydrolysis. Slime growths are a particular problem because microorganisms prefer to grow on surfaces, and the RO system is ideal for this purpose. Together these problems mean that a carefully controlled, sterile feed stream is highly desirable.

Ultrafiltration

Ultrafiltration is quite similar in concept to RO except that pore sizes in the membrane are large enough to eliminate osmotic pressure as a factor. The product water flux then becomes

$$F_W = K_u \Delta P \qquad (3\text{-}25)$$

For convenience, the mass-transfer coefficient K_u is often given in terms of the membrane resistance ψ_m. This is particularly useful because the buildup of separated material on the membrane surface as a gel imposes a second resistance to the water flux ψ_g. Water flux can be expressed in an analogous fashion to electric current.

$$F_W = \frac{\Delta P}{\psi_m + \psi_g} \qquad (3\text{-}26)$$

Ultrafiltration is useful in separating solutes with molecular weights greater than 500. This, as was noted, allows use of high-permeability membranes that eliminate osmotic pressure effects. A second result is that materials building up on the surface do not diffuse away rapidly because their diffusion coefficients are much lower than salts such as NaCl.

In practice, ultrafiltration is used for removal or concentration of macromolecules such as proteins, enzymes, starches, and other organic polymers. Many of these compounds have molecular weights greater than 1 million and thus are well suited to the process. Obviously, many of the same operating problems exist in both cases.

Process Design

Until 1970, cellulose acetate membranes were almost exclusively used for RO systems treating either water or wastewater. In 1970 a new polymer was introduced by E. I. du Pont de Nemours under the trade name Permasep. This membrane is an aromatic polyamide hollow fiber and used in the configuration shown in Fig. 3-17. Properties of this membrane are very suitable for water and wastewater treatment processes, and Permasep systems are now marketed by a number of equipment manufacturers. These systems are sold as modules on a turn-key basis. Expansion of the treatment component of the system is done by adding modules.

Since 1970, a number of applications of hollow-fiber RO systems have been made. Greenfield, Iowa,[11] has installed a 570 m^3/d (150,000 gpd) RO plant to treat a brackish water with a TDS concentration of 2200 mg/l and a hardness of 600 mg/l. Product water has a TDS value below 200 mg/l and a hardness of less than 20 mg/l. Total cost of the Greenfield system has been about $0.20/m^3$ ($0.77/1000 gal) based on $0.01/kWh power costs, and the capital cost of the membranes and auxiliaries was $94,300.

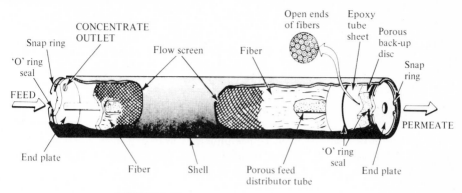

FIGURE 3-17
Cutaway drawing of Permasep Permeator. (*E. I. du Pont de Nemours & Co.*)

Plating-rinse treatment by hollow-fiber RO units has also proven successful.[12] The Permasep membrane withstands continued operation at pH values in the 10 to 11 range and gives high rejection and permeate production values. Typical values reported are presented in Table 3-4.

Use of RO Systems as Concentrators

The point that RO systems produce a brine waste stream has already been made. In some cases, this brine stream will contain material concentrations great enough for economic recovery. Industrial wastewaters from plating or

Table 3-4 PERMEATE FRACTION AND REJECTION FOR VARIOUS SOLUTES WITH PERMASEP MEMBRANE†

Solute	Feed concentration, mg/l	pH	Rejection	Ref.
Copper	793	10–11	0.99	12
Zinc	144	10–11	0.99	12
Sodium cyanide	1000	10–11	0.92	12
Nitrate ion	1500	7.0	0.80	13
Nitrate ion	1500	11.1	0.86	13
Ammonium ion	500	6.9	0.85	13
Ammonium ion	500	3.2	0.84	13
Formic acid	500	3.2	0.50	14
Acetic acid	500	3.7	0.40	14
Sodium acetate	680	8.1	0.98	14
Phenol	2000	7–9	0.55	15
Phenol	2000	11	0.95	15

† Standard operating conditions were 2.76 MPa (400 psi) and 75 percent influent going to permeate.

electronics manufacturing operations are possible applications of RO with materials recovery. Combining water or wastewater treatment with an economically justifiable process is obviously beneficial from many viewpoints.

3-4 CHEMICAL OXIDATION

Chemical oxidation of dissolved compounds has application in both water and wastewater treatment. Examples include oxidation of taste- and odor-causing compounds in water with permanganate, chlorine, or ozone, oxidation of iron and manganese to insoluble ferric and manganic ions in water treatment, and breakpoint chlorination for nitrogen removal from wastewater.

Chemical Oxidation of Organic Materials

Chemical oxidation or organic compounds is relatively expensive. The method is generally limited to the removal of low concentrations of organics or to contaminants that cannot be removed by less expensive methods. As noted above, taste and odor problems in water supplies can often be controlled by the addition of permanganate (MnO_4^-), chlorine, or ozone. Chlorine is often used for the pretreatment of wastewaters also. Prechlorination reduces organic concentration somewhat but also retards biooxidation. The latter effect is the more important.

Ozone can be used in a manner similar to chlorine. Because ozone is unstable, the gas must be generated on site. Ozone generators utilize a corona discharge, which occurs when a high-voltage ac current is imposed across a discharge gap (Fig. 3-18). The method is inefficient with only about 10 percent

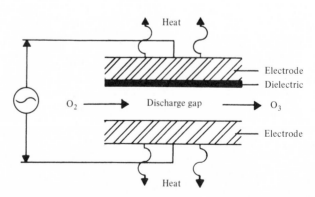

FIGURE 3-18
Schematic diagram of basic ozone-generator configuration. (*W. R. Grace and Co.*)

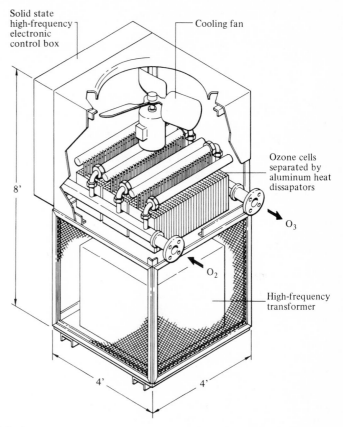

FIGURE 3-19
Commercial ozone generator with capacity of approximately 650 kg/d ozone.
(*W. R. Grace and Co.*)

of the applied energy going into the ozone. Heat produced must be removed because decomposition of ozone to oxygen is accelerated at higher temperatures. Commercial ozone generators can be purchased in various sizes from a number of manufacturers. A sketch of a typical commercial unit is shown in Fig. 3-19.

Chemical oxidation can also be used to partially break down difficult-to-degrade organics such as phenolic compounds. After chemical oxidation, these compounds may be more amenable to conventional treatment procedures. Ozonation of secondary (biological) treatment plant effluents has been considered as a method of reducing final effluent organic concentrations.

In most cases, chemical oxidation of organic compounds is an alternative to carbon adsorption. Reported costs are similar,[16] and choices between alternatives must be based on characteristics of the particular problem.

Oxidation of Inorganic Compounds

Chemical oxidation of inorganic compounds covers a broad range of reactions and objectives. Examples include removal of iron and manganese by oxidation from a soluble to an insoluble state, oxidation of ammonium ion to molecular nitrogen, and oxidation of odor-causing compounds such as sulfides.

Oxidation of iron and manganese is of particular importance in domestic water treatment because these compounds result in both taste and color problems. Both iron and manganese are unstable in their $+2$ states in the presence of oxygen. Simple aeration at neutral pH values for iron and pH values above 9.5 for manganese results in oxidation at high rates.

$$2Fe^{2+} + \tfrac{1}{2}O_2 + 5H_2O \rightarrow 2Fe(OH)_3 + 4H^+$$
$$Mn^{2+} + \tfrac{1}{2}O_2 + H_2O \rightarrow MnO_2 + 2H^+$$

Permanganate also acts as an oxidant and is often used rather than aeration.

$$3Fe^{2+} + MnO_4^- \rightarrow 3Fe^{3+} + MnO_2$$
$$3MN^{2+} + 2MnO_4^- \rightarrow 5MnO_2$$

Ferric hydroxide and manganic oxide precipitates are often removed on sand or multimedia filters. This procedure has an added benefit in that the precipitates adsorb unoxidized Fe^{2+} and Mn^{2+} ions. Thus, as precipitates accumulate in a filter, removal of iron and manganese improves.

Nitrogen Removal by Breakpoint Chlorination

Hypochlorous acid reacts with ammonia to form the chloramines:

$$NH_3 + HOCl \rightarrow NH_2Cl + H_2O \qquad \text{monochloramine}$$
$$NH_2Cl + HOCl \rightarrow NHCl_2 + H_2O \qquad \text{dichloramine}$$
$$NHCl_2 + HOCl \rightarrow NCl_3 + H_2O \qquad \text{trichloramine}$$

Addition of chlorine in amounts up to 1 mol/mol ammonia initially present results in the formation of both monochloramine and dichloramine. Increasing the molar ratio results in formation of some trichloramine and oxidation of some chloramine to nitrogen gas (N_2). When a molar ratio of approximately $2:1$ is reached, both nitrogen and chlorine are nearly eliminated from solution. Adding chlorine to this "breakpoint" is, therefore, a method of removing ammonia from solution.

3-5 PRECIPITATION

Removal of inorganic ions from water and wastewater by precipitation is a commonly used procedure. Precipitation of iron and manganese was discussed in the previous section on chemical oxidation. In this section, discussion of precipitation will be based on the law of mass action [Eq. (3-27)] for the general reversible reaction.

$$v_1 A + v_2 B \rightleftharpoons v_3 C + v_4 D$$

$$\frac{[C]^{v_3}[D]^{v_4}}{[A]^{v_1}[B]^{v_2}} = K_e \qquad (3\text{-}27)$$

where K_e is the equilibrium constant, A, B, C, and D are reactant species, the brackets, [ ], signify molar concentration, and the v_i values are stoichiometric coefficients.

Ionic equilibrium in water is extremely complex. Individual ions are in equilibrium with many different materials. The objective of precipitation is to produce a situation in which the dominant form is insoluble. Perhaps the most common example is calcium carbonate equilibria.

Calcium Carbonate Equilibria

Calcium and carbonate exist in complex equilibria even when they are the only solutes in water. The pertinent equations are as follows.

	Dissociation constant, 20°C	
$CO_2 + H_2O \rightleftharpoons H_2CO_3$	3.89×10^{-2}	(3-28)
$H_2CO_3 \rightleftharpoons HCO_3^- + H^+$	4.17×10^{-7}	(3-29)
$HCO_3^- \rightleftharpoons CO_3^{2-} + H^+$	4.17×10^{-11}	(3-30)
$CaCO_3 \rightleftharpoons Ca^{2+} + CO_3^{2-}$	5.25×10^{-9}	(3-31)
$CaCO_3 \rightleftharpoons Ca^{2+} + HCO_3^-$	1.26×10^2	(3-32)
$CaCO_3 + H_2CO_3 \rightleftharpoons Ca^{2+} + 2HCO_3^-$	5.25×10^{-5}	(3-33)
$H_2O \rightleftharpoons H^+ + OH^-$	10^{-14}	(3-34)

Calcium carbonate has a very low solubility; therefore precipitation of calcium is favored by producing a situation in which calcium exists primarily as $CaCO_3$. This can be accomplished by increasing the carbonate concentration and forcing the equilibrium of Eq. (3-31) to the left. In a given water a specific quantity of carbonate exists and is distributed among the various forms: CO_2, H_2CO_3, HCO_3^-, and CO_3^-. The relative amount of each form present is dependent upon pH. These values can be computed for suspensions of $CaCO_3$ in pure distilled water,[17] and the computed values have some application in predicting stabilization of waters. A stable water is one that does not tend to dissolve solid surface coatings of $CaCO_3$ (and thus expose surfaces to corrosion) or lay down a precipitate that may clog pipes or equipment.

Accurate equilibrium calculations are dependent on the extent of knowledge of water constituents. Such calculations are useful in predicting long-term effects such as may occur in lakes or in the ocean but are not particularly useful in process control.

The conclusion that can be drawn from observation of the equilibria equations and their dissociation constants is that raising the pH pushes the carbonate equilibrium more strongly toward CO_3^{2-}. For example, the approximate distribution of carbonate forms can be calculated from Eqs. (3-29) and (3-30), assuming that H_2CO_3 incorporates all CO_2 in solution. A selected range of values is presented in Table 3-5.

Precipitation of Ca^{2+} can, therefore, be accomplished by raising the pH to values above 10. At these pH values, CO_3^{2-} will be available to combine with Ca^{2+} and form a precipitate. Because of the small dissociation constant of $CaCO_3$, the solid form is extremely dominant.

Lime–Soda Ash Water Softening

An excellent example of precipitation in water treatment is the lime–soda ash water-softening process. Hardness in water is due to the presence of divalent cations. The principal hardness cations are Ca^{2+} and Mg^{2+}, although Sr^{2+}, Fe^{2+}, and Mn^{2+} are significant in some cases. Commonly associated anions are HCO_3^-, SO_4^{2-}, Cl^-, NO_3^-, and SiO_3^{2-}.

Hardness is an important factor in both municipal and industrial water treatment, and a standard nomenclature has been developed for hardness and water softening. Because the objective of this section is to use softening as an example of precipitation, the conventional nomenclature will not be used.

Table 3-5 DISTRIBUTION OF H_2CO_3, HCO_3^-, AND CO_3^{2-} AS A FUNCTION OF pH AT 20°C

pH	Distribution fraction		
	H CO_3	HCO_3^-	CO_3^{2-}
5	9.6×10^{-1}	4.0×10^{-2}	1.7×10^{-7}
6	7.1×10^{-1}	2.9×10^{-1}	1.2×10^{-5}
7	1.9×10^{-1}	8.1×10^{-1}	3.4×10^{-4}
8	2.3×10^{-2}	9.7×10^{-1}	4.1×10^{-3}
9	2.3×10^{-3}	9.6×10^{-1}	3.4×10^{-2}
10	1.7×10^{-4}	7.1×10^{-1}	2.9×10^{-1}
11	4.6×10^{-6}	1.9×10^{-1}	8.1×10^{-1}
12	5.6×10^{-8}	2.3×10^{-2}	9.8×10^{-1}

In removing hardness the equations for calcium carbonate equilibria can be used. Three additional relationships must be included that account for the removal of magnesium hardness.

$$Mg^{2+} + CO_3^{2-} + Ca(OH)_2 \rightleftharpoons CaCO_3 + Mg(OH)_2 \qquad (3\text{-}35)$$

$$Mg^{2+} + SO_4^{2-} + Ca(OH)_2 \rightleftharpoons Ca^{2+} + SO_4^{2-} + Mg(OH)_2 \qquad (3\text{-}36)$$

$$Ca^{2+} + SO_4^{2-} + Na_2CO_3 \rightleftharpoons CaCO_3 + 2Na^+ + SO_4^{2-} \qquad (3\text{-}37)$$

where the SO_4^{2-} is a common co-ion but not a necessary one.

Equation (3-37) is necessary (i.e., the addition of $NaCO_3$) because of the soluble calcium added when Mg^{2+} is precipitated.

EXAMPLE 3-4   A water with characteristics listed in the following table is to be softened by the lime–soda ash method.

Constituent	Concentration	
	mg/l	meq/l
Ca^{2+}	120	6.00
Mg^{2+}	30	2.47
Na^+	10	0.43
HCO_3^-	428	7.03
SO_4^{2-}	20	0.20
Cl^-	58	1.64
Total TDS	666	
pH = 8.0		

An ionic balance can be made by adding up the positive and negative equivalent weights. These two sums are equal within satisfactory limits, and therefore the list of constituents is satisfactory.

To precipitate the Ca^{2+} and Mg^{2+}, a base must be added to raise the pH. The least expensive base is lime, CaO, which must be hydrated or slaked to $Ca(OH)_2$ before application. The quantity of Ca^{2+} and Mg^{2+} that can be precipitated by lime addition alone can be calculated as follows:

$$Ca^{2+} + 2HCO_3^- + Ca(OH)_2 \rightarrow 2CaCO_3 + 2H_2O$$
$$Mg^{2+} + 2HCO_3^- + 2Ca(OH)_2 \rightarrow 2CaCO_3 + Mg(OH)_2 + 2H_2O$$
$$Mg^{2+} + Ca(OH)_2 \rightarrow Mg(OH)_2 + Ca^{2+} + 2H_2O$$

For the reaction forming $Mg(OH)_2$ to go to completion (i.e., to the point where nearly all the Mg^{2+} is in the form of magnesium hydroxide precipitate), a high concentration of OH^- must exist. This requires a pH of 10.8 or above. To achieve this, in practice an excess of lime of 35 mg/l is added.

The lime (as CaO) necessary can be estimated as follows:

1   Calcium precipitation consumes 6 meq/l Ca^{2+}, HCO_3^- and CaO.
2   Lime addition required for magnesium precipitation is 2.06 meq/l for the first reaction and 1.44 meq/l for the second reaction.
3   Excess lime (as CaO) added of 35 mg/l = 1.25 meq/l.
4   Total lime added = 10.8 meq/l = 302 mg/l.

The excess lime added can be removed in a second stage by recarbonation with CO_2.

$$Ca^{2+} + 2OH^- + CO_2 \rightarrow CaCO_3 + H_2O$$

Maintaining the pH above 9.5 will prevent substantial amounts of soluble calcium bicarbonate being formed. Carbon dioxide can be added by sparging with the rate controlled by a pH meter with feedback control. The quantity of CO_2 necessary in this example would be $\frac{44}{56}(35) = 27.5$ mg/l.

At this point in the softening process, calcium remains in solution as the result of the second magnesium-precipitation reaction. In addition, some calcium will not have precipitated or settled out in an actual process and $Mg(OH)_2$ is soluble to about 9 mg/l. Calcium can be removed by adding soda ash (Na_2CO_3). Soda ash is expensive, and usually only enough is added to bring the total hardness down to about 50 to 80 mg/l as $CaCO_3$ (the conventional units for hardness). In this example, 1.44 meq/l (29 mg/l) Ca^{2+} is in solution as the result of the second magnesium-precipitation reaction. This corresponds to 73 mg/l of hardness as $CaCO_3$. Solubility of calcium carbonate is about 17 mg/l, and therefore removable $CaCO_3$ is approximately 44 mg/l. To achieve this removal, all the Ca^{2+} must be converted to $CaCO_3$, which requires 1.44 meq/l (60 mg/l) of Na_2CO_3.

The solubility of $Mg(OH)_2$ is about 9 mg/l, which corresponds to 15 mg/l hardness as $CaCO_3$. With the additional $CaCO_3$ in solution and allowing for incomplete precipitation, the final hardness will be over 50 mg/l.

Because the pH of the water is above 9.5, recarbonation must be used to produce a product in an acceptable pH range. The most important reason for this pH reduction is to prevent precipitation of $CaCO_3$ downstream. Reduction of the pH to a value of 8.6 is usually satisfactory. At this pH value, the solubility product of $CaCO_3$ is usually greater than $[Ca^{2+}][CO_3^{2-}]$.   ////

A schematic diagram of the lime–soda ash softening process is shown in Fig. 3-20. Filtration is necessary because of residual, unsettled precipitate in the water. The system is approximately twice as expensive as a conventional water-filtration plant. Operating costs are considerably higher because of the amount of chemical addition necessary and the increased sludge-disposal problem.

Lime–soda ash water softening provides an excellent example of precipitation in water and wastewater treatment because of the type of operational steps

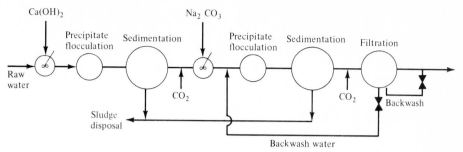

FIGURE 3-20
Schematic diagram of water-softening system using the lime–soda ash process.

involved. Many ions can be precipitated by modifying the pH of the water. Interaction of the constituents is also important. For example, cyanide can be precipitated at high pH values, and chromium can be precipitated at low pH values. The two species must be removed in separate steps where they occur together.

PROBLEMS

3-1 An industrial waste has an average total organic carbon concentration of 100 mg/l and an average flow rate of 50 l/s. The pH of the wastewater is 4.4, and therefore no adjustment is necessary for activated-carbon adsorption. Three carbons have been studied. The results of batch-equilibrium experiments are presented in Table 3-6, and the results of rate studies are presented in Table 3-7.

Table 3-6 EQUILIBRIUM DATA FROM 1-LITER BATCH
SYSTEM USING 1 GRAM OF CARBON

Initial TOC, mg/l	Carbon A	Carbon B	Carbon C
10	0.523	0.003	0.498
20	1.161	0.045	1.047
40	2.944	0.93	2.321
80	11.09	10.83	5.887
160	66.95	58.18	22.17
320	222.2	186.7	133.90
640	540.9	474.5	444.40
1280	1180.	1080.5	1081.81
2560	2460.	2322.5	2360.8

Table 3-7 CARBON-ADSORPTION RATE
DATA FROM BATCH EXPERI-
MENTS
Unit volume = 1 l
Mass of activated carbon = 1 g
Initial TOC concentration = 320
mg/l

Time, s	TOC, mg/l		
	A	B	C
1	313	312	311
2	307	305	302
4	296	292	286
8	278	274	258
16	254	239	216
32	233	207	170

Compare the three carbons with respect to:
(a) Equilibrium coefficients
(b) Mass-transfer coefficients
(c) Fractional capacity
(d) Time to breakthrough in a 10-m column with a mass liquid flux of 1 kg/m^2-s and assuming a dry packed density of the carbon to be 300 kg/m^3 in all cases

3-2 Many carbon-adsorption processes are designed as pairs of columns operating in series. The first column produces an effluent with an organic concentration greater than breakthrough but less than the influent concentration, while the second column produces the final product. A third column must be included to allow one column to be undergoing regeneration at all times.
(a) What limitations relating to column effectiveness, adsorption-zone length, and channeling are important in determining the effectiveness of this procedure?
(b) Would this procedure have any advantage over a fluidized bed with continuous regeneration?

Table 3-8 CARBON-TOC EQUILIBRIUM
DATA
Liquid volume = 1 l
Total carbon mass = 1 g

C_i, mg/l	C_e, mg/l
10	0.10
20	0.20
30	0.31
40	0.41
50	0.52
60	0.62

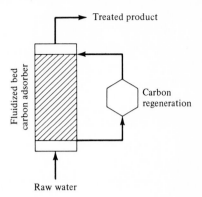

FIGURE 3-21
Proposed adsorption system for Prob. 3-3.

3-3 A continuous-flow, continuous-carbon regeneration process is to be designed for a wastewater with a flow rate of 50 l/s and a carbon (TOC) concentration of 350 mg/l. Effluent requirements state that effluent TOC concentrations must be below 3 mg/l. Data on the adsorption process are shown in Table 3-8 and a process schematic is shown in Fig. 3-21. Determine the regeneration rate (g carbon/min).

3-4 An activated-carbon-adsorption process pilot plant has been designed from batch data. The pilot plant is 5 m long and has a diameter of 30 cm. Flow rates for these studies were 0.71 m³/s and 1.42 m³/s. Breakthrough occurs when an effluent concentration of 5 mg/l TOC is attained. Influent TOC concentration was 125 mg/l. Breakthrough volume predicted from the batch experiments was 23,000 m³. The pilot plant data are presented in Table 3-9. Give two possible explanations for the results and a method of determining the validity of the explanations.

3-5 Would the volume V_0 in Fig. 3-12*b* be expected to equal the true pore volume or some fraction of the true pore volume?

Table 3-9

Time, min	TOC, mg/l	
	0.71 m³/s	1.42 m³/s
50	0.0	4.7
100	0.0	23.4
150	0.3	61.2
200	0.5	91.5
250	4.0	109.3
300	17.0	115.4
350	37.0	116.5
400	85.7	118.1
450	106.9	120.1

3-6 An ion-exchange pilot plant 2 m long and having a diameter of 0.1 m was set up to remove calcium from a wastewater and has been run at four flow rates: 5, 10, 20, and 40 m^3/min-m^2. Bulk dry density of the exchange resin is 750 g/l, and the batch-tested capacity is 0.8 meq/g. Wastewater calcium chloride concentration is 375 mg/l. Using the continuous-flow data in Table 3-10, select an operating flow rate and justify your reasoning.

3-7 Derive Eq. (3-24).

3-8 A wastewater contains a sucrose concentration of 900 mg/l and has a flow rate of 50 l/min. Allowable-product sucrose concentration is 5 mg/l, and if the brine is above 3000 mg/l, recovery becomes profitable. Determine the operating region with respect to K, W, and A. Assume that a single RO unit will be used and that the concentration

Table 3-10 EFFLUENT-ION CONCENTRATIONS, mg/l

Time, min	Loading rate, m^3/m^2-min							
	5		10		20		40	
	Cl^-	Ca^{2+}	Cl^-	Ca^{2+}	Cl^-	Ca^{2+}	Cl^-	Ca^{2+}
0	0	0	0	0	0	0	0	0
5	0	0	15	0	196	0	217	0
10	20	0	175	0	239	0	239	0
15	90	0	240	0	239	0	239	2
20	160	0	239	0	240	0	240	51
25	220	0	240	0	239	0		100
30	237	0	239	0		2		125
60	238	0	239	0		14		130
90	239	0		0		57		131
120	240	0		0		102		132
150						133		134
180						135		135
210						234		135
240						135		135
270				0				135
300				5				
330				10				
360				75				
390				118				
420				128				
450				134				
				135				
630		0						
660		5						
690		10						
720		20						
750		40						
780		80						
810		120						
840		135						

averages used in the derivation of Eq. (3-24) are valid here. Use an applied pressure differential of 5.52×10^{-7} dyn/cm^2 (800 psig) and determine the relationship between pressure and the other variables (for example, K_i, ΔP^x).

GENERAL READING

1. STUMM, W., and J. J. MORGAN: "Aquatic Chemistry," Wiley-Interscience, New York, 1970.
2. WEBER, W.: "Physicochemical Treatment of Wastewaters," John Wiley & Sons, Inc., New York, 1973.
3. SAWYER, C. N., and P. L. MCCARTY: "Chemistry for Sanitary Engineers," 2d ed., McGraw-Hill Book Company, New York, 1967.
4. COULSON, J. M., and J. R. RICHARDSON: "Chemical Engineering," vols. I, II, Pergamon Press, New York, 1968.

REFERENCES

1. WEBER, W.: "Physicochemical Treatment of Wastewaters," John Wiley & Sons, Inc., New York, 1973.
2. CULP, R. L., and G. L. CULP: "Advanced Wastewater Treatment," Van Nostrand Reinhold Company, New York, 1971.
3. DORFNER, K.: "Ion Exchangers," Ann Arbor Science Publishers, Inc., Ann Arbor, Mich., 1972.
4. ANDERSON, R. E.: A Contour Map of Anion Exchange Resin Properties, *J. Ind. Eng. Chem.*, Prod. Res. Dev., vol. 3, p. 85, 1964.
5. BECKER-BOOST, E. H.: Process Technology of Ion Exchange, *Chem.-Ing. Tech.*, vol. 28, p. 579, 1955.
6. KRONE, R. B.: Predicted Suspended Sediment Inflows to the San Francisco Bay System, prepared for Central Pacific Basins Comprehensive Water Pollution Control Project, FWPCA Southwest Region, 1966.
7. SAWYER, C. N., and P. L. MCCARTY: "Chemistry for Sanitary Engineers," 2d ed., McGraw-Hill Book Company, New York, 1967.
8. TCHOBANOGLOUS, G.: Notes on Water and Wastewater Treatment, limited circulation document, University of California, Davis, 1973.
9. KESTING, R. E.: "Synthetic Polymeric Membranes," McGraw-Hill Book Company, New York, 1971.
10. SOURIRAGAN, S.: "Reverse Osmosis," Academic Press, Inc., New York, 1970.
11. MOORE, D. H.: Operation of a Reverse Osmosis Desalting Plant at Greenfield, Iowa, presented at a panel discussion of Operation Experiences with Desalting Plants, 92d Annual Conference, AWWA, Chicago, Ill., June 4–8, 1972.
12. MATTAIR, R., and J. B. KELLER: Closed Loop Recovery of Nickel Plating Rise, presented at the 75th National Meeting, AIChE, Detroit, Mich., June 4, 1973.

13.   PERMASEP PERMEATORS, *Du Pont Tech. Bull.* 403, May 8, 1972.

14.   PERMASEP PERMEATORS, *Du Pont Tech. Bull.* 401, Feb. 1, 1972.

15.   PERMASEP PERMEATORS, *Du Pont Tech. Bull.* 405, Nov. 13, 1972.

16.   ROSEN, H. M.: Ozone Generation and Its Economical Application to Wastewater Treatment, *Water Sewage Works*, vol. 119, pp. 9, 114, 1972.

17.   FAIR, G. M., J. C. GEYER, and D. A. OKUN: "Water and Wastewater Engineering," John Wiley & Sons, Inc., New York, 1968.

4

GAS TRANSFER

Interphase gas transfer is of importance in aerobic biological wastewater treatment processes and in gas stripping. Biological treatment is the most common method of treating domestic and other organic-containing wastewaters, and because of this aeration, the transfer of oxygen from the gas to the liquid phase is of primary interest here. Gas stripping, typified by ammonia removal, is of interest in a limited number of cases and will also be discussed.

Two types of aerobic biological treatment processes are in widespread use: fluidized cultures and film-flow reactors. In the fluidized-culture processes, air or in some cases oxygen is distributed through the liquid volume as small bubbles, thus providing a large amount of interphase surface contact area throughout the tank. Tank components are mixed by the air bubbles or a combination of air and mechanical mixing mechanisms (Fig. 4-1a). In recent years, mechanical surface aerators (Fig. 4-1b,c) have become increasingly common because of relatively low capital costs, ease of maintenance, and the move toward design of stirred-tank reactors. Similar geometries can also be used for gas stripping.

Film-flow reactors can be visualized as shown in Fig. 4-2. Oxygen dissolves in the liquid film and is transported to the reactive microbial film. Reactions

(a)

(b)

(c)

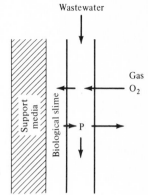

FIGURE 4-2
Gas transfer of oxygen and reaction products (P) in film-flow reactor.

take place in the microbial film, and products are released into the liquid. Gaseous end products are released back into the gas phase from the liquid. In the case of gas stripping, the biological slime does not exist, and the only gas movement of interest is the transfer of gases from the liquid film to the gas phase.

4-1 MASS-TRANSFER MODELS

Three gas-liquid mass-transfer models are in common use: two-film, penetration, and surface renewal. Two-film theory is both the oldest and the simplest, and the other two theories can be partially coupled to the two-film theory approach. Hence two-film theory is by far the most widely used model for describing gas transfer in water and wastewater treatment.

Two-film Theory

Three assumptions are fundamental to two-film theory development: (1) laminar flow along both sides of the gas-liquid interface, (2) steady-state conditions, and (3) instantaneous establishment of equilibrium conditions between the gas and liquid phases *at the interface*.[1] A model of the two-film concept is shown in Fig. 4-3.

At steady state, the flux of oxygen (or any other gaseous component i) through the gas film (N_{O_g}) must be the same as the flux through the liquid

FIGURE 4-1
Commonly used aeration devices. (*a*) Swing-diffuser apparatus. (*Chicago Pump Co.*) (*b*) Cutaway of floating-surface aerator. (*Peabody-Welles, Inc.*) (*c*) Brush (or cage rotor) aerator. (*Culp, Wesner and Culp, El Dorado Hills, Calif.*)

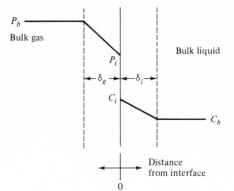

FIGURE 4-3
Schematic diagram of two-film gas-
transfer model.

film N_{O_l}. The concentration (or partial-pressure) gradient is caused by oxygen-utilizing reactions in the bulk liquid (or at least primarily in the bulk liquid as the film thickness δ_l is assumed small), and therefore the concentration profiles will be nearly linear. We can write

$$N_{O_g} = N_{O_l}$$

$$N_{O_g} = k_g(P_b - P_i) \qquad (4\text{-}1)$$

$$N_{O_g} = k_l(C_i - C_b) \qquad (4\text{-}2)$$

where k_g and k_l are mass-transfer coefficients for the respective films and correspond ideally to $\mathscr{D}/\delta$, where $\mathscr{D}$ = diffusivity, cm^2/s. Flux is given here as mol (or mg)/cm^2-s; thus the coefficients have units of cm/s. From Henry's law, we know that the interface concentration C_i is at equilibrium with the gas-phase partial pressure

$$C_i = HP_i \qquad (4\text{-}3)$$

where H is the Henry's law constant. Introducing C^* and P^*, which correspond to the equilibrium concentrations that would be associated with the bulk-gas partial pressure P_b and bulk-liquid concentration C_b, respectively, we can write

$$H(P_b - P_i) = C^* - C_i$$

$$H(P_i - P^*) = C_i - C_b$$

Substituting these relationships into Eqs. (4-1) and (4-2) and eliminating the interface concentrations results in the following development:

$$k_g(P_b - P_i) = \frac{k_g}{H}(C^* - C_i)$$

$$= k_l(C_i - C_b)$$

Hence

$$C_i = \frac{k_l C_b + (k_g/H)C^*}{k_l + k_g/H}$$

Therefore, the total flux can be expressed as

$$N_O = \frac{k_g}{H}\left[C^* - \frac{k_l C_b + (k_g/H)C^*}{k_l + k_g/H}\right]$$

or

$$N_O = \frac{k_g}{H}\frac{(k_l C^* - k_l C_b)}{k_l + k_g/H}$$

$$= \frac{k_g k_l}{Hk_l + k_g}(C^* - C_b)$$

$$= K_L(C^* - C_b)$$

where K_L is an overall mass-transfer coefficient. A similar expression can be developed for the gas phase.

$$N_O = K_g(P_b - P^*) \qquad (4\text{-}4)$$

Because the interface concentrations are impossible to determine experimentally, Eqs. (4-3) and (4-4) are very useful. In the case of oxygen transfer, the liquid-film resistance $1/k_l$ is considerably greater than the gas-film resistance $1/k_g$, and the liquid film usually controls the oxygen-transfer rate across the interface.

Penetration Theory

Higbie[2] considered the system shown in Fig. 4-3 for unsteady-state and liquid-film-controlling conditions. A mass balance on a differential section gives

$$\text{In} + \text{generation} = \text{out} + \text{accumulation}$$

$$AO_{O_z} + A\,\Delta z\,r_0 = AN_{O_{z+\Delta z}} + A\,\Delta z\,\frac{\partial C}{\partial t}$$

Rearranging the expression, allowing Δz to approach zero, and assuming no chemical reaction gives

$$\frac{\partial C}{\partial t} = \mathcal{D}\frac{\partial^2 C}{\partial z^2} \qquad (4\text{-}5)$$

Higbie was interested in the initial steps of interphase mass transfer in which the boundary conditions below apply.

$$
\begin{array}{lll}
t = 0 & z > 0 & C = C_b \\
t > 0 & z = 0 & C = C^* \\
t > 0 & z = \infty & C = C_b
\end{array}
$$

Solving Eq. (4-5) is simpler if the difference variable $C' = C - C_b$ is used.

$$\frac{\partial C'}{\partial t} = \mathcal{D}\frac{\partial^2 C'}{\partial z^2} \qquad (4\text{-}6)$$

Applying the Laplace transform $\mathscr{L}\{C'\} = \int_0^\infty C' \exp(-\jmath t)\,dt = \overline{C}'$ to both sides of Eq. (4-6),

$$\mathscr{L}\left\{\frac{\partial C'}{\partial t}\right\} = [C' \exp(-\jmath t)]_0^\infty + \jmath \int_0^\infty C' \exp(-\jmath t)\,dt$$

where
$$C' = 0 \qquad \text{at } t = 0$$
$$\exp(-\jmath t) = 0 \qquad \text{at } t \to \infty$$

Therefore
$$\mathscr{L}\left\{\frac{\partial C'}{\partial t}\right\} = \jmath \overline{C}'$$

Because the Laplace transform is independent of z, the transform of the right-hand side of Eq. (4-6) is equal to the second derivative of the transformed function $\overline{C}'$.

$$\mathscr{L}\left\{\frac{\partial^2 C'}{\partial z^2}\right\} = \frac{\partial^2 \overline{C}'}{\partial z^2}$$

and
$$\jmath \overline{C}' = \mathscr{D}\frac{\partial^2 \overline{C}'}{\partial z^2}$$

Solving by conventional procedures gives

$$\overline{C}' = B_1 \exp\left(z\sqrt{\jmath/\mathscr{D}}\right) + B_2 \exp\left(-z\sqrt{\jmath/\mathscr{D}}\right)$$

Applying the second boundary condition $C' = C^* - C_b$ at $z = 0$ yields

$$\overline{C}_0' = \int_0^\infty C_0' \exp(-\jmath t)\,dt = \frac{C_0'}{\jmath}$$

and applying the third boundary condition $C_\infty' = 0$ gives

$$\overline{C}_\infty' = 0$$

Inserting these conditions into the transformed solution Eq. (4-6) gives

$$\overline{C}' = \frac{C_0'}{\jmath} \exp\left(-z\sqrt{\frac{\jmath}{\mathscr{D}}}\right) \qquad (4\text{-}7)$$

The inverse Laplace transform of Eq. (4-7) is

$$C' = C_0' \operatorname{erfc}\frac{z}{2\sqrt{\mathscr{D}t}}$$

where
$$\operatorname{erfc}\alpha = \frac{2}{\sqrt{\pi}}\int_\alpha^\infty e^{-f^2}\,df$$

thus
$$\frac{C'}{C_0'} = \frac{C - C_b}{C^* - C_b} = \frac{2}{\sqrt{\pi}}\int_{z/(2\sqrt{\mathscr{D}t})}^\infty \exp\left(-\frac{z^2}{4\mathscr{D}t}\right)d\frac{z}{2\sqrt{\mathscr{D}t}}$$

We are interested in the concentration gradient because of its relation to the flux; thus

$$\frac{\partial C}{\partial t} = \frac{C^* - C_b}{\sqrt{\pi \mathscr{D} t}} \left[-\exp\left(-\frac{z^2}{4\mathscr{D} t} \right) \right] \qquad (4\text{-}8)$$

Now the flux may be determined at the interface as

$$N_0 = -\mathscr{D}\left(\frac{\partial C}{\partial z} \right)_{z=0}$$

and
$$N_0 = -\sqrt{\frac{\mathscr{D}}{\pi t}}\,(C^* - C_b) \qquad (4\text{-}9)$$

Considering a time of contact t_c of a small volume at the interface, we can determine the average flux over the period.

$$\overline{N}_0 = -(C^* - C_b)\sqrt{\frac{\mathscr{D}}{\pi}}\,\frac{1}{t_c}\int_0^{t_c} \frac{dt}{t^{1/2}}$$

$$= -2\sqrt{\frac{\mathscr{D}}{t_c}}\,(C^* - C_b) \qquad (4\text{-}10)$$

Equation (4-10) can be interpreted to mean the shorter the contact time, the higher the rate of mass transfer. (Note that a high rate over a small time interval is still a small number.) This has little meaning in film-flow reactors where the contact time is limited by geometry and flow rates but can be useful in designing diffused-air (i.e., bubble-type) processes, and to an extent in designing turbine aerators. In the latter two cases, it is desirable to form interfaces that are large in number, small in size, and have a short lifetime. In diffused aeration, the lifetime may be the time of travel of a bubble from the release point to the surface, while in the case of a surface aerator, the time t_c might be the time between formation of a liquid droplet until it reenters the bulk of the liquid. Maximizing oxygen flux would be expensive in both cases. Decreasing contact time for diffused aeration requires increasing mixing at the gas-liquid interface. This means shallow tanks and high volumetric flow rates of air in diffused-air systems, and for surface aerators, decreased time means more turbulence; i.e., increased energy input is required. Cost considerations, therefore, put practical limits on the transfer rate in both cases. Consideration of increasing turbulence or decreasing contact time can often lead to improved process geometry with little cost increase, however.

Surface Renewal Theory

Danckwerts[3] extended the penetration model by considering the case where liquid elements would be at the interface for finite time periods. Because of turbulence, the time of contact of the liquid elements would be randomly distributed, and the concept has become known as the *random surface renewal theory* or more commonly the *surface renewal theory*. In developing the theory, Danckwerts assumed that there existed a rate of production of "fresh" surface per unit total surface available, r_s, and that this rate was independent of the age of the element in question. He then defined an area of surface, $A(t)\,\Delta t$, with age between t and $t + dt$, and made what is essentially a balance on the area.

$$A(t)\,\Delta t = A(t - \Delta t)\,\Delta t - [A(t - \Delta t)\,\Delta t]r_s\,\Delta t \qquad (4\text{-}11)$$

where $[A(t - \Delta t)\,\Delta t]r_s\,\Delta t$ is the quantity of the area renewed in time Δt. This expression can be rewritten in differential form by rearranging the terms and allowing Δt to approach zero in the limit; thus

$$\frac{dA(t)}{dt} = -r_s\,A(t) \qquad (4\text{-}12)$$

which upon integration yields

$$(\text{constant})e^{-r_s t} = A(t) \qquad (4\text{-}13)$$

The total area under consideration was unity; therefore

$$\int_0^\infty A(t)\,dt = 1$$

and

$$1 = \int_0^\infty (\text{constant})e^{-r_s t}\,dt$$

since

$$1 = \frac{\text{constant}}{r_s}$$

$$A(t) = r_s\,e^{-r_s t} \qquad (4\text{-}14)$$

From penetration theory we found that

$$N_0 = -(C^* - C_b)\sqrt{\frac{\mathscr{D}}{\pi t}} \qquad (4\text{-}9)$$

Including the age-distribution term in Eq. (4-9) gives

$$N_0 = -(C^* - C_b)\int_0^\infty \sqrt{\frac{\mathscr{D}}{\pi t}}\, r_s\,e^{-r_s t}\,dt \qquad (4\text{-}15)$$

Substituting β^2 for $r_s\,t$,

$$N_0 = 2(C^* - C_b)r_s^{1/2}\left(\frac{\mathscr{D}}{\pi}\right)^{1/2}\int_0^\infty e^{-\beta^2}\,d\beta$$

$$= 2(C^* - C_b)\left(\frac{r_s\mathscr{D}}{\pi}\right)^{1/2}\frac{\pi^{1/2}}{2}$$

$$= (C^* - C_b)\sqrt{\mathscr{D}r_s} \qquad (4\text{-}16)$$

FIGURE 4-4
Activated-sludge process using porous diffusers.

The surface renewal rate r_s, like the contact time in Eq. (4-10), is difficult to determine. The three theories do fit together quite well as all three have the function $C^* - C_b$, and the coefficients of the penetration theory and the surface renewal theory can be used in conceptual interpretation of K_L. Appearance of the diffusivity in the models [Eqs. (4-10), (4-16)] should not be surprising because this is the hypothesized mechanism of transport in the laminar film. The reader should remember that the surface renewal theory assumes turbulent flow while the penetration model does not. This is of interest in discussing oxygen transfer in trickling filters and will be of importance later in this chapter.

4-2 BUBBLE AERATION

Many aeration systems in use today are of the diffused-air or bubble-aeration type. Introduction of the bubbles into the liquid is accomplished in a number of ways, but most commonly a porous tube or plate is used (Fig. 4-4). The major constraints on diffuser design are to produce a minimum-size bubble without plugging of the pores of the apparatus. Common causes of plugging are biological growth and oil and dirt in the air lines.

Bubbles are released from the diffuser in a cloud. On collision they tend to coalesce and form larger bubbles that, of course, are undesirable but also are unavoidable. Thus analysis of bubble aeration depends on average values and experience. Oxygen transfer rate depends on the average surface area of the bubbles and thus on the mean bubble diameter d_b. Eckenfelder[4] reported that an empirical relationship was satisfactory to describe the typical bubble diameter associated with the airflow rates used in activated-sludge processes. He suggested

$$d_b \propto Q_g^n \qquad (4\text{-}17)$$

where Q_g is the air volumetric flow rate and n is an empirical coefficient ranging from $0.2 < n < 1.0$. Bubble shape is dependent upon the Reynolds number and the characteristics of the liquid.[5-7] At Reynolds numbers less than 300, air bubbles are generally spherical; between 300 and 4000, they are ellipsoidal; and above 4000, they form spherical caps.

Correlations between three dimensionless groups—the Sherwood number $(K_L d_b/\mathscr{D})$, which is the molar rate of transfer per unit area for a unit molar concentration difference, the Reynolds number $(v_b d_b \rho_w/\mu)$, which is the ratio of inertial forces to the viscous forces in the liquid, and the Schmidt number $(\mu/\rho_w \mathscr{D})$, which is the ratio of the kinematic viscosity to the diffusivity—are commonly used in mass-transfer studies. Eckenfelder[4] gave such a correlation based on data reported in the literature [Eq. (4-18), Fig. 4-5].

$$\frac{K_L d_b}{\mathscr{D}} = \beta \frac{d_b v_b \rho_w}{\mu} \left(\frac{\mu}{\rho \mathscr{D}} \right)^{1/2} \qquad (4\text{-}18)$$

As one can see from the log-log plot, data scatter are fairly great. This would be expected from the variety of materials and apparatus used in the various studies analyzed by Eckenfelder. However, there is a consistency to the data pattern, and an expression of the form of Eq. (4-18) can be obtained by dimensional analysis.[8] For these reasons, Eq. (4-18) and Fig. 4-5 seem to be an acceptable approach for design purposes. Eckenfelder found that the coefficient β was equal to $\beta'/H_L^{1/3}$, where β' was a constant and H_L was the tank depth.

In practice the oxygen-transfer coefficient K_L is rarely used by itself. Instead, an overall coefficient (or perhaps an overall-overall coefficient because K_L is actually an overall coefficient) $K_L a$, which combines both the rate and the interface area available, is used. This coefficient can be estimated from the following development:

$$a = \frac{Q_g}{(\pi/6)d_b^3} \frac{\pi d_b^2}{V} t_c \qquad (4\text{-}19)$$

where a = area of interface per unit of tank volume V
t_c = contact time of bubble with liquid

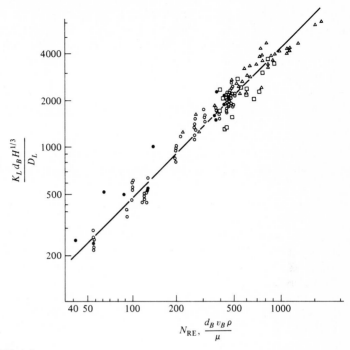

FIGURE 4-5
Correlation used by Eckenfelder[4] for aeration parameter evaluation and process
scale-up.

The contact time is dependent upon the path of the bubble through the liquid
and will be given here in terms of the average bubble velocity v_b and the liquid
depth H_L:

$$t_c = \frac{H_L}{v_b} \qquad (4\text{-}20)$$

The area of bubble interface per unit of tank volume is then

$$a = \frac{6Q_g H_L}{d_b v_b V} \qquad (4\text{-}21)$$

Equation (4-21) should contain a term for the free surface because oxygen transfer
occurs across this boundary. Total transfer across the free surface is small relative
to that across the bubble-liquid interfaces and can be neglected, however.[9]
Combining Eq. (4-19) and (4-20) and solving for the overall transfer rate $K_L a$,

using Eckenfelder's coefficient $\beta'/H_2^{1/3}$, gives

$$K_L a = \frac{6\beta' H_L^{2/3} Q_g \rho_w^{1/2} \mathscr{D}^{1/2}}{d_b V (\mu)^{1/2}}$$

$$= \frac{6\beta' H_L^{2/3} Q_g}{d_b V N_{sc}^{1/2}} \qquad (4\text{-}22)$$

The units of the overall rate term are t^{-1} as noted earlier. Checking the units forces the conclusion that β' must have units of $l^{1/3}$ as can also be concluded from the fact that it is part of the unitless proportionality factor β of Eq. (4-18).

Oxygen mass transfer rate per unit tank volume can now be written in terms of the overall rate coefficient $K_L a$.

$$M_{O_2} = K_L a (C^* - C_b) \qquad (4\text{-}23)$$

Equation (4-23) is used in most aeration studies to describe the oxygen-transfer process, even though it was developed for bubble aeration only. Because the derivation of Eq. (4-22) depends on the empirical coefficient β', theoretically obtained values of $K_L a$ are not possible. Direct determination of $K_L a$ is far easier and a bit more honest than determining β' and using Eq. (4-22) to find values for $K_L a$. Thus tables of β' for various conditions are not available. The usefulness of Eq. (4-22) is in scale-up or in predicting the effect of conditions that have not been experimentally studied, e.g., estimates of the effect of tank depth or gas flow rate change on oxygen transfer.

EXAMPLE 4-1   An aeration system is being designed for a municipal waste-water treatment process. The tank volume needed is 10^6 1. A test system using a 30-cm diameter cylinder with a liquid depth of 200 cm is to be used. Air is introduced in fine bubbles through a porous bottom plate, and average bubble sizes are determined through photographs. Oxygen-transfer data[†] were taken at three airflow rates and are presented in Table 4-1. Estimate the value of K_L for the same environmental conditions for a tank 366 cm deep and for a tank 458 cm deep. The saturation concentration of oxygen was found to be 7.8 mg/l, and the diffusivity of oxygen in water is approximately 2.1×10^{-5} cm^2/s.

$$K_L a_1 = \frac{6\beta'_1 (200)^{2/3} (125)(1)^{1/2} (2.1 \times 10^{-5})^{1/2}}{0.2(1.41 \times 10^5)(0.01)^{1/2}}$$

$$= 4.17 \times 10^{-2} \beta'_1$$

$$K_L a_2 = \frac{6\beta'_2 (200)^{2/3} (100)(1)^{1/2} (2.1 \times 10^{-5})^{1/2}}{0.18(1.41 \times 10^5)(0.01)^{1/2}}$$

$$= 3.71 \times 10^{-2} \beta'_2$$

---

† A number of methods of determining $K_L a$ are available. See Refs. 10 to 12.

$$K_L a_3 = \frac{6\beta'_3(200)^{2/3}(75)(1)^{1/2}(2.1 \times 10^{-5})}{0.16(1.41 \times 10^5)(0.01)^{1/2}}$$

$$= 3.13 \times 10^{-2}\beta'_3$$

Determination of $K_L a$ values from plots of the time-concentration data gives

$$K_L a_1 = 0.0025 \text{ s}^{-1}$$
$$K_L a_2 = 0.0022 \text{ s}^{-1}$$
$$K_L a_3 = 0.0018 \text{ s}^{-1}$$

Corresponding β' values are

$$\beta'_1 = 0.0600$$
$$\beta'_2 = 0.0593$$
$$\beta'_3 = 0.0586 \qquad ////$$

The β' values obtained seem to show a pattern (increasing with Q_g), but with only three data points, it is very difficult to determine whether or not this is true. Variation of β' is small (~ 2 percent), while variation of Q_g is large (~ 67 percent), and, therefore, assuming $\beta' = 0.0593$ is probably fairly safe.

Determination of the overall $K_L a$ for the full-size (10^6 l) tank will be dependent on airflow rate and the assumption that aeration devices are the same in both systems.

Table 4-1

Time, s	O_2 concentration, mg/l		
	Q_{g1}	Q_{g2}	Q_{g3}
	7.5 SLM†	6.0 SLM	4.5 SLM
	Average bubble diameter, cm		
	0.2	0.18	0.16
0	0	0	0
100	1.59	1.40	1.21
200	2.83	2.53	2.21
300	3.80	3.44	3.05
600	5.60	5.24	4.80
900	6.44	6.17	5.82
1200	6.84	6.67	6.40
1500	7.03	6.92	6.74
2000	7.15	7.11	7.02
2500	7.19	7.17	7.13
3000	7.20	7.19	7.17

† SLM = standard liter/min.

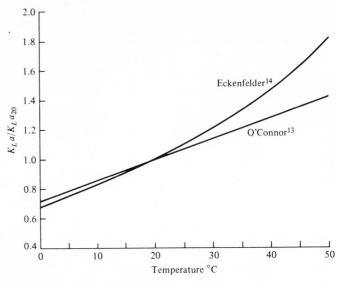

FIGURE 4-6
The effect of temperature on oxygen transfer rate.

4-3   TEMPERATURE EFFECTS

Temperature affects both equilibrium values for oxygen concentration (see Appendix B) and the rate at which transfer occurs. Equilibrium-concentration values have been established for water over a range of temperature and salinity values, but similar work for the rate coefficient is less abundant. O'Connor[13] developed the semitheoretical equation

$$\frac{K_L a_1}{K_L a_2} = \sqrt{\frac{T_1 \mu_2}{T_2 \mu_1}} \qquad (4\text{-}24)$$

where μ is the viscosity at the respective temperatures and T is in kelvins.

Eckenfelder[14] suggested an exponential expression

$$\frac{K_L a}{K_L a_{20}} = 1.02^{(T-20)} \qquad (4\text{-}25)$$

where temperature is in degrees Celsius.

A comparison of the results of Eqs. (4-24) and (4-25) is given in Fig. 4-6. As can be seen from the curves, the two expressions give quite similar results below 30° where the preponderance of field data is. Eckenfelder's empirical expression was developed from experimental data and comparisons of his work with that of Streeter,[15] Howland,[16] and Wilke[17] to yield Eq. (4-25).

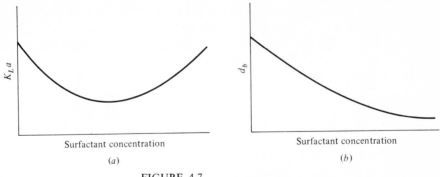

FIGURE 4-7
Effects of surfactant concentration on mass-transfer characteristics.

4-4 THE EFFECT OF SURFACE-ACTIVE AGENTS

Film thickness and surface tension are affected by the presence of materials such as proteins, detergents, oils, and other surface-active agents. Increased film thickness δ or perhaps changes in film characteristics due to the presence of dissolved organics decrease the transfer rate. Aiba and Toda[18] suggested that the effect of surface-active agents is to make the interface more quiescent, i.e., decreasing the rate of surface renewal. Experiments with pure surfactants (Eckenfelder and Barnhart,[19] Mancy and Okun,[20] McKeown and Okun,[21] Timson and Dunn,[22] Aiba and Toda[18]) have typically resulted in curves such as shown in Fig. 4-7. The presence of a minimum transfer rate is unexplained but may be due to the increased interface area as the bubble diameters decrease.

The effect of surfactant concentrations on mass transfer rates is far more complicated in real systems where there are effects of an interaction with suspended solids and dissolved materials. Thus one can expect real process transfer rates (or $K_L a$ values) to be lower than those obtained with pure-water experimental systems. Perhaps a better way of saying this is that $K_L a$ values given for aeration equipment are meaningful only if they were derived from a system with constituents similar to the one you plan to use. Attempts have been made to develop correlations between $K_L a$ values in particular wastes and in pure water. These correlations, in general, are not very useful. As Busch[23] noted, it is a considerably better idea to obtain data for individual wastewater-aerator combinations rather than rely on data from systems which may have many chemical and, hence, mass-transfer-limiting differences.

4-5 AGITATION RESULTING FROM AERATION

Aeration serves to mix the tank contents as well as to transfer oxygen to the liquid for biooxidation reactions. This is particularly important in activated sludge because the biological culture is flocculant, i.e., made up of particles or clumps of cells, and will settle out if the tank is quiescent. Air bubbles are usually in the 0.1- to 1.0-cm diameter-size range and can be expected to have rise velocities of the order of 20 to 50 cm/s and corresponding Reynolds numbers of 300 to 5000. Hence most of the bubbles are in the Newton's law range where the drag coefficient is nearly constant. The Reynolds numbers given above are based upon the bubble diameter and therefore do not give an indication of the state of turbulence in the tank. This can be estimated by determining the power transferred to the tank contents by the gas. Assuming spherical bubbles (actually they are more elliptical in shape) rising at terminal velocity, the buoyant force is nearly equal to the drag force. Therefore the power transfer per bubble is

$$P_b = \frac{\pi d_b^3 \rho_w g}{6} v_b \qquad (4\text{-}26)$$

The number of bubbles in the tank at any time is

$$n_b = \frac{Q_g t_c}{V_b} = \frac{Q_g H_L}{V_b v_b} \qquad (4\text{-}27)$$

Multiplying Eq. (4-26) by Eq. (4-27) gives the total power input to the liquid:

$$P_L = Q_g H_L \rho_w g \qquad (4\text{-}28)$$

Power transmission alone does not give a complete indication of mixing. For example, if all the gas were introduced at one point, there would undoubtedly be dead spots, remaining unmixed, in the tank. Assuming uniform bubble distribution, we can estimate liquid velocities resulting from the aeration, however.

$$P_L = \frac{\pi Z^2 g \rho_w}{4} \left(\frac{Z}{4} + H_L \right) v_l \qquad (4\text{-}29)$$

where Z = tank diameter or other characteristic dimension
v_l = average liquid velocity

Equation (4-29) is based on a force balance over the liquid and the Fanning friction factor

$$f' = \frac{1}{4} \frac{Z}{H_L} \frac{(P_1 - P_2)}{\frac{1}{2} \rho v_l^2}$$

where P_1 and P_2 are the entrance and exit pressures. Substituting Eq. (4-28) into Eq. (4-29) and solving for the liquid velocity gives

$$v_l = \frac{4 Q_g H_L}{\pi Z^2 (Z/4 + H_L)} \qquad (4\text{-}30)$$

Because mixing is directly related to turbulence, we can obtain an approximation of the extent of convective mixing from the Reynolds number for the liquid, which, of course, is easily calculated once a velocity term is obtained.

$$N_{\text{Re}} = \frac{vZ\rho}{\mu} = \frac{4Q_g H_L \rho_w}{\pi\mu Z(Z/4 + H_L)} \qquad (4\text{-}31)$$

The average liquid velocity can be used directly in determining if particles will be held in suspension. Activated-sludge particles or flocs are of the order of 0.5 to 1.5 cm in size and have a wet specific gravity of approximately 1.05. Velocities of the order of 0.5 cm/s are necessary to keep them in suspension.

Mixing and Transfer Efficiency

The relative importance of mixing and oxygen transfer in determining the air-flow rate now becomes of interest. We can estimate the "efficiency" of oxygen transfer by estimating the quantity of oxygen transferred per bubble and dividing this quantity by the amount of oxygen available per bubble.

$$m_b = \text{mass of oxygen transferred per bubble}$$
$$= \frac{V K_L a(C^* - C_b)}{n_b}$$
$$= \frac{V_b V K_L a(C^* - C_b)}{Q_g} \qquad (4\text{-}32)$$

Therefore the efficiency of oxygen transfer is

$$E_{O_2} = \frac{m_b}{V_b \rho_{O_2}} \qquad (4\text{-}33)$$

$$E_{O_2} = \frac{V K_L a(C^* - C_b)}{Q_g \rho_{O_2}} \qquad (4\text{-}34)$$

From Eq. (4-22) we see that the mass transfer rate coefficient is proportional to the gas flow rate, and we can write

$$E_{O_2} = \frac{6\beta' H_L^{2/3} \rho_w^{1/2} \mathscr{D}^{1/2}(C^* - C_b)}{d_b \mu^{1/2} \rho_{O_2}} \qquad (4\text{-}35)$$

For a given set of conditions, the controlling variables will be the liquid depth and the bulk-liquid-oxygen concentration. Bubble diameter will vary somewhat with gas flow rate, and therefore, the effect of gas flow rate on efficiency is not completely eliminated, although effects are considerably decreased from those apparently shown in Eq. (4-34).

An extremely important point to be remembered in discussing oxygen-transfer efficiency is that the liquid-phase concentration enters into the determination. Thus, unless oxygen is being removed from the liquid phase by reaction, adsorption or some other mechanism, C_b will quickly approach C^*, and the efficiency of transfer will drop to zero. Thus efficiency by itself gives little insight into the effectiveness of a particular system. Liquid-phase oxygen concentration is inversely related to the gas flow rate ($C_b \propto Q_g^{-x}$) for a given situation, and so efficiency is again indirectly related to gas flow rate.

EXAMPLE 4-2   A biological wastewater treatment process consumes oxygen at the rate of 0.017 mg/s-l. The tank depth is 4.58 m, and the tank volume is 1.439×10^3 m³. Air bubbles from the diffused-air aeration system have been observed to average 0.2 cm in diameter. The saturation concentration of oxygen in the wastewater is 7.8 mg/l, and the system is maintained at a bulk-liquid-oxygen concentration of 1 mg/l. Assuming an average oxygen density in the bubbles of 3.85×10^{-4} g/cm³, calculate (a) the oxygen-transfer efficiency to be expected, (b) the oxygen and air input rates needed, and (c) the air input rate necessary to keep 0.05-cm-diameter floc particles in suspension.

Oxygen-transfer efficiency

$$E_{O_2} = \frac{6\beta' H_L^{2/3} \rho_w^{1/2} \mathscr{D}^{1/2}(C^* - C_b)}{d_b \mu^{1/2} \rho_{O_2}}$$

The only factors not furnished in the problem statement are β', μ, and $\mathscr{D}$. We will assume that these are the same as in Example 4-1. Certainly the viscosity and diffusivity will not change greatly, and their roots will change less.

$$E_{O_2} = \frac{6(0.06)(458)^{2/3}(1)^{1/2}(2.1 \times 10^{-5})^{1/2}(6.8 \times 10^{-6})}{0.2(0.01)^{1/2}(3.85 \times 10^{-4})}$$

$$= 0.087 \text{ or } 8.7 \text{ percent}$$

Oxygen and airflow rate required

$$Q_{O_2} = \frac{V(0.017 \text{ mg/s-l})}{E_{O_2} \rho_{O_2}}$$

$$= \frac{(1.439 \times 10^6)(0.017 \times 10^{-3})}{0.087(3.85 \times 10^{-1})}$$

$$= 0.730 \text{ m}^3/\text{s}$$

$$Q_g = \frac{1}{0.24} Q_{O_2}$$

$$= 3.043 \text{ m}^3/\text{s air}$$

Air input rate to maintain scour   The necessary velocity can be estimated from the terminal settling rate of the particle. A force balance on a falling particle (assumed spherical) gives

$$m_p \frac{dv}{dt} = \frac{\pi}{6} d_p g^3 (\rho_p - \rho_l) - C_D A_p \rho_l \frac{v^2}{2}$$

When terminal velocity is reached, $dv/dt \to 0$. By trial and error, using $\rho_p = 1.05$, we find that the Reynolds number is of the order of 3, and thus the system is out of the Stokes' law region. The drag coefficient is given by

$$c_D = \frac{18.5}{N_{Re}^{0.6}}$$

Terminal velocity is therefore

$$v_D = \frac{0.072 d^{1.6} g (\rho_p - \rho_l)}{\mu^{0.6} \rho_l^{0.4}}$$

$$= \frac{0.072 (0.05)^{1.6} (981)(0.05)}{(0.01)^{0.6} (1)^{0.4}} = 0.464 \text{ cm/s}$$

$$N_{Re} = \frac{0.464 (0.05)(1)}{0.01} = 2.3$$

which is close to the predicted value. Assuming a circular tank and using Eq. (4-30), the gas flow rate for mixing is calculated.

$$Q_g = \frac{v_l \pi Z^2 (Z/4 + H_L)}{4 H_L}$$

$$= \frac{0.464 (3.14)(1000)^2 (\frac{1000}{4} + 458)}{4(458)}$$

$$= 563,406 \text{ cm}^3/\text{s}$$

$$= 0.563 \text{ m}^3/\text{s}$$

Therefore, oxygen transfer evidently controls the air input rate in this case.

////

4-6   AERATION BY MECHANICAL MIXERS

Mechanical agitation may be used alone or with gas addition. When gas addition is used, submerged turbines are almost always used also. In wastewater treatment, the most common mechanical aeration device is the surface turbine (Fig. 4-1b). This system comes in a variety of forms as shown by Fig. 4-8 but is basically a system for contacting the liquid with the air above it at a high rate.

As might be expected, the influence of tank geometry on the results achieved in a given situation is very great. In order to compare results of two aeration devices, tank geometry and liquid characteristics must be identical. This fact makes prediction of aeration capability difficult. For this reason, mechanical aeration devices are usually specified with respect to the volume, depth, and area they are expected to mix and aerate. In addition, several aerators distributed around a large tank are preferable to one large aerator of the same net power

(a)

(b)

(c)

FIGURE 4-8
Types of mechanical aerators in use. (a) Floating aerator with simplex cone. (*Aqua Aerobics, Inc.*) (b) Fixed-surface aerator with submerged turbine for additional deep mixing. (c) Brush aerators in operation.

input. Aspects of this problem have been discussed in detail by Busch[23] and Conway.[24]

Mechanical aerators are often described in manufacturers' literature by transfer-coefficient $(K_L a)$ values or by transfer rates (mass transferred/unit power-time). These values while useful are representative of the test systems only, and the design engineer must either have additional information concerning the relationship of the system he is designing to the system the reported values were developed from, or he must conduct tests himself. Transfer-coefficient values are developed in much the same way as for dispersed-air systems (see Eckenfelder and O'Connor[25]), but the physical system is much more difficult to conceptualize or describe mathematically.

Eckenfelder[4] reported that oxygen transfer rates associated with surface turbine aerators may be as high as 3.6 kg O_2/kWh (6.0 lb O_2/hp-h) under ideal conditions (i.e., zero oxygen in liquid). Experience with conventionally designed activated-sludge processes has resulted in the use of design values between 1.5 kg O_2/kWh (2.5 lb O_2/hp-h) and 2.4 kg O_2/kWh (4.0 lb O_2/hp-h), however.

There is no reason why transfer rate correlations of the form of Eq. (4-36) could not be developed, but their applicability is doubtful because of the role of geometry in determining actual rates. One scale-up expression commonly used is

$$V K_L a = C H_L^\alpha N_t^\beta Z_i^\gamma N_t^\beta Z_i^\gamma \qquad (4\text{-}36)$$

where N_t = impeller tip speed,
Z_i = impeller diameter

The major problem with the use of equations such as (4-36) is that a correlation developed for one tank will not, in general, be applicable to another tank with different dimensions. In practice, nearly all the studies have been done in a narrow band of tip speeds (N_t) in the vicinity of 70 r/min, which effectively eliminates tip speed from consideration as a variable. In addition, impeller diameter sizes are limited, primarily due to structural limitations. Thus experience is the best source of information at the present time.

Considerable work has been done on submerged turbine aerators with air addition because of the wide use of this system in the fermentation industry. Richards et al.[26] suggested Eq. (4-37) for this type of system.

$$K_L a \propto \left(\frac{P_L}{V}\right)^{0.4} \bar{v}_b^{0.5} N_t^{0.5} \qquad (4\text{-}37)$$

where v_b is the nominal gas velocity determined by the flow rate of the gas divided by the cross-sectional area of the tank. Eckenfelder[14] has presented a similar expression.

$$K_L a \propto \frac{Q_g^\alpha Z_i^\beta N_i^\gamma}{V} \qquad (4\text{-}38)$$

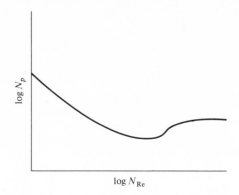

FIGURE 4-9
Typical power curve for turbine agitator.

where the exponents α, β, and γ range from 1.2 to 2.4, 0.9 to 1.6, and 0.6 to 0.9, respectively, and N_i is the impeller rotational speed, t^{-1}.

Power requirements for turbine aerators and their relationship to mixing effectiveness have also been studied to a considerable degree. Results are generally quite empirical but have been very useful in the fermentation industry. Quite possibly direct applicability in wastewater treatment will be greatest in systems such as anaerobic digestion where mass transfer is not a limiting process. In most cases mixing effectiveness is estimated from correlations of the power number N_p, a ratio of applied force to inertial force, and Reynolds number.

$$N_p = \frac{P_L}{N_i^3 D_i^5 \rho} \qquad (4\text{-}39)$$

The Froude number might be expected to enter the correlation also but does so only in vortexing systems. Vortexing is inhibitory to mixing and thus is eliminated as much as possible. A correlation is developed for a particular mixing device and a particular tank geometry, and this curve is then used for scale-up purposes. Correlations are usually of the form shown in Fig. 4-9.

4-7 MAXIMUM OXYGEN TRANSFER RATES AND COMPARISON OF OPERATING SYSTEMS

Perhaps the simplest and most important fact in considering oxygen-transfer systems is that oxygen cannot be transferred faster than it is being utilized. This fact must be remembered when comparing operating or experimental data. Normally, oxygen-transfer coefficients are developed for systems with deoxygenated water. The water normally is low in dissolved salts and has a high oxygen-saturation value and, of course, the maximum gradient because the bulk-liquid-oxygen concentration is zero. If data are reported as mass oxygen transferred per unit energy input (kg O_2/kWh), the results are obviously quite misleading.

A more serious error is to assume that more oxygen can be transferred by increasing the oxygen flow rate, using more impellers or larger motors. If a satisfactory amount of oxygen is being transferred, i.e., if the reactions in the liquid are not dependent on oxygen concentration, the actual rate of mass transfer cannot be increased. This fact can be used to explain many low transfer rate values reported in the literature.[23] A possible reason for the misuse of oxygen transfer rates in practice is a lack of complete understanding of the method of their determination. In most cases, liquid-phase oxygen concentrations are maintained at zero by an extremely fast-reacting sulfite–cobalt catalyst reaction system. This means that effectively the reaction rate never becomes limiting. Increasing power input (either as mixing energy or increased gas flow rate) results in increased surface area for mass transfer and thus an increase in overall transfer rate. This is not the case in biological processes where the oxygen uptake rate is generally less than the potential transfer rates. Therefore, it is likely that many existing units are overdesigned.

4-8   FILM-FLOW OXYGEN TRANSFER

Transfer of oxygen into liquid films, as in the case of the trickling filter, presents a new set of problems. Several physical situations must be considered: (1) reaction only in the semisolid phase, (2) reaction in both the liquid and semisolid phase, (3) mass-transport controlling, (4) reaction-rate controlling, (5) laminar, and (6) turbulent flow. With respect to the trickling filter, case (1) would be best approximated by a standard-rate filter with insignificant amounts of cells in the liquid phase and, by definition, no recycling. Case (2) would be approximated by a high-rate filter, i.e., with recycle of effluent, which has a significant concentration of cells suspended in the liquid phase. Figure 4-10*a* and *c* are representations of these two conditions. The situation shown in Fig. 4-10*b* in which modular growth extends through most of the liquid film would probably give results similar to case (2). Either condition will be referred to as the pseudohomogeneous case. Cases (3) and (4) were discussed to some extent in Chap. 2 and will be discussed again in Chap. 9. Here we will point out that rate limitations can occur within the film as well as in the liquid phase. Thus consideration of the film depth and reactions within the film is necessary to obtain an accurate picture of what is occurring. A general statement can be made on the effect of oxygen transfer rate on the system, however. A maximum rate of oxygen transfer will exist for any given system, and as long as the rate of conversion of reactants does not exceed this maximum transfer rate, there will not be an oxygen deficiency. A sound method of predicting whether or not oxygen transfer will be limiting is to determine the organic conversion rate in a system known to be independent of oxygen concentration and compare this value with the possible rate of oxygen transfer. This procedure was proposed by Busch[27] as a method for setting standards on rivers and lakes. Prediction of

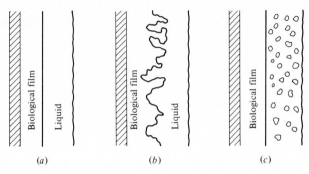

FIGURE 4-10
Possible slime conditions in film-flow reactors.

the maximum oxygen-transfer capacity is dependent on knowledge of the hydraulic conditions, i.e., laminar or turbulent flow. We will begin our analyses by considering flow characteristics.

Determination of Condition of Flow

Film-flow reactors used in wastewater treatment are actually porous beds. The solid media on which biological slimes grow and over which the wastewater flows has traditionally been rock (2 to 4 in nominal size), but virtually any material which withstands abrasion and is nonbactericidal would be satisfactory. In recent years, there has been an increasing use of plastic media because of the larger porosity, surface area, and lower weight per unit volume. Rock and plastic media usually have open areas of approximately 0.22 and 0.97 m^2/m^2, respectively. Typical hydraulic loading rates used range from 5×10^{-5} m^3/m^2-s (100 gal/ft²-d), which would be for a standard-rate rock trickling filter, to 4.5×10^{-4} m^3/m^2-s (950 gal/ft²-d) for a high-rate plastic-media trickling filter. Both loading rates were determined on the basis of open area rather than total cross section.

Assuming that wastewater is uniformly distributed over the solid surface, we can use the equation of motion to determine the velocity profile *if* laminar conditions exist.

$$0 = -\frac{\partial P}{\partial x} + \mu \frac{\partial^2 v_x}{\partial y^2} + \rho g_x \qquad (4\text{-}40)$$

For the purposes of determining the state of flow, it is practical to assume that the liquid depth δ is constant and, therefore, that the pressure gradient $\partial P/\partial x$

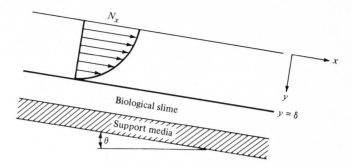

FIGURE 4-11
Open-channel flow model of film-flow reactor segment.

is zero. Integrating with the boundary conditions below gives Eq. (4-41).

$$y = \begin{cases} 0 & \dfrac{\partial v_x}{\partial y} = 0 \\[2ex] \delta & v_x = 0 \end{cases}$$

$$v_x = \frac{\rho \delta^2 g_x}{2\mu}\left(1 - \frac{y^2}{\delta^2}\right) \qquad (4\text{-}41)$$

Wastewaters are, in general, dilute solutions and can be considered newtonian in nearly all cases. The shear stress τ_{xy} is then

$$\tau_{xy} = \mu \frac{dv_x}{dy} \qquad (4\text{-}42)$$

Substituting Eq. (4-41) into Eq. (4-42) and determining the shear stress of the solid-liquid interface gives

$$\tau_\delta = -\rho g_x \delta = -\rho g \delta \sin\theta \qquad (4\text{-}43)$$

Assumption of turbulent-flow conditions also leads to Eq. (4-43). Using a macroscopic force balance as shown in Fig. 4-12 and determining the shear stress at the wall for a wide flat channel (one in which the shear stress at the walls is negligible) leads to Eq. (4-44).

$$W \sin\theta = -pl\tau_\delta \qquad (4\text{-}44)$$

where W = weight of liquid in segment
p = wetted perimeter
P = pressure
A = cross-sectional area
l = length of segment

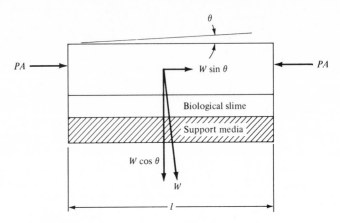

FIGURE 4-12
Open-channel flow model for turbulent flow in film-flow reactor.

The wetted area multiplied by the shear stress must equal the weight component in the direction of flow under steady, uniform-flow conditions. For a wide channel the wetted area is approximately bl, where b is the width. Thus we can write

$$\tau_\delta = \frac{-W \sin \theta}{bl} = \frac{-\rho g b l \delta \sin \theta}{bl} = -\rho g \delta \sin \theta \quad (4\text{-}43)$$

The depth of flow δ can be determined for laminar conditions as follows:

$$\frac{dQ}{dy} = v_x \frac{dA}{dy}$$

$$dQ = v_x b \, dy$$

$$Q = b \int_0^\delta \frac{\rho g \sin \theta}{2\mu} (\delta^2 - y^2) \, dy$$

$$= \frac{\rho b g \sin \theta}{3\mu} \delta^3$$

$$\delta_l = \left(\frac{3\mu Q}{\rho b g \sin \theta}\right)^{1/3} \quad (4\text{-}45)$$

If the flow is turbulent, a partially empirical expression including a friction factor must be used.

$$\rho g \delta \sin \theta = f \rho \frac{v_x^2}{8} \quad (4\text{-}46)$$

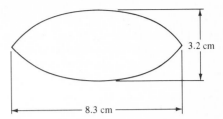

FIGURE 4-13
Typical cross section of plastic trickling-
filter media.

$$\delta = \frac{f v_x^2}{8g \sin \theta}$$

$$= \frac{f Q^2}{8g \delta^2 b^2 \sin \theta}$$

$$\delta_t = \left(\frac{f Q}{8 b^2 g \sin \theta} \right)^{1/3} \qquad (4\text{-}47)$$

Consideration of flow through one section or pore of a plastic-media trickling filter will allow us to estimate both film depths and whether the flow is laminar or turbulent. A typical plastic-media cross section is shown in Fig. 4-13. The area of the section is approximately 21 cm², and the perimeter is approximately 20.3 cm. Using Eq. (4-45) (laminar flow) and the maximum hydraulic loading rate $(4.5 \times 10^{-6} \text{ 1/cm}^2\text{-s})$, we can estimate the film thickness δ for a vertical orientation.

$$\delta_1 = \left(\frac{3(0.01)(4.5 \times 10^{-3})(21)}{1(20.3)(981)(1)} \right)^{1/3}$$

$$= 0.005 \text{ cm}$$

The Reynolds number can be calculated using the hydraulic radius as a parameter.

$$r_H = \frac{\text{cross-sectional area}}{\text{wetted perimeter}}$$

$$\approx \frac{b\delta}{b + 2\delta} \approx \delta$$

$$N_{Re} = \frac{v \, d\rho}{\mu}$$

$$= \frac{(Q/A)4 r_H \, \rho}{\mu}$$

$$= \frac{\dfrac{(4.5 \times 10^{-3})(21)}{2.03(0.005)}(4)(0.005)(1)}{0.01}$$

$$= 1.9$$

This would seem to mean that the flow regime in trickling filters is laminar as turbulence does not occur in liquid films until the Reynolds number approaches 1000. Howell[28] has pointed out that uniform flow over the media surface is quite unlikely and that one would expect considerable channelization. Assuming this is true and taking an arbitrary value (which on the surface seems conservative) of $b_{\text{actual}} = b/10$, we estimate

$$\delta^* = 0.011 \text{ cm} \qquad N_{\text{Re}}^* = 18.6$$

where the starred variables indicate the assumption that channelization has reduced the flow area by a factor of 10. Thus laminar flow probably exists in the trickling filter under most conditions. It is of interest that the velocities calculated here on a purely theoretical basis are close to those reported by Eckenfelder and Barnhart.[19]

The flow situation is considerably more complex than simple laminar flow, however. Rivulets flowing over a flat surface nearly always flow together and pull apart as they move downward. At points where two rivulets meet, mixing must occur due to the need to dissipate lateral momentum, and thus some sort of turbulence exists on a local basis at least. Additionally, experimental studies on flow over biological slimes by Atkinson and his coworkers[29–31] have shown that rippling is normal even some distance from the entrance. We can, therefore, assume that pure laminar flow does not exist, but neither does pronounced turbulence (except on a local basis).

Flux of oxygen into the liquid can be defined by Eq. (4-3) with the mass-transfer coefficient K_L being defined by either the penetration theory expression [Eq. (4-10)] or the surface renewal expression [Eq. (4-16)]. Three distinct situations must at least be considered: (1) mass transfer through the liquid phase and into the reacting biological slime (Fig. 4-10a), (2) mass transfer into the pseudohomogeneous-reaction region and then into the slime (Fig. 4-10b, c), and (3) mass transfer into the rivulets for either case 1 or 2 and directly into the reacting slime. Case 3 is real because some reacting material would be stored in the slime, and therefore reaction will continue even if the liquid is temporarily removed. This brings a very unwelcome additional complexity to the problem which will be eliminated here by assuming that direct transfer into the slime is faster than transfer through the liquid. We thus reduce our problem to cases 1 and 2 for unchanneled and channeled situations.

Considering a conservative situation, case 2, with the liquid film covering the entire surface, an estimate of the mass transfer rate can be made using Higbie's penetration model. Corrugations in plastic media and corners on rock media are of the order of 1 cm apart. Using a velocity of 0.9 cm/s, a contact time of approximately 1.1 s can be used in estimating a mass-transfer coefficient K_L value of 4.9×10^{-3} cm/s (note this is K_L, not $K_L a$). The flux of oxygen, N_O, across the gas-liquid interface is then

$$N_O = 4.9 \times 10^{-3}(C^* - C_b) \qquad (4\text{-}48)$$

Assuming a minimum bulk-liquid concentration of 1 mg/l and a saturation value C^* of 7.8 mg/l gives an estimated transfer rate of 3.3×10^{-5} mg/cm^2-s. As will be developed in Chap. 9, the maximum flux necessary in a trickling filter can be expected to be considerably below this value. Thus for this case at least, oxygen transfer rate does not seem to be a constraint. The fact that this is a conservative estimate is attested to by the "diffusion velocity" $\mathscr{D}/\delta$ magnitude of 4.2×10^{-3} being larger than the calculated K_L and by the fact that most trickling-filter effluents have a dissolved-oxygen concentration above 1.0 mg/l except when badly overloaded.

An oxygen-transfer limitation can occur in trickling filters if the organic concentration is very high or if the movement of air through the pores is not great enough. In most trickling filters, the draft of air is caused by temperature differences between the air inside and outside the filter, and stagnation can occur. If airflow rates are not sufficient, the filter can become anaerobic. This situation will be discussed in Chap. 9.

4-9 GAS STRIPPING

Dissolved gases often must be removed from wastewaters instead of being transferred into the liquid phase. Depending on the solubility of the particular gas, this may be a simple or difficult process. The mechanism is relatively straightforward. A clean gas is brought in contact with the liquid. Because the clean gas initially contains little or none of the pollutant gas, there is a net transfer from the liquid to the gas phase. Air is the normal clean gas used for stripping because of its availability and the large quantities that normally must be used. Diffused-aeration or film-flow processes are both used, although film-flow systems are more common.

The discussion here will be limited to film-flow processes, and the example of ammonia stripping will be used to provide numbers for calculation and an introduction to some of the problems that can occur with a stripping process. A schematic diagram of a stripping system is shown in Fig. 4-14. In practice, the support media are usually redwood strips, and the stripper operates much like a cooling tower. Analysis of the system is based on the assumption that the air leaving the stripper is in equilibrium with the incoming liquid. This assumption is incorrect but in most situations is satisfactory. Henry's law can now be used to predict the final air concentration of the pollutant. The statement of Henry's law given below [Eq. (4-49)] is in terms of mole fractions and partial pressures rather than concentrations, as used earlier in the chapter. This latter version is preferable, and Henry's law constants are normally reported in these terms.

$$P_i = H_i X_i \qquad (4\text{-}49)$$

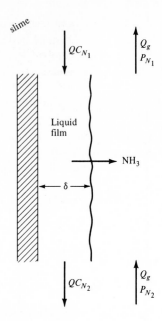

FIGURE 4-14
Schematic diagram of stripping process.

where P_i = partial pressure of component i
     X_i = mole fraction of component i in liquid phase

Relationships of interest concerning partial pressures are as follows.

$$P = \text{pressure of gas over the liquid}$$

$$= \sum_{i=1}^{n} P_i$$

$$y_i \propto P_i, \quad \text{where } y_i = \text{gas-phase mole fraction of } i$$

$$y_j = \frac{P_j}{\sum\limits_{i=1}^{n} P_i}$$

$$X_i = \frac{C_i}{\sum\limits_{i=1}^{n} C_i}$$

$$\sum_{i=1}^{n} X_i = 1$$

$$\sum_{i=1}^{n} C_i = C$$

Temperature and pH Effects

Equilibrium between gas- and liquid-phase composition is strongly affected by both pH and temperature. Ammonia is an excellent example of these effects. In water, ammonia is in equilibrium with ammonium ion.

$$NH_3 + H_2O \rightleftharpoons NH_4^+ + OH^- \qquad (4\text{-}50)$$

A convenient method of noting the relationship is to calculate the percentage of ammonia nitrogen in the ammonia form (on a molar basis):

$$NH_3 \text{ percent} = \frac{[NH_3]100}{[NH_3] + [NH_4^+]}$$

where the brackets indicate molar concentration. We can also write an expression in terms of the equilibrium constant for ammonia, K_b.

$$\frac{[NH_4^+]}{[NH_3]} = \frac{K_b}{[OH^-]}$$

The hydroxal-ion concentration is related to the hydrogen-ion concentration by the solubility product for water.

$$[H^+][OH^-] = K_w$$

Substituting these last two relationships into the ammonia percentage expression gives

$$NH_3 \text{ percent} = \frac{100}{1 + K_b[H^+]/K_w} \qquad (4\text{-}51)$$

The effect of pH on the ammonia concentration can be readily seen from Eq. (4-51). As pH decreases, $[H^+]$ increases, and the ammonia concentration decreases. While the solubility product of water has a constant value of 10^{-14}, the equilibrium product of ammonia-ammonium ion varies with temperature as shown in Table 4-2. The combined effect of temperature and pH on the liquid-

Table 4-2 K_b FOR AMMONIA

T, °C	$K_b \times 10^5$
0	1.374
10	1.570
20	1.710
30	1.820
40	1.862
50	1.892

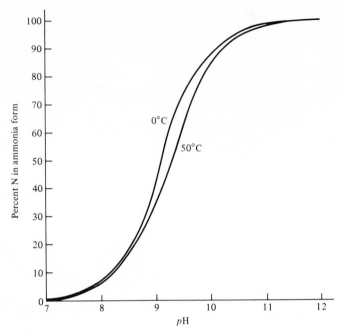

FIGURE 4-15
Percentage of N in ammonia form as a function of pH and temperature.

phase ammonia concentration is shown in Fig. 4-15. Because most wastes are nearly neutral with respect to pH, addition of a base is normally necessary. For economic reasons, lime is the preferred base. The amount of lime added is dependent upon the chemical makeup of the particular wastewater. This quantity can be predicted from knowledge of the alkalinity but in practical terms is usually determined experimentally. In most cases, the ammonia percentage required must be greater than 98 percent; thus pH values must be greater than 11.

Temperature also affects the gas-liquid ammonia equilibrium. The effect of temperature on the Henry's law constant for ammonia is shown in Fig. 4-16. For convenience in calculation, the gas-phase ammonia content has been given in mole fractions rather than partial pressure. Conversion to partial pressure is given by Eq. (4-52).

$$P_i = PY_i \qquad (4\text{-}52)$$

Calculation of the Henry's law constants from the curves is then possible by multiplying the slope of any of the curves by the total pressure.

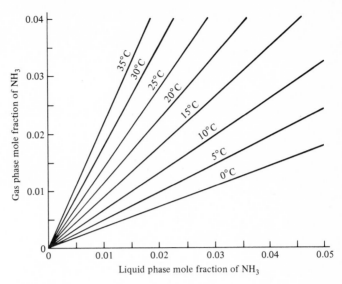

FIGURE 4-16
Liquid- and gas-phase NH_3 mole fractions as functions of temperature.

4-10 AIR REQUIREMENTS FOR GAS STRIPPING

If we consider a countercurrent process such as a film-flow scrubber and make a macroscopic mass balance (Fig. 4-17) on the pollutant gas entering and leaving, we obtain Eq. (4-53).

$$G(Y_2 - Y_1) = L(X_2 - X_1) \qquad (4\text{-}53)$$

where G = gas molar flow rate
L = liquid molar flow rate

EXAMPLE 4-3   A wastewater containing 40 mg/l of ammonia nitrogen (as N) has been raised to pH 12 by lime addition. The nitrogen is to be stripped out using a tower-type scrubber. If the liquid flow rate is 4000 l/s, determine the gas flow rate necessary to produce an effluent liquid-nitrogen concentration of 1 mg/l. Determine the gas flow rate necessary at 0, 10, and 20 °C.

Necessary removal   At pH_{12}, NH_3 percent > 99.8 percent.

$$\text{Effluent concentration} = 1 \text{ mg/l N}$$
$$= C_e(\tfrac{14}{17}) + (1 - 0.998)(40)$$
$$C_e = 0.76 \text{ mg/l}$$
$$\text{Removal necessary} = 0.998(40) - 0.76$$
$$= 39.16 \text{ mg/l}$$

FIGURE 4-17
Schematic diagram of countercurrent scrubber.

Determine influent and effluent mole fractions

$$X_2 = \frac{(40 \times 10^{-3})/17}{55.5 + (40 \times 10^{-3})} = 4.24 \times 10^{-5}$$

$$X_1 = \frac{(0.76 \times 10^{-3})/17}{55.5 + (0.76 \times 10^{-3}/17)} = 8.05 \times 10^{-7}$$

$$Y_1 = 0 \quad \text{(assumed clean air)}$$

$$Y_2 = H'X_2$$

where H' is the slope of the appropriate curve in Fig. 4-16.

$$H'_0 = 0.36$$
$$H'_{10} = 0.67$$
$$H'_{20} = 1.11$$

Determine gas flow rates necessary

$$G = \frac{L(X_2 - X_1)}{Y_2}$$

$$L = (4000 \text{ l/s})(55.5 \text{ mol/l}) = 222,000 \text{ mol/s}$$

$$G = 2.22 \times 10^5 \frac{4.16 \times 10^{-5}}{Y_2}$$

$$G_0 = 6.05 \times 10^5 \text{ mol/s}$$
$$= (6.05 \times 10^5 \text{ mol/s})(22.4)(\tfrac{273}{273})$$
$$= 13.6 \times 10^6 \text{ l/s}$$

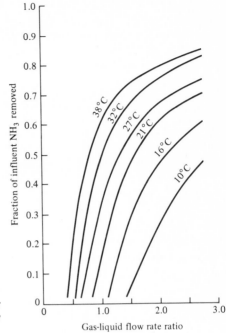

FIGURE 4-18
Fraction removed as a function of air temperature and gas-liquid molar flow rate ratio.

$$G_{10} = 3.25 \times 10^5 \text{ mol/s}$$
$$= (3.25 \times 10^5)(22.4)(\tfrac{283}{273})$$
$$= 7.55 \times 10^6 \text{ l/s}$$
$$G_{20} = 1.85 \times 10^5 \text{ mol/s}$$
$$= (1.85 \times 10^5)(22.4)(\tfrac{294}{273})$$
$$= 4.46 \times 10^6 \text{ l/s} \qquad \text{////}$$

A corollary to the above calculations is that removal efficiency drops with temperature if flow rate remains constant. Thus if the effluent ammonia liquid fraction C_1/C_2 is calculated as a function of temperature for a constant flow rate value, curves such as shown in Fig. 4-18 are obtained.

The problems illustrated in Fig. 4-18 must be carefully considered in design around. Culp and Culp,[32] presenting data from the Lake Tahoe, Calif., advanced waste-treatment plant, have emphasized this point. If a plant is to operate successfully under all weather conditions, it must be designed for the lowest temperatures to be experienced, and additional consideration must be given to the cooling effect of evaporation within the tower.

A more important limitation on the use of ammonia stripping is that the airstream containing the ammonia must not cause a pollution problem. For example, if the ammonia stripper is designed to protect a lake or river, there is little sense in allowing the airstream containing highly soluble ammonia to come in contact with the receiving water. Thus ammonia stripping is limited to cases where the wastewater is to be used for some purpose that requires ammonia removal and where the airstream will not come in contact with receiving waters. This situation can be modified to an extent by putting an acid adsorber on the the stripping unit. Here low-pH water is used to adsorb the ammonia from the airstream. Because of the low pH, solubility is high, and the entire process effectively becomes a concentration system. Some method of disposal of the final effluent stream must, of course, be provided. Extraction of the concentrated nitrogen for use as fertilizers is one possibility.

The equilibrium model presented here does not consider rate of interphase transfer and, therefore, does not include any terms which indicate the time of contact needed, the tower depth, or any other physical design parameters. Clearly, the process is simply the reverse of the oxygen-transfer systems discussed in this chapter, and similar models can be constructed. If, as is probable, the liquid-phase resistance is controlling, the models would be virtually identical. The ratio of gas-to-liquid flow rates (G/L) might be used as a mixing parameter such that a correlation between fraction-removed G/L and tower depth would be useful. Slecta and Culp,[33] reporting on pilot studies of ammonia stripping using various packing materials, a range of G/L ratios between 0 and 9 mol air/mol wastewater (0 to 1500 ft^3 air/gal), and a tower depth of 7.3 m (24 ft), found that at 20°C removal increased rapidly up to a value of $G/L = 1.5$. Removal efficiency at $G/L = 1.5$ was approximately 0.90 and gradually increased to nearly 1.0 with increasing airflow. From Fig. 4-18 we can see that at 20°C (68°F) the 0.90 to 1.0 fraction-removed region should be reached between G/L values of 0.83 and 1.10. Thus mass transfer rates are evidently of importance and need to be investigated if proper design procedures are to be developed.

Cross-flow or bubble-type ammonia-removal processes are of interest also. Because of the lack of suitable data, discussion of these processes will not be considered.

PROBLEMS

4-1 Scale-up of aeration equipment is a difficult problem in practice. Discuss why the factor $H_L^{2/3}$ in Eq. (4-22) does not satisfactorily account for all depth effects. What other scale-up factors would be important?

4-2 A batch-oxygen-transfer experiment is run using a full-scale aeration basin (4 m deep and 20 m square). The wastewater was initially deoxygenated with excess sodium sulfite, and oxygen transfer was measured by the increase in sulfate concentration. Aeration was through a set of porous plates placed in the tank bottom. Data from the experiment are given in Table 4-3. Assuming that the tank behaves as an approximate ideal CFSTR, determine the proper air input rate to maintain 1 mg/l oxygen in the system. Expected flow rate is 50 l/s and the expected influent and effluent COD concentrations are 350 and 60 mg/l, respectively. The O_2-utilized/COD-removed ratio is 0.4. Saturation concentration of oxygen in the wastewater is 7.4 mg/l.

Table 4-3 SULFATE CONCENTRATION, mg/l

Time, min	540 l/s	1080 l/s	2160 l/s	4320 l/s
0	26.0	26.0	26.0	26.0
1	28.9	32.1	40.4	60.9
2	31.9	38.3	54.8	95.8
3	34.8	44.4	69.2	130.7
4	37.7	50.5	83.5	165.6
5	40.6	56.6	97.9	200.5
7	46.5	68.9	126.7	270.3
10	55.3	87.3	169.9	375.0

4-3 Scale-up studies for diffused-aeration processes have resulted in the data presented in Table 4-4. The experimental system consisted of a cylindrical tube (30-cm ID) with a porous plate for a bottom. Water was added to a given height, de-oxygenated by nitrogen stripping, and then a non-steady-state batch-aeration experiment was run. Bubble sizes were estimated by scaling values off the photo-

Table 4-4 OXYGEN CONCENTRATION, mg/l

	Depth of water, cm								
	60			120			240		
	Q_g, SLM								
	4	8	12	4	8	12	4	8	12
	$\bar{d}_b$, cm								
Time, s	0.18	0.22	0.25	0.19	0.23	0.26	0.21	0.26	0.29
0	0	0	0	0	0	0	0	0	0
50	0.82	1.29	1.66	0.62	1.00	1.30	0.45	0.72	0.45
100	1.55	2.38	2.98	1.20	1.88	2.39	0.88	1.37	1.79
200	2.80	4.05	4.85	2.22	3.32	4.07	1.66	2.51	3.18
400	4.62	6.05	8.78	3.82	5.26	6.07	2.98	4.23	5.09
800	6.57	7.52	7.81	5.81	7.06	7.53	4.85	6.22	6.94
1600	7.75	7.97	7.99	7.40	7.89	7.97	6.76	7.61	7.86

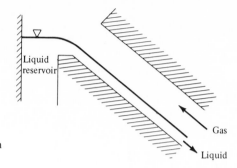

FIGURE 4-19
Experimental system used to develop in
Table 4-5.

graphs taken with a high-speed camera. All oxygen-concentration values are averages of samples taken at two or more depths. Estimate bubble velocities and make a plot similar to Fig. 4-5. Determine whether or not Eckenfelder's model for β holds for this data. Determine whether or not Eq. (4-22) holds.

4-4 Using the data presented in Prob. 4-3, determine the airflow rate necessary and the oxygen-transfer efficiency expected for a CFSTR activated-sludge system used to treat a flow rate of 40×10^6 l/d (10.6 mgd) with an incoming COD concentration of 650 mg/l, an effluent COD concentration of 60 mg/l, a yield factor of 0.34, a hydraulic residence time of 0.2 d, and a tank depth of 5 m. Assume an oxygen concentration of 1 mg/l must be maintained.

4-5 Discuss the effect of placement of surface aerators in tanks. Consider the dimensions of tanks using a single aerator, distances between aerators in tanks using multiple aerators, and effects of depth.

4-6 Plot the gas-to-liquid flow rate ratios necessary for 98 percent nitrogen removal at pH values of 11 and 12 and temperatures of 0 to 30°C.

4-7 Estimate the air requirements for a process that is to reduce the ammonia-nitrogen concentration of a wastewater from 58 to 3 mg/l. The pH of the wastewater entering the treatment unit will be 11.0. Assume design temperature is 20°C.

4-8 Experiments with a model stripping unit (see Fig. 4-19) have resulted in the data presented in Table 4-5. Evaluate the data in the table and explain the implications that can be derived from your evaluation.

Table 4-5

P, mm Hg	760	760	760	760	760
Q_g, l/min	25	25	25	25	25
C_{N_i}, mg/l	60	60	60	60	60
C_{N_o}, mg/l	49.5	39.0	31.0	22.0	6.0
P_{N_i}, mm Hg	0	0	0	0	0
P_{N_o}, mm Hg	0.039	0.067	0.078	0.081	0.084
Q_L, l/min	0.070	0.060	0.050	0.040	0.030
pH	11	11	11	11	11
T, °C	26.7	26.7	26.7	26.7	26.7

REFERENCES

1. LEWIS, W. K., and W. C. WHITMAN: Principles of Gas Adsorption, *J. Ind. Eng. Chem.*, vol. 16, 1924.
2. HIGBIE, R.: The Rate of Absorption of Pure Gas into a Still Liquid during Short Periods of Exposure, *Trans. Am. Inst. Chem. Eng.*, vol. 31, p. 365, 1935.
3. DANCKWERTZ, P. V.: Significance of Liquid Film Coefficients in Gas Absorption, *J. Ind. Eng. Chem.*, vol. 43, p. 1460, 1951.
4. ECKENFELDER, W. W., JR.: "Industrial Water Pollution Control," McGraw-Hill Book Company, New York, 1966.
5. WEBB, F. C.: "Biochemical Engineering," Van Nostrand Reinhold Company, New York, 1964.
6. AIBA, S., et al.: "Biochemical Engineering," 2d ed., Academic Press, Inc., New York, 1973.
7. LEMLICH, R.: "Adsorptive Bubble Separation Techniques," Academic Press, Inc., New York, 1972.
8. COULSON, J. M., and J. F. RICHARDSON: "Chemical Engineering," vol. II, 2d ed., Pergamon Press, New York, 1968.
9. CAMP, T. R.: Discussion of Mechanics of Stream Reaeration, by D. J. O'Connor and W. Dobbins, *J. Sanit. Eng. Div.*, ASCE, vol. 81, 1955.
10. ECKENFELDER, W. W., JR.: Absorption of Oxygen from Air Bubbles in Water, *J. Sanit. Eng. Div.*, ASCE, vol. 85, SA4, p. 89, 1959.
11. IPPEN, A. T., and C. E. CARVER, JR.: Basic Factors of Oxygen Transfer in Aeration Systems, *Sewage Ind. Wastes*, vol. 26, p. 813, 1954.
12. KING, H.: Mechanics of Oxygen Absorption in Spiral Flow Tanks, *Sewage Ind. Wastes*, vol. 27, p. 894, 1955.
13. O'CONNOR, D. J., and W. DOBBINS: The Mechanics of Reaeration in Natural Streams, *J. Sanit. Eng. Div.*, ASCE, vol. 82, SA6, 1956.
14. ECKENFELDER, W. W., JR., and D. J. O'CONNOR: "Biological Waste Treatment," Pergamon Press, New York, 1966.
15. STREETER, H. W., ET AL.: Measures of Natural Oxidation in Polluted Streams, *Sewage Works J.*, vol. 8, 1936.
16. HOWLAND, W. E.: Effect of Temperature on Sewage Treatment Processes, *Sewage Ind. Wastes*, vol. 25, p. 161, 1953.
17. WILKE, C. R., and P. CHANG: Correlation of Diffusion Coefficients in Dilute Solutions, *AIChE J.*, vol. 1, p. 264, 1955.
18. AIBA, S., and K. TODA: The Effect of Surface Active Agents on Oxygen Absorption in Bubble Aeration, *Gen. Appl. Microbiol.*, vol. 9, p. 443, 1963.
19. ECKENFELDER, W. W., JR., and E. L. BARNHART: Performance of a High Rate Trickling Filter Using Selected Media, *Water Pollut. Control Fed.*, vol. 35, p. 1838, 1963.
20. MANCY, K. H., and D. A. OKUN: The Effects of Surface Active Agents on Bubble Aeration, *Water Pollut. Control Fed.*, vol. 32, p. 351, 1960.
21. MCKEOWN, J. J., and D. A. OKUN: Effects of Surface Active Agents on Oxygen Bubble

Characteristics, in W. W. Eckenfelder and B. J. McCabe (eds.), "Advances in Biological Wastewater Treatment," Pergamon Press, New York, 1963.

22. TIMSON, W. J., and C. G. DUNN: Mechanism of Gas Absorption from Bubbles under Shear, *J. Ind. Eng. Chem.*, vol. 52, p. 799, 1960.

23. BUSCH, A. W.: "Aerobic Biological Treatment of Wastewaters," Olygodynamics Press, Houston, Tex., 1968.

24. PRICE, K. S., et al: Surface Aerator Interactions, *J. J. Environ. Eng. Div.*, *ASCE*, vol. 99, p. 263, 1973.

25. ECKENFELDER, W. W., and D. J. O'CONNOR: "Biological Waste Treatment," Pergamon Press, New York, 1961.

26. RICHARDS, J. W.: Studies in Aeration and Agitation, *Prog. Ind. Microbiol.*, vol. 3, p. 143, 1961.

27. BUSCH, A. W.: A Five Minute Solution for Stream Assimilative Capacity, *Proc. 26th Ind. Waste Conf.*, 1971.

28. HOWELL, J.: Personal communication, Swansea, Wales, Nov. 8, 1973.

29. ATKINSON, B., and E. L. SWILLEY: A Mathematical Model for the Trickling Filter, *Proc. 18th Ind. Waste Conf.*, 1963.

30. ATKINSON, B., and D. WILLIAMS: The Performance Characteristics of Trickling Filter with Hold Up of Microbial Mass Controlled by Periodic Washing, *Trans. Inst. Chem. Eng.*, vol. 49, p. 215, 1971.

31. ATKINSON, B.: "Biochemical Reactors," Pion Press, London, 1974.

32. CULP, R. L., and G. L. CULP: "Advanced Wastewater Treatment," Van Nostrand Reinhold Company, New York, 1971.

33. SLECTA, A. F., and G. L. CULP: Water Reclamation Studies at the South Lake Tahoe Public Utilities District, *Water Pollut. Control Fed.*, vol. 39, p. 787, 1967.

5

SOLIDS REMOVAL

Removal of solids from suspension in water and wastewaters is one of the most common treatment operations. Most waters contain solids to some degree, and in cases where biological treatment of wastewater is used, additional solids are created during treatment. Because of the high volumetric flow rates associated with water supplies and wastewater discharges, there are economic constraints on the capital and operating costs which can be incurred. These constraints limit removal methods to gravity sedimentation and thickening, in most cases. In some instances, the addition of coagulant aids (chemicals that increase the flocculation rate) and flocculation may be feasible. Where effluent suspended-solids standards are stringent (such as for domestic water supplies) filtration of the final effluent may be necessary, regardless of the additional cost. Finally, concentration of the sludges produced from these processes is sometimes necessary prior to final disposal. Both centrifugation (a form of sedimentation) and vacuum filtration are used and will be discussed separately because of certain peculiarities associated with these two processes.

5-1 IDEAL SEDIMENTATION

Out of respect for tradition, we will start our discussion by considering a rigid sphere falling through a newtonian fluid. A force balance on the sphere gives Eq. (5-1).

$$\frac{\pi}{6} d_p^3 \rho_p \frac{dv_p}{dt} = \frac{\pi}{6} d_p^3 g(\rho_p - \rho_l) - C_D A_p \rho_l \frac{v_p^2}{2} \qquad (5\text{-}1)$$

where ρ_p, ρ_1 = particle and liquid densities, respectively
  d_p = particle diameter
  v_p = particle velocity
  A_p = cross-sectional area of the particle
  C_D = coefficient of drag

As noted in Chap. 4, the drag coefficient for a rigid sphere falling through a liquid is

N_{Re}	C_D
$N_{Re} < 0.1$	$24/N_{Re}$
$0.1 < N_{Re} < 1000$	$18.5/N_{Re}^{3/5}$
$N_{Re} > 1000$	0.44

where N_{Re} is the Reynolds number.

Our interest is primarily in the terminal velocity of the particles, and therefore dv_p/dt is assumed to be zero and v_p is assumed to be the terminal particle velocity from this point on. Because sedimentation is a removal process (i.e., the purpose is to remove as many particles as possible), the slowest-settling particles are of interest rather than the faster ones. These critical particles have densities close to that of water and diameters (or effective diameters in the case of nonspheres) of less than 1 mm and can be expected to settle according to Stokes' law.

$$v_p = \frac{d_p{}^2 g(\rho_p - \rho_l)}{18\mu} \qquad (5\text{-}2)$$

where g = gravitational constant
  μ = viscosity of liquid

Equation (5-2) is useful in considering why coagulation and flocculation are of interest in water and wastewater treatment. Particle velocity is a function of particle diameter, particle density, liquid density, and liquid viscosity. The latter two variables cannot be changed in most treatment systems, but both the particle diameter and particle density can be changed through the aggregation of small particles into larger ones. In general, bulk-particle density of the aggregated particles decreases as the overall diameter increases due to inclusion of water in the floc. The decrease in density is extremely important as indicated by Eq. (5-2). Thus, for example, a decrease in density from 1.1 to 1.09 is 0.9 percent decrease in density of the particle but a 10 percent decrease in the function $\rho_p - \rho_l$. This difference is normally more than made up for by the fact that particle diameters are squared. The following example demonstrates this concept.

EXAMPLE 5-1   Particles of diameter 0.02 cm and density 1.1 g/cm^3 are settling in water (viscosity 0.01 P or 10^{-3} Pa-s). Determine the terminal velocities of single particles, groups of 8, and groups of 64 such particles. Assume the packing arrangement is such that lines drawn from particle center to particle center are at right angles to each other.

Single-particle

$$v_{p1} = \frac{(0.02)^2(1.1 - 1.0)}{18(0.01)}$$

$$= 2.2 \times 10^{-4} \text{ cm/s}$$

Aggregate of 8 particles with maximum packed volume

$$\rho_8 = \rho_p \frac{\pi}{6} + \left(1 - \frac{\pi}{6}\right)\rho_1 = 1.05$$

$$d_8 = \left[\frac{6}{\pi}(2d)^3\right]^{1/3} = 0.05$$

$$v_{p8} = \frac{1}{18}\frac{(0.05)^2}{0.01}(1.05 - 1.0) = 6.9 \times 10^{-4} \text{ cm/s}$$

Aggregate of 64 particles with maximum packed volume

$$\rho_{64} = 1.05$$

$$d_{64} = \left[\frac{6}{\pi}(4d)^3\right]^{1/3}$$

$$= 0.099 \text{ cm}$$

$$v_{p64} = \frac{1}{18}\frac{(0.099)^2}{0.01}(1.05 - 1.0)$$

$$= 2.77 \times 10^{-3} \text{ cm/s} \qquad ////$$

5-2 COAGULATION

Increasing particle size is advantageous, particularly in cases where a significant fraction of particles are quite small. A problem exists in the fact that particles of colloidal size usually are charged, and in a water or wastewater suspension, charges

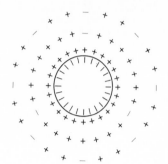

FIGURE 5-1
Ionic double layer.

on particles are usually the same. Thus the particles repel each other, and the colloidal suspension is termed *stable*. Coagulation is the process by which these particle suspensions are destabilized and particles are allowed to grow or flocculate to sizes that settle at satisfactory velocities. There are four major mechanisms of coagulation: double-layer compression, adsorption and charge neutralization, enmeshment in a precipitate and adsorption by polymers, and interparticle bridging. In practice, all four mechanisms are probably involved, but the latter two, enmeshment in a precipitate and adsorption and interparticle bridging, are the principal mechanisms for particle-size growth. An understanding of all four is necessary for an understanding of the phenomena of coagulation, however.

As previously stated, colloidal- and near-colloidal-size particles usually have an "apparent" surface charge where the magnitude and sign are a function of pH. Because the particle-liquid system is neutral (i.e., does not have a net charge), a *double ionic layer*[1] develops as illustrated in Fig. 5-1. Ions with the same charge as the particle are rare near the particle surface but gradually increase in number as the distance from the particle increases. Unlike ions, or counterions,

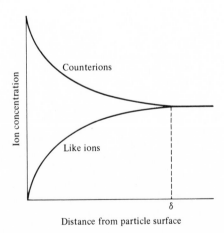

FIGURE 5-2
Ion-concentration profile near colloid surface.

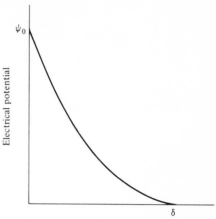

FIGURE 5-3
Electrical potential variation with distance
from colloid surface.

predominate near the particle surface and gradually decrease in number
concentration with increasing distance (Fig. 5-2). This predominance of one
type of ion near the particle surface results in a potential energy gradient, which
is graphically illustrated in Fig. 5-3. The shape of the curves in Figs. 5-2 and 5-3
is affected by the ionic strength of the water. Because of the ion-concentration
gradient existing, there is a tendency for counterions to move away from the
particle surface by molecular diffusion. When the ionic strength of the solution
is high, the *diffuse* layer, i.e., the layer in which an ion-concentration gradient
exists, is decreased in depth and the potential gradient becomes steeper. However,
the net potential difference between the particle surface and the bulk liquid (Ψ_0)
does not change because this value is solely determined by the particle-surface
charge. A potential termed the *zeta* potential $\varkappa$ has traditionally been used in
water and waste treatment. The zeta potential is based upon the diffuse-layer
thickness δ.

$$\varkappa = \frac{4\pi\delta q}{D^*} \qquad (5\text{-}3)$$

where q = net (apparent) particle charge
D^* = dielectric constant of water

The zeta potential is useful in a qualitative sense because it can be measured,[1]
and because (for a given experimental system) the colloidal suspension stability
decreases as the zeta potential decreases. Prediction of the critical value (the
value when destabilization occurs) has not been possible, and the use of this
parameter has decreased in recent years.

Attractive forces (van der Waals forces) analogous to gravitational forces
also exist between particles. The strength of these forces is as for other body

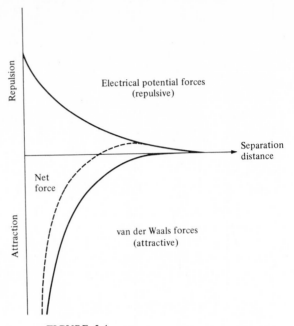

FIGURE 5-4
Net force between colloidal particles of like charge.

forces related to the distance between particles.[1] A qualitative picture of the two types of forces, repulsion and attraction, and their net effect is illustrated in Fig. 5-4. Clearly, the depth of the diffuse layer is important in determining whether or not attractive forces predominate and whether or not particles of like charge can be brought together. As solution ionic strength increases, the effective distance of repulsive forces is decreased toward the particle surface. Eventually, the net-charge curve drops entirely into the attractive-force region, and the particles are attracted to each other by the van der Waals forces. As indicated in Eq. (5-2), the zeta potential decreases under these conditions because the diffuse-layer thickness δ between particles decreases.

Increasing ionic strength by adding electrolytes, termed *double-layer compression*, is not believed to be a major factor in the coagulation processes used in water and wastewater treatment. There is some basis for believing that it is significant in the coagulation of colloidal clay particles moving from freshwater rivers into estuaries, however.

Coagulation can also be induced by simple electrostatic adsorption of counterions that effectively neutralize the particles and decrease the surface potential Ψ_0. This effect is dependent upon the particular ion and follows the Schulze-Hardy rule, which states that the effectiveness of a counterion in bringing about coagulation increases markedly with charge. Thus the effectiveness of Na^+, Ca^{2+}, and Al^{3+} as measured by ionic concentrations necessary to bring about coagulation is usually approximately $1 : 10^{-2} : 10^{-3}$. In the case where counterions are simple cations or anions, double-layer compression and adsorption and charge neutralization would be parallel operations. When large, complex molecules are used as coagulants, ordinary adsorption becomes important also. Overdosage with coagulant can result in particle-charge reversal due to adsorption of excess ions, and the result is a stable colloid.

Commonly used coagulants include metal salts such as $Al_2(SO_4)_3$(alum), $FeCl_3$(ferric chloride), or metal oxides and hydroxides such as lime [CaO, $Ca(OH)_2$]. Hydrated-metal hydroxides are formed which have extremely low solubility limits, and hence, precipitates of these hydroxides form rapidly at low concentrations. For an explanation of the chemistry of the precipitate formation, the reader should investigate the work of Stumm and his coworkers.[2-4]

Precipitates may be adsorbed on particle surfaces and act through charge neutralization. If this mechanism is significant, overdosing is possible, and pH effects are also important. When the pH of the system is below the isoelectric point of the metal hydroxide, the precipitate will exhibit a positive charge. Above the isoelectric point, it will have a negative charge. Therefore, if adsorption and charge neutralization is an important mechanism, careful matching of coagulant dosage and particle charges is necessary for effective particle destabilization.

In most cases of coagulating wastewaters, coagulant dosages are high enough to produce large, loose-matrix, precipitate particles. These particles collect small particles such as colloids as they settle. Effectiveness of this process, enmeshment in the precipitate, is dependent on colloid concentration as well as coagulant dosage. At higher colloid concentrations, lower coagulant dosages are possible. For this reason, bentonite or activated silica is sometimes added to increase the probability of particle collision.

The metal hydroxides formed when coagulants are added to water are acidic; i.e., they are hydrogen donors. Solubility of the metal ions and the metal hydroxides is strongly related to pH. Alkalinity of the water is, therefore, an important factor in determining the effectiveness of a coagulant. In cases where alkalinity is low, lime addition may be necessary to produce a satisfactory precipitate.

Precipitates formed by metal hydroxides are actually polymers, and the process of enmeshment can, to a certain extent, be considered a process of adsorption and bridging between particles. Organic polymers provide a much more explicit example of this phenomenon, however. Cationic, anionic, and nonionic polymers are available for use as coagulants, and all three are extensively used in water and wastewater treatment. Choice of the most suitable polymer for a given situation depends on both the type of particle being removed and the chemical characteristics of the water or wastewater. Ions present in the water interact with the polymer and, in some cases, may result in the surprising situation where an anionic polymer is the most suitable coagulant for a negatively charged colloid.

Selection of Coagulant Type and Dosage

Determination of the need for coagulant addition and the type and amounts of coagulant to be used should be based upon laboratory and if possible pilot plant studies. A significant expense, both in capital outlay and operational costs, is incurred whenever coagulation is added to a treatment system. Careful analysis of the improvement in product quality provided by coagulant addition should, therefore, be made.

Flash mixing is necessary to provide uniform coagulant species in the water. Many colloidal particles are relatively fragile (e.g., biological flocs) and cannot withstand the high shearing rates desirable for flash mixing of Fe^{3+} and Al^{3+} ions. When slower mixing rates are used, higher coagulant dosages may be necessary due to the range of metal-complex species produced. At high mixing rates, the increased surface area produced by particle disintegration increases coagulant dosage requirements. In general, the complexity of the water, colloid, and chemical mixture requires experimental data for design.

Ideally, data on residual turbidity (turbidity of the settled, coagulated suspension) should be measured as a function of coagulant dosage, pH, Reynolds number, and temperature (where temperature varies significantly). Remembering that the coagulant-dosage variable often includes more than one coagulant (for example, $FeCl_3 + CaO$) presents us with a large number of these variables at the start. Many can be eliminated by inexpensive jar tests. Pilot plant testing of the final set of coagulants and Reynolds numbers is recommended, however.

5-3   FLOCCULATION

Particulate material in wastewater must be brought together in order for aggregation or particle growth to occur. This process is called *flocculation*. Two principal mechanisms of flocculation may be considered, perikinetic and orthokinetic.[5] Perikinetic flocculation is due to the random motion (brownian motion) of particles resulting from collisions with fluid molecules and is only significant for very small particles. Not surprisingly, the process is very slow, and consequently perikinetic flocculation is not useful in water treatment processes, although it is undoubtedly of great importance in nature. An analysis of the kinetics of perikinetic flocculation was given by Swift and Friedlander.[6]

Orthokinetic flocculation is induced by the presence of velocity gradients in the fluid. Because of the velocity gradients, particles travel at different velocities, and collisions result. Particles that collide may aggregate as previously described, and depending upon particle and fluid characteristics (e.g., colloid stability), an aggregation efficiency ε will be established for each combination. An expression giving the frequency of collision between particles of various sizes in suspension was presented by von Smoluchowski[7] in 1918.

$$J_{ij} = \frac{4}{3} n_i n_j (z_{ij})^3 \frac{du}{dz} \qquad (5\text{-}4)$$

where J_{ij} = number of collisions per unit time between ith and jth particles, $t^{-1}l^{-3}$
$\quad n_i, n_j$ = number concentrations of ith and jth particles, respectively, l^{-3}
$\quad\quad z_{ij}$ = collision distance
$\quad\quad\quad$ = $Z_i + Z_j$, sum of particle radii
$\quad du/dz$ = local velocity gradient, t^{-1}

Harris[5] used the Smoluchowski[7] equation to develop a rate of aggregation expression for flocculating systems with a range of particle sizes. His development assumes that when two particles collide and aggregate into a larger particle, the new particle's volume is the sum of the volumes of the smaller particles. This is the same as saying that particles fit perfectly together, which while undoubtedly overoptimistic, allows particle density to be considered constant. We have previously seen (Example 5-1) that particle density is roughly constant when particles aggregate on some fixed packing scheme, and Krone[8] has shown that this is probably the case in particle flocculation. Thus, Harris' assumption is probably quite satisfactory and, at the most, will result in the introduction of a constant to the expression. Harris[5] also introduced Krone's[8] concept of maximum particle size into his model. Maximum particle size is controlled by shear strength. As particles grow, their shear resistance decreases. At some limiting-particle size, the shearing strength of the particle is exceeded by the shearing stress due to the local velocity gradients, and the particle ceases to grow.

The smallest aggregating particle is termed the *primary* particle, and its number concentration is n_1. Larger particles are made up of i, j, or k primary particles and have number concentrations of n_i, n_j, and n_k. The radii of these larger particles are $i^{1/3}Z_1$, $j^{1/3}Z_1$, and $k^{1/3}Z_1$. Harris' expressions for the rate of change of the concentration of primary particles and the total particle concentration are given here without derivation by Eqs. (5-5) and (5-6), respectively.

$$\frac{dn_1}{dt} = -\varepsilon \alpha a^3 \frac{du}{dz} \phi n_1 \qquad (5\text{-}5)$$

where ϕ = volume fraction of particles in suspension
ε = collision aggregation efficiency

$$\phi = \frac{4}{3}\pi Z_1^3 \sum_{i=1}^{p} in_i$$

$$\alpha = \frac{\sum\limits_{i=0}^{p} n_i(i^{1/3} + 1)^3}{\sum\limits_{i=0}^{p} in_i}$$

$$a = \frac{z_{ij}}{r_i(i^{1/3} + j^{1/3})}$$

$$\frac{dn_t}{dt} = \frac{-\varepsilon\gamma a^3}{\pi}\frac{du}{dz}\phi n_t \qquad (5\text{-}6)$$

$$\gamma = \frac{\sum\limits_{j=0}^{p}\sum\limits_{i=0}^{p-1} n_i n_j(i^{1/3} + j^{1/3})^3}{\sum\limits_{j=0}^{p}\sum\limits_{i-0}^{p} n_i n_j(i + j)}$$

where n_t is the total number concentration of the particles, and the pth particle is the maximum size.

In any removal process the rate is the parameter of interest. Because of the approximations used and the difficulty of measuring parameters such as n_i and n_j, Eqs. (5-5) and (5-6) are of primarily conceptual importance. We can see that increasing the shearing rate increases the number of contacts, and therefore, high shearing rates seem desirable. The rate is also proportional to the number of particles present, which means that high concentrations and high flocculation rates go together. As a consequence, we see that as particle sizes increase, the number concentration n_t must decrease, and therefore the overall rate of aggregation will slow down with time. Maximum particle size cannot be determined from the model, and, in fact, a method of predicting the maximum particle size is not available. This parameter effectively limits the values of the velocity gradient that can be used, however. Larger flocs give higher settling rates [Eq. (5-2)], and thus some economic balance between shearing rate and settling velocity desired must be determined for good design.

In water and wastewater treatment the major controllable flocculation variable is the velocity gradient. Particle concentration can be modified by the addition of clay or other particulates, but this is rarely done. In mixed systems the gradient is a function of the power input to the liquid. Camp and Stein[9]

FIGURE 5-5
Stake and stator flocculation unit at Cincinnati, Ohio, water-treatment plant.

gave the following equation for the average velocity gradient.

$$\frac{\overline{du}}{dz} = \left(\frac{P_L}{V\mu}\right)^{1/2} \qquad (5\text{-}7)$$

where V = tank volume, l^3
    μ = viscosity, P or Pa-s
    P_L = power input, W
    u = velocity

Obviously, the range of velocity-gradient values is extremely important. Flocculators such as the stake and stator units (Fig. 5-5) would be expected to have a high shearing rate near the stakes and stators, a lower shearing in the bulk field, and an intermediate value near the tank walls. Dead spaces undoubtedly exist in these units and would be expected to decrease the efficiency of flocculation. The flocculation equipment shown in Fig. 5-5 leads one to the conclusion that the average velocity gradient du/dt, as determined by Eq. (5-7),

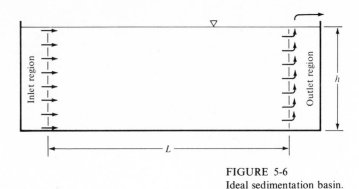

FIGURE 5-6
Ideal sedimentation basin.

is a qualitative parameter at best. Thus, a velocity gradient found satisfactory at one wastewater flow rate in a given unit may not be satisfactory at another flow rate and is quite probably irrelevant in a quantitative sense with respect to another type of unit. Values reported in the literature usually vary between 10 and 100 s^{-1}.

5-4 DISCRETE PARTICLE SEDIMENTATION

When particles settle independently through a quiescent fluid, their motion can be described by Eq. (5-1). Particles of concern in wastewater treatment are usually small and have densities near that of water, allowing the use of Stokes' law [Eq. (5-2)] to describe their terminal velocities. If we consider an ideal sedimentation tank (Fig. 5-6) in which the wastewater flow is horizontal throughout the settling region and the particles are uniformly distributed at the entrance, several useful relationships can be developed.

Particles entering the ideal tank are assumed to differ in size and density, but each will have a characteristic terminal settling velocity as defined by Eq. (5-2). We can, therefore, classify the particles by settling velocity. For example, there will be a weight fraction y_i, which has a velocity equal to or less than v_i. A critical velocity can be defined for a given tank as the liquid depth h divided by the hydraulic residence time Θ_H.

$$v_c = \frac{h}{\Theta_H}$$

$$= \frac{h}{V/Q} \qquad (5\text{-}8)$$

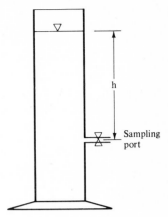

FIGURE 5-7
Discrete particle-settling apparatus.

$$v_c = \frac{Q}{Lb} = \frac{Q}{A_{surf}} \qquad (5\text{-}9)$$

where b is the tank width. The critical velocity is often given in terms of volume per unit surface area of the tank per unit time (e.g., m³/m²-d or gal/ft²-d) and is referred to as the *overflow rate*. A particle that settles with a velocity equal to or greater than the critical velocity will be removed no matter at what height (distance from the bottom) it enters the sedimentation region. Particles having a velocity less than v_c will be removed only if they enter at a height $v_i h/v_c$. For example, if a particle has a characteristic settling velocity of $v_c/2$ and enters anywhere above $h/2$, it will pass into the exit region and be carried out of the tank. Because we assume each type of particle is uniformly distributed at the entrance, the fraction of particles having a velocity v_i (which will be removed) is 1.0 if $v_i > v_c$ and v_i/v_c if $v_i < v_c$.

A batch-settling test can be used to experimentally determine the settling characteristics of a suspension that exhibits discrete particle settling. Equipment necessary is a column with a sampling port (Fig. 5-7) and filters and balances necessary for a suspended-solids determination. Column diameter need not be great because wall effects are negligible for discrete particle settling when $d_{cyl}/d_p > 100$. At the initiation of an experiment, the wastewater sample is poured into the cylinder and stirred while an initial ($t = 0$) sample is taken. Stirring is stopped, and samples are then taken at appropriate time intervals. In most cases, intervals of 30 to 60 s for the first 5 min and 120 to 300 s thereafter is satisfactory. If the particles do settle discretely, each sample will contain only particles having a velocity v_i less than h/t. All particles with a given velocity v_i in a given sample will be in the same concentration as at $t = 0$, however.

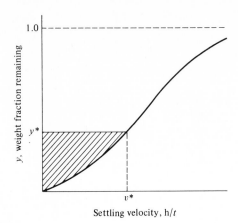

FIGURE 5-8
Discrete particle-settling curve.

Plotting the weight fraction of particles remaining y versus the velocity h/t produces a curve similar to that shown in Fig. 5-8.

Entering Fig. 5-8 at a chosen value of v, designated as v^*, a corresponding value of the weight fraction remaining y^* is obtained. We know that a fraction $1 - y^*$ has a velocity greater than v^* and will be removed in an *ideal* system which has an overflow rate equal to v^*. The fraction of particles with a lower velocity that will be removed is dependent on the size-density distribution. For an incremental weight fraction Δy the weight fraction removed is $(v/v^*)\,\Delta y$. Summing these increments and applying the limit as $\Delta y \to 0$ allows us to write an expression for the total weight fraction removed y_R.

$$y_R = 1 - y^* + \frac{1}{v^*} \int_0^{y^*} v \, dy \qquad (5\text{-}10)$$

The integral is equal to the shaded area in Fig. 5-8. Because the curve in Fig. 5-8 is developed from a relatively small number of data points, graphical integration is a suitable means of evaluation.

Equation (5-10) is useful in predicting the removal of suspended solids by a settling basin. Several constraints must be considered, however. Flow patterns in real basins do not correspond to those assumed for the ideal basin. Sedimentation-tank effluent is removed at several points on the surface of rectangular tanks and on the periphery of circular tanks. There is no separate exit region, and therefore there is an upward component to the wastewater velocity that partially counteracts the settling velocity of the particle. Currents are often present because of density differences between incoming wastewater and that in the tank or because of wind, and these must be considered in choosing tank dimensions. As a rule of thumb, the overflow rates associated with a fraction removed in Eq. (5-10) should be multiplied by a factor between $\frac{2}{3}$ and $\frac{4}{5}$ for tanks using conventional length-to-width ratios.

Analysis of the ideal sedimentation tank assumes that a particle that reaches the tank bottom is removed. This is a good approximation as long as the wastewater velocity is less than scour velocity. When scour occurs, particles are resuspended and carried out of the tank. The critical scour velocity can be estimated from Eq. (5-11).

$$u_s = \left(\frac{8}{f}\right)^{1/2} v_c \qquad (5\text{-}11)$$

where u_s = average horizontal velocity
f' = Fanning friction factor
v_c = critical settling velocity

5-5 FLOCCULANT SEDIMENTATION

In water treatment the solids are flocculant, and most of the suspended material in municipal wastewater, other than grit, is organic in nature and tends to flocculate or aggregate rather easily. Because there is a distribution of particle sizes, larger particles catch up with smaller ones, and the contacts result in particles that are larger and settle faster than either of the parent particles. Models such as presented for discrete particle settling do not apply to these systems, and the range of particle sizes, shapes, and densities precludes developing a general analytical approach. Prediction of removals (or effluent quality) in this type of system is based upon experimental data.[10] As in the case of discrete particle settling, a batch experimental system is used. The settling column differs in that a number of sampling ports are necessary (Fig. 5-9), however. Experimental procedure is the same as in the previous case except that samples are taken at each sampling port of each time interval. The results are plotted as weight fraction removed at a given time and depth. Smooth constant curves are then plotted by eye.

Estimation of the fraction removed at a given overflow rate is begun by entering the diagram at a time t^* corresponding to $v^* = h/t^*$. For convenience, this time is usually chosen at a point where a constant removal curve crosses the abscissa. The suspended solids can then be considered in increments. For example, using Fig. 5-10, there is a fraction $y_D - y_C$ which has an *average* settling

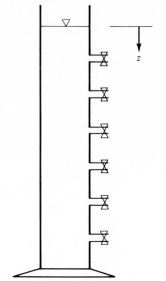

FIGURE 5-9
Settling column for flocculant-settling test.

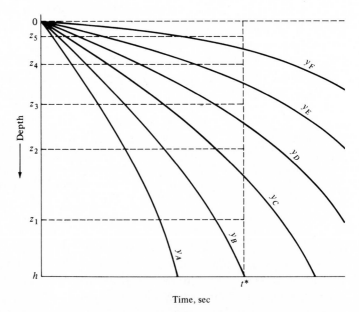

Time, sec

FIGURE 5-10
Typical flocculant-sedimentation contours.

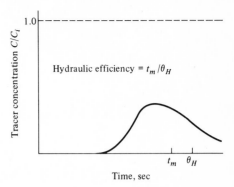

FIGURE 5-11
Typical tracer curve showing hydraulic
residence time and time to center of mass
of the tracer, t_m.

velocity of Z_2/t^*, and so on. Making the same assumptions as previously with respect to fraction removed as a function of velocity results in Eq. (5-12).

$$y_R = y^* + \frac{v_1}{v^*}(y_1 - y^*) + \frac{v_2}{v^*}(y_2 - y_1) + \cdots + \frac{v_{n-1}}{v^*}(y_{n-1} - y_{n-2})$$

$$+ \frac{v_n}{v^*}(y_n - y_{n-1}) \qquad (5\text{-}12)$$

where $y^* =$ fraction-removed value associated on curve with t^*
$y_1, y_2, \ldots, y_n =$ succeeding fraction-removed values

The average settling values are most easily determined by assuming Z_1 to be the midpoint between two curves. This approximation is easily as accurate as the curves themselves, and there is little use in applying a more sophisticated procedure.

As in the case of discrete particle settling, consideration must be given to currents, scour, and other factors that make the actual sedimentation tank behave differently than the batch experimental system. The same safety factors (0.66 to 0.8) are applied to the overflow rates developed for flocculant particles as for nonflocculant suspensions. The hydraulic efficiency of the tank can be used as a safety factor if prototype settling data are unavailable. Hydraulic efficiency is defined as the ratio of the time associated with the center mass for a dye tracer study to the theoretical hydraulic residence time Θ_H (Fig. 5-11).

5-6 HINDERED SETTLING AND THICKENING

When suspended-solids concentrations are high, particles do not settle independently. Flow around a particle affects the flow around neighboring particles under those circumstances. Kynch[11] proposed a model for hindered settling which assumes that settling velocity is proportional to suspended-solids concentration as long as there is no mechanical interaction between particles. This model

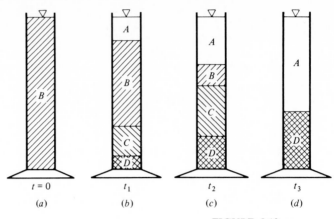

FIGURE 5-12
Sludge-settling experiment.

has been widely applied to sludge sedimentation and thickening and has been demonstrated to be a satisfactory representation of the systems.

A conceptual understanding of both the Kynch model and the physical system can be developed by considering the settling of a sludge in a column (in this case a 1-l graduated cylinder is a satisfactory experimental system). Initially, the solids concentration is uniform throughout the column (Fig. 5-12a), but as time passes, solids settle to the bottom, and a four-layered system develops (Fig. 5-12b) that includes a region in which particles have mechanical interaction D, a region of variable concentration C, a region with the initial concentration B, and a clear-water region A. As time progresses, the relative volumes of the regions change until the sludge is all contained in the compaction zone and the rest of the column of water is solids-free (Fig. 5-12c, d). According to the Kynch model, a plot of interface (between the sludge and the clear water) height vs time would have a constant slope until the layer of constant concentration (B) disappeared (Fig. 5-12d). Interface velocity would then be a function of the interface concentration until zone C disappeared. Because particles mechanically interact in the compaction zone D, the Kynch model does not apply after zone C disappears. A typical curve of interface height vs time for activated sludge is shown in Fig. 5-13.

As in the previous two cases, discrete and flocculant particle settling, we are interested in determining an overflow rate for use in process design. Characteristics of this process are different from the previous two systems, however. Because the solids settle in zones with a clear-water zone (essentially zero suspended solids) appearing at the top, we do not consider removal efficiency. Instead our concern is for choosing an overflow rate that will provide a maximum sludge (underflow solids) concentration, a minimum underflow rate, and a minimum tank area for a continuous-flow system to thicken solids to the desired level

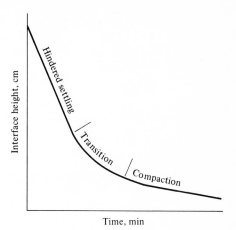

FIGURE 5-13
Curve of sludge-interface height vs time.

(Fig. 5-14). Sludge concentration is not considered in dilute suspensions because the volume of solids is relatively small and concentration is very difficult to modify or control.

A process operating at steady state will have a constant flux of solids (m/l^2t) moving downward. The maximum flux of solids possible is the sum of the flux due to the hindered settling of the solids plus the flux due to the underflow.[12]

$$N_s = Xv_H + \frac{XQ_u}{A} \qquad (5\text{-}13)$$

where N_s = solids flux, m/l^2t
X = solids concentration entering tank, m/l^3
v_H = hindered settling velocity, l/t, at concentration X
Q = incoming volumetric flow rate, l^3/t
Q_u = underflow rate, l^3/t
A = cross-sectional area of unit, l^2

Hindered settling velocity is a function of solids concentration. Dick[12] has reported that unit depth also affects this parameter. Thus flux values developed through batch experiments must be carried out over a range of solids concentrations and column depths. For a given depth, a typical batch-flux curve is

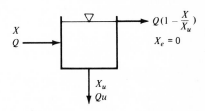

FIGURE 5-14
Schematic diagram of sludge concentrator.

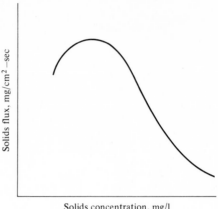

FIGURE 5-15
Batch solids-flux curve.

presented in Fig. 5-15. The curve does not extend back to zero solids concentration because hindered settling does not occur below concentrations of approximately 500 mg/l.

Adding underflow flux (based on a hypothetical underflow velocity Qu/A) to the maximum batch flux gives the curves in Fig. 5-16. Choosing the value of Q_u/A is somewhat arbitrary. Experience with biological sludges has led to the use of values in the range 7.1×10^{-3} to 1.4×10^{-2} cm/s (150 to 300 gal/ft²-d) for

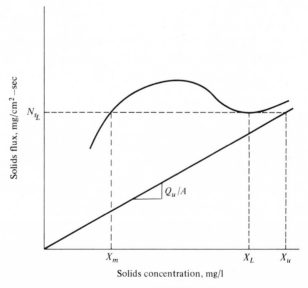

FIGURE 5-16
Continuous-flow solids-flux curve.

activated-sludge processes. In a design situation, more than one value should be used in order to develop some idea of the sensitivity of the resulting design parameters to influent variations. The limiting value of solids flux N_{S_L} in Fig. 5-16 is the maximum solids flux that can be used for any influent suspended-solids value less than X_L. It should be noted again that depth of column affects the values on the batch-flux curve,[12] and thus the data only reflect systems with the same depth. The primary reason for this is probably associated with interaction between the supposedly independently settling interface layer and the solids farther down in the column. In any case, if the limiting solids-flux value is exceeded, accumulation and eventual overflow of solids can be expected.

A mass balance on solids entering the tank and the solids passing any given section gives

$$AN_{S_L} \geq QX \qquad (5\text{-}14)$$

or

$$A \geq \frac{QX}{N_{S_L}} \qquad (5\text{-}15)$$

Because the slope of the underflow-flux line can be modified, the limiting flux value can also be modified. While this allows additional flexibility in choosing the tank area, increasing the flux will always result in lower compaction ratios, lower sludge-solids concentrations, and higher underflow rates.

EXAMPLE 5-2 Thickening tests made on a mineral sludge resulted in the data presented in Table 5-1. Settling velocities are the initial velocities at each sludge concentration. Determine the area-required underflow rate and solids concentrations for a feed with a flow rate of 3.785×10^6 l/d (1 mgd), and a solids, concentration of 10 g/l. Use underflow velocities of 500, 1000, and 2000 cm/d.

Table 5-1 **HINDERED SETTLING
DATA FOR EXAMPLE 5-2**

X, g/cm^3	v_i, cm/s
0.005	0.039
0.010	0.030
0.015	0.022
0.020	0.011
0.025	0.0046
0.030	0.0023
0.035	0.0015
0.040	0.0018
0.045	0.0008
0.050	0.0007

(a) $v_u = 500$ cm/d

$N_{s_{L1}} = 20.8$ g/cm^2-d

$$A \geq \frac{(3.785 \times 10^9 \text{ cm}^3/\text{d})(10^{-2} \text{ g/cm}^3)}{20.8 \text{ g/cm}^2\text{-d}}$$

$$\geq 1.82 \times 10^6 \text{ cm}^2 = 182 \text{ m}^2$$

$$\frac{Q}{Q_u} = \frac{X_u}{X} = \frac{3785 \text{ m}^3/\text{d}}{(5 \text{ m/d})(182 \text{ m}^2)}$$

$$= 4.16$$

$X_u = 41.6$ g/l

$Q_u = 910$ m^3/d

(b) $v_u = 1000$ cm/d

$N_{s_{L2}} = 35$ g/cm^2-d

$A_2 N_{s_{L2}} \geq QX$

$$A_2 \geq 1.08 \times 10^6 \text{ cm}^2 = 108 \text{ m}^2$$

$$\frac{Q}{Q_u} = \frac{X_u}{X} = \frac{3785 \text{ m}^3/\text{d}}{(10 \text{ m/d})(108 \text{ m}^2)}$$

$$= 3.50$$

$X_u = 35$ g/l

$Q_u = 1080$ m^3/d

(c) $v_u = 2000$ cm/d

Table 5-2 SOLIDS-FLUX DATA FOR EXAMPLE 5-2

X, g/l	v_i, cm/d	N_{batch}, g/cm-d	Total flux† v_u, cm/d		
			500	1000	2000
5	3400	17.0	19.5	22.0	27.0
10	2600	26.0	31.0	36.0	56.0
15	1870	28.1	35.6	43.1	58.1
20	950	19.0	29.0	39.0	59.0
25	400	10.0	22.5	35.0	60.0
30	200	6.0	21.0	36.0	66.0
35	130	4.6	22.1	39.6	74.6
40	90	3.6	23.6	43.6	83.6
45	70	3.2	25.7	48.2	93.2
50	60	3.0	28.0	53.0	103.0

† Curves for these three conditions are plotted in Fig. 5-17.

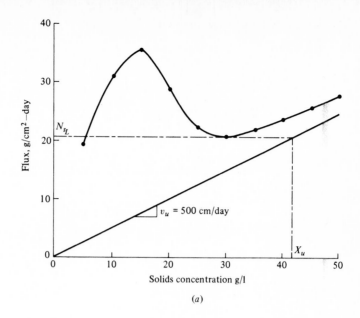

(a)

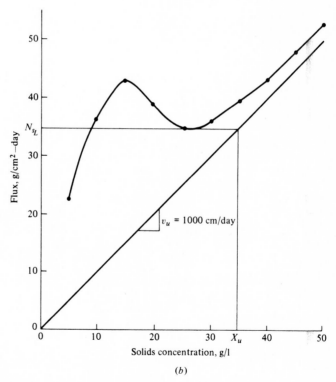

(b)

FIGURE 5-17
Solids-flux curves in Example 5-2 for underflow velocities of (a) 500 cm/d, (b) 1000 cm/d, and (c) 2000 cm/d.

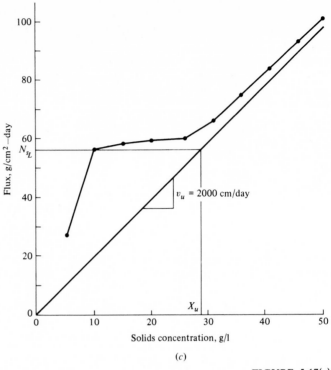

FIGURE 5-17(c)

Here (Fig. 5-17c) there is no actual minimum flux. Increasing feed-solids concentration limits the concentration factor X_u/X, however. A practical maximum flux is given by the breakpoint, which coincidentally in this problem is 10 g/l.

$$N_{s_{L3}} \approx 56 \text{ g/cm}^2\text{-day}$$

$$A_3 \geq 6.75 \times 10^5 \text{ cm}^2 = 67.6 \text{ m}^2$$

$$\frac{Q}{Q_u} = \frac{X_u}{X} = \frac{3785 \text{ m}^3/\text{d}}{(20 \text{ m/d})(67.6 \text{ m}^2)}$$

$$= 2.8$$

$$X_u = 28 \text{ g/l}$$

$$Q_u = 1350 \text{ m}^3/\text{d} \qquad \qquad ////$$

The fact that the calculated values of the underflow solids concentration were the same as the X_u values developed graphically in Example 5-2 results from using the maximum flux value. If a more conservative value had been chosen, a higher required area would have resulted together with lower values of X_u/X and underflow solids concentration and a higher value of underflow rate. Thus, the limiting flux rate provides the maximum suspended-solids concentration in the underflow stream. This is true without regard to influent solids concentration as long as $X_m < X < X_L$. For example, using the data for case (a) in Example 5-2 and a feed concentration of 20 g/l, the required minimum area is 364 m^2, the concentration factor X_u/X is 2.1, but the underflow solids concentration is 42 g/l as before.

If the influent suspended-solids concentration X is greater than X_L, higher fluxes are possible, but concentration factors will be small, and thickening will, in general, be uneconomic.

Process Control

Howell[13] has pointed out that the major drawback of the methods for predicting or designing sedimentation and thickening basins presented here is that no provision is included that can be used in automatic process control. This problem is particularly important in secondary settling units used in the activated-sludge wastewater treatment process where solids are recycled to the aeration-reaction tank. The problem is also important in discrete- or flocculant-settling processes that must meet an effluent standard. Four approaches may be used to overcome this difficulty: (1) Use photoelectric cells to detect when the sludge reaches a particular height, (2) provide the operator with a set of instructions for manual control based on pilot plant analysis, (3) develop a function based on the experimental curves that can be used to develop transfer functions for the units, and (4) assume or develop a new model for the process based upon traditional control variables or models and backfit the data to it. Approaches (1) and (2) are both used presently. In case (3) real-operator understanding seems to be conveyed with the instructions only rarely. In the case of primary clarification, the matter is usually not dealt with to any great extent as the solids carried over are simply passed to the secondary treatment unit. Recognition of this factor is evident in that some plants (e.g., Sacramento County, Calif.–Arden Wastewater Treatment Plant) have been designed without primary clarifiers.

Approaches (3) and (4) are oriented toward automatic process control. In this case, effluent solids, sludge solids, or both concentrations would be monitored by an automatic device of some sort. Howell[13] suggests using a series of N CFSTRs with or without backmixing (as necessary) as a model for thickeners. This approach relates well to the Kynch model, which assumes a continuum of solids with increasing concentrations as depth increases. Continuous parameter measurement is the major problem in such a model.

Design Flow Rates

In most treatment situations, flow rate fluctuates with time. A question is then raised with respect to the design flow rate which must be used. Clearly, the maximum rate is the design flow rate because the maximum overflow rate will be associated with the maximum flow rate once a unit cross-sectional area is chosen. Remembering that overflow rate is really an imposed settling velocity explains this relationship.

Underflow rates are, in some cases, limited by solids and operational characteristics. Biological solids continue to react and metabolize in the sludge layer, and the reaction products include gases that cause rising sludge which may carry over the effluent weir. Thus underflow rates must be great enough to remove sludges in a reasonable time. This time has traditionally been given in terms of hydraulic residence time in the tank rather than solids residence time. Values in use are in the range of 1 to 3 h for primary municipal sludges and $\frac{1}{2}$ to $1\frac{1}{2}$ h for secondary treatment sludges. Hydraulic and solids residence times vary because solids are often stored on the tank bottom for periods of 15 min to 1 h or longer in order to make removal easier.

5-7 CONFIGURATIONS USED IN SEDIMENTATION TANKS

Sedimentation tanks can be constructed in any desired configuration. In common use are circular, square, and rectangular units. Square units are really modifications of the circular- or radial-flow configuration. Features of importance in any configuration are the inlet structure, sludge-collection system, and outlet structure. The inlet structure should distribute the incoming flow evenly over the tank cross section. Consideration must be given to possible density variations that may occur with changing temperatures.

Sludge collection in radial-flow tanks is accomplished by revolving rakes that move the solids toward a center well to be pumped off. In rectangular tanks, endless chain rakes are used that scrape the solids to one end (usually the inlet end) of the tank. Where biological sludges are involved or where the sludge is particularly light and fluffy, solids can be pumped off the bottom of the tank rather than from a center or end well. This is accomplished by placing collection arms in radial-flow tanks or across the width of the tank on a traveling bridge in rectangular tanks. In the latter case, the suction orifices are placed at the bottom of a V-notched rake so that all the solids are pushed into the orifice as the rake travels the length of the tank.

Sludge-collection apparatus controls the width or radial dimension of sedimentation tanks. In most cases, standard sizes are chosen rather than ordering specially constructed equipment. Rectangular sedimentation-tank equipment is readily available up to 9 m (30 ft) in width. For circular tanks, units up to 60 m (200 ft) in diameter are available.

Sedimentation-tank bottoms are sloped to aid in cleaning. Radial-tank slopes are of the order of 8 percent, and rectangular tanks have slopes of approximately 1 percent.

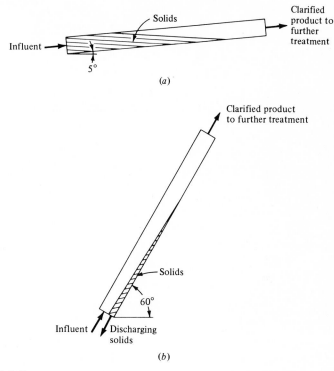

FIGURE 5-18
Tube-settler configurations. (*a*) Schematic diagram of tube settler with low slope;
(*b*) schematic diagram of tube settler with steep slope.

Tube Settlers

A recent innovation in sedimentation-unit configuration has been the tube settler
developed by Microfloc Corporation.[14–16] In these units, large numbers of small-
diameter (2 to 5 cm) tubes are nestled together to act as a single unit. When
the angle is low (~5°), the tube gradually fills with solids. When the tube is full,
the flow is turned off and the solids are allowed to drain out. Inclining the units
at 60° induces continual sludge discharge. A schematic diagram of the tube
settler is shown in Fig. 5-18.

Overflow rates as high as 290 m³/m²-d (7200 gal/ft²-d) have been
reported[15] for tube settlers in water treatment applications. Some installations
have been made in conventional sedimentation tanks with the objective of
increasing capacity.[16, 17] This modification is relatively straightforward because
the tube settler is constructed of PVC plastic in modules approximately 3 m
long, 1 m wide, and $\frac{2}{3}$ m high.

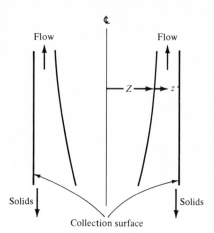

FIGURE 5-19
Schematic diagram of continuous-flow centrifuge.

5-8   CENTRIFUGATION

Centrifuges are commonly used for thickening previously concentrated sludges. The principles of centrifugation are the same as for gravity sedimentation, but an additional force is involved due to the radial acceleration imposed on the particles by the rotating centrifuge basket (Fig. 5-19). A force balance on a particle results in Eq. (5-16).

$$\frac{\pi}{6} d_p^3 \rho_p \frac{d^2z}{dt^2} = \frac{\pi d_p}{6}\left(\rho_p - \rho_1\right)(Z + z)\omega^2 + g_z) - C_D \frac{\pi}{4} d_p \rho_1 \left(\frac{dz}{dt}\right)^2 \qquad (5\text{-}16)$$

where ω = radial velocity, rad/s
$\quad C_D$ = drag coefficient
$\quad Z + z$ = particle position in bowl

For the laminar-flow case, C_D is $24/N_{Re}$, and assuming acceleration and gravitational effects are negligible, Eq. (5-16) becomes

$$t = \frac{18\mu}{d_p^2 \omega^2} \ln \frac{Z + z_2}{Z + z_1} \qquad (5\text{-}17)$$

where t is the time necessary for a particle to move from z_1 to z_2.
If the flow is turbulent with respect to the particle, the drag coefficient is

approximately 0.44, and Eq. (5-16) becomes

$$t = \frac{2}{\{3d_p \omega^2[(\rho_p - \rho_1)/\rho_1]\}^{1/2}} [(Z + z_2)^{1/2} - (Z + z_1)^{1/2}] \qquad (5\text{-}18)$$

The assumption of constant velocity is not correct because the radial acceleration changes with position, and thus the radial force changes as the particles move toward the wall. This change is directly proportional to the distance moved, and in most cases this is a relatively small value, making the assumption approximately correct.

Equations (5-16) to (5-18) are nothing more than a restatement of the discrete-particle-settling model presented in Eq. (5-1). If clarification is of interest, a relationship similar to Eq. (5-10) could be developed also. In most cases, thickening is of far more interest to the treatment process designer than clarification because the supernatant will be recycled back through the plant, however. Satisfactory theoretical relationships are not available for thickening in centrifuges, and consequently, empirical relationships are used.

The most useful way of correlating thickening data for centrifuges is by relating the cake moisture content and the fraction of incoming solids in the cake to the relative centrifugal force and the hydraulic flow rate. Relative centrifugal force (rcf) can be and is calculated as the ratio of the average

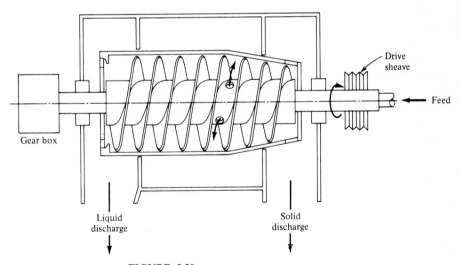

FIGURE 5-20
Imperforate-bowl-conveyor discharge centrifuge. (*Penwalt Corporation.*)

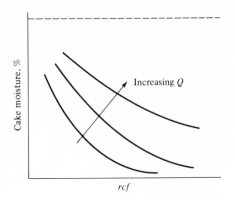

FIGURE 5-21
Cake moisture as a function of relative
centrifugal force.

centrifugal force to the gravitational force.

$$\text{rcf} = \frac{\overline{(Z + z)}\omega^2}{g} \qquad (5\text{-}19a)$$

where $\overline{Z + z}$ is the average radial distance from the axis of rotation to the
particles being separated. In scroll-type centrifuges (Fig. 5-20), which are the
most common type in water and wastewater treatment, the liquid layer is thin
and $\overline{Z + z}$ is approximately equal to the radius Z^*. Therefore, the rcf value is
given by Eq. (5-19b) for scroll centrifuges.

$$\text{rcf} = \frac{Z^*\omega^2}{g} \qquad (5\text{-}19b)$$

As the rcf value increases, cake moisture content decreases and the fraction
of solids removed increases. Increasing flow rate increases cake moisture content
and decreases fraction removed (Figs. 5-21, 5-22). Thus an experimentally
determined optimum must be found for each sludge type and variation.

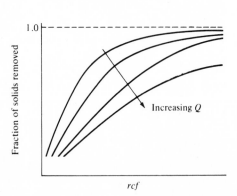

FIGURE 5-22
Fraction of solids removed as a function
of relative centrifugal force.

Commercial centrifuges used for sludge dewatering are usually of the scroll type (Fig. 5-20). Continuous discharge of solids and product liquid is possible with this machine. Typical diameters and rotational speeds are 15 to 75 cm and 1000 to 6000 r/min, respectively.

5-9 FILTRATION

Filtration has been used for water and wastewater treatment since the first days of pollution control. Significant problems associated with the use of filters have been that high suspended-solids concentrations in some water supplies and most wastewaters result in short filter runs before backwashing or cleaning is required. Filtration of organic-containing wastes may result in the growth of biological slimes that not only plug the filter but are very difficult to remove. These problems can be greatly alleviated by using the process only for tertiary treatment effluents.

Filtration processes can be loosely classified as precoat filters or depth filters. Precoat filters are those that use a hydraulically applied particulate coating on thin support media such as cloth or finely woven wire. The coating acts as a filter medium or as the base for the filter medium. In some cases, the solids in the filtered stream act as the precoating material. This is the case in vacuum filtration of wastewater sludges and microscreens.

Depth filters utilize a permanent, relatively thick medium that is either cleaned and reused or disposed of and replaced when clogging occurs. Depth filters used in water and wastewater treatment incorporate a granular medium of varying porosity and density supported by gravel layers and/or porous underdrains (Fig. 5-23). In most cases, the medium is graded sand, but in recent years, multimedia filters using sand, garnet, and anthracite coal or resins have been used. A primary factor in choosing material is resistance to abrasion. Nearly all installations constructed in the last 40 years have been "high-rate" filters that require regular backwashing for cleaning purposes. This involves fluidizing the bed and results in considerable abrasion. Fluidization also sorts the particles by density and diameter. In a single-media filter, the larger particles gravitate toward the bottom and the smaller particles "float" to the top (Fig. 5-23b), while in a dual-media filter, a similar process evolves over two layers (Fig. 5-23a).

While the particle-size distribution that evolves has a definite benefit in that the underdrainage system does not have to be designed to screen out the smallest filter-media particles, there is the obvious problem of having the particles in exactly the wrong order. To make optimal use of the filter, the larger granules should be at the top, allowing smaller particles to penetrate the filter to some depth and screen out the larger particles near the top. Plugging would be slower and filtration more efficient. Even without backwashing this would be a

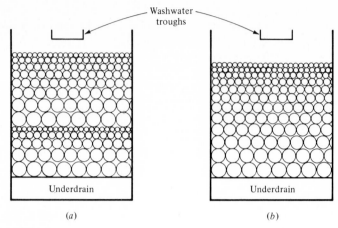

FIGURE 5-23
Particle-size distribution in depth filters. (*a*) Dual media, (*b*) single media.

difficult problem with a single-media filter, and this is the reason the multimedia filter was developed. The less dense media are also coarser, thus providing removal of large particles near the top of the filter.

Removal Mechanisms

Sand filters use grain sizes between 0.3 and 0.5 mm, yet nearly all the bacteria and a significant percentage of the viruses entering a filter are removed, even at the beginning of a filter run. Thus straining is not the only and perhaps not the major mechanism of removal. Tchobanoglous[18, 19] suggested that the major mechanisms of removal in wastewater filtration were straining, inertial impaction, gravity settling, interception, diffusion, adsorption, adhesion, flocculation, and biological activity. Straining involves both sieve action and the chance lodging of particles in crevices. Gravity settling is similar to straining in that lodging of particles in crevices or on surfaces to which they adhere to is an important factor. Inertial impaction occurs where particles leave the fluid streamline and collide with and stick on media granules, while interception occurs in situations where particles following fluid streamlines are large enough to strike media granules and be removed. Diffusion is important only in the case of very small particles such as viruses. Because of the small distances involved in pores and the fact that laminar flow exists, diffusion is an important parameter. Adhesion of particles to granule surfaces may be due to a number of forces: electrostatic, chemical reaction, or electrokinetic. Physical adsorption may also be involved. Particles moving through the pores will, on occasion, collide and aggregate; thus flocculation is a mechanism that couples with other mechanisms in the removal process.

As filter clogging proceeds, stress on adhered particles increases. Eventually, particles begin to shear off the surface, and effluent solids concentration begins to rise. Breakthrough of solids may occur as a result of long filter runs at a single head or as the result of increasing the pressure head on an operating filter. The problem is a hydraulic phenomenon.

Which mechanism is paramount is open to question, but there is little doubt that all the mechanisms are significant. Dominance of one mechanism over the others may change as clogging proceeds, and therefore the conceptual understanding is more important than a quantitative knowledge at any time.

Precoat filters, particularly those using filtered particles as the media, evidently have straining as the primary mechanism. These filters are characterized by poor-quality product water at the beginning of a filter run with improved quality as the run and volume of water filtered progresses.

Hydraulics of Filtration

Flow in filters has been demonstrated to follow Darcy's law.

$$v = KS \qquad (5\text{-}20)$$

where v = apparent velocity, Q/cross-sectional area
K = coefficient of permeability
S = pressure gradient

Kozeny[20] proposed an expression for flow through a uniform porous media in 1927.

$$\frac{h_L}{H} = \frac{k\mu v}{g\rho} \frac{(1 - \phi)^2}{\phi^3} a_v \qquad (5\text{-}21)$$

where h_L = head loss through bed of depth H
ϕ = bed porosity
a_v = average grain surface-area-to-volume ratio
k = dimensionless coefficient
≈ 5 in water and wastewater filtration

Because filters do not have uniform media, Fair and Hatch[21] modified the Kozeny equation to include layers of average sieve size d_i and weight fractions y_i.

$$\frac{h_L}{H} = \frac{k\mu v}{g\rho} \frac{(1 - \phi)^2}{\phi^3} \left(\frac{6}{\Psi}\right)^2 \sum_{i=1}^{n} \frac{y_i}{d_i^2} \qquad (5\text{-}22)$$

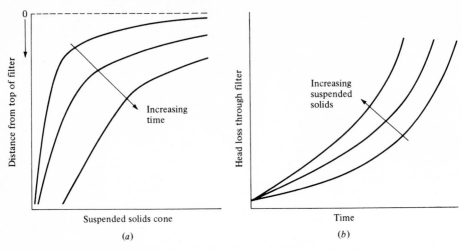

FIGURE 5-24
Head loss and solids removal in depth filters. (*a*) Suspended solids removed as a function of depth; (*b*) head loss vs time.

where Ψ is the average particle sphericity, a parameter that measures the closeness to spherical shape of a particle.

$$\Psi = \frac{d_p}{d_s} \qquad (5\text{-}23)$$

where d_p = particle effective diameter
d_s = diameter of an equivalent sphere

Equation (5-18) is useful in predicting the initial, or clean, flow properties of a filter and therefore is useful in design. As clogging progresses, the porosity and possibly the shape factor changes. The effect of these changes is not predictable because filtration does not occur uniformly throughout the filter.

Most of the filtration action takes place within a few centimeters of the top of the bed. As time progresses during a filter run, suspended-solids concentrations in the water increases at any depth as does head loss (Fig. 5-24). Tchobanoglous,[18] working with activated-sludge-process effluent, concluded that straining was the primary effect for the flocculant biological particles in the wastewater. He proposed an expression for solids removal rate as a function of depth and solids concentration.

$$\frac{dX}{dH} = \frac{1}{(1 + aH)^n} r_0 X \qquad (5\text{-}24)$$

where r_0 = initial ($t = 0$) removal rate
a, n = constants

Backwashing of Granular Filters

Backwashing of filters involves fluidizing the bed. Head loss incurred during fluidization is equal to the buoyant weight of the bed.

$$\frac{h_L}{H_e} = (1 - \phi) \frac{\rho_p - \rho_\omega}{\rho_\omega} \qquad (5\text{-}25)$$

where H_e is the unexpanded bed depth. The extent of expansion is dependent upon the relative velocity of the smallest media granules with respect to the water. This relationship is a function of the local porosity in the expanded bed because actual fluid velocity is related to the true cross-sectional area. Thus fluid velocity approaches the apparent velocity as the bed expands. Richardson and Zaki[17] proposed the following expressions to describe these relationships:

$$(\phi_e)^{n_e} = \frac{v_a}{v_t} \qquad (5\text{-}26)$$

$$\frac{H_e}{H} = \frac{1 - \phi}{1 - \phi_e} \qquad (5\text{-}27)$$

$$n_e = 4.45 N_{Re}^{-0.1} \qquad (5\text{-}28)$$

where the subscript e denotes the expanded bed, v_t is the terminal particle-settling velocity, and v_a is the apparent backwash velocity Q/A. Clearly, the Richardson-Zaki expression would have to be applied in a layered approach as was done with the Fair-Hatch modification of the Kozeny equation.

Process Operation

The simplest method of operation is to control head loss. Filter runs are then limited by the amount of head loss that is acceptable for a given circumstance. Head-loss acceptability may be based on hydraulic limitations or on solids breakthrough, whichever controls.

A filter may be operated with a constant-head–variable-flow rate, variable-head–constant-flow rate, or a combination of the two. Storage limitations often limit the variability of flow allowable in water treatment, but this would not ordinarily be the case in wastewater treatment. Thus efficiency of operation would be the controlling factor in most cases.

Applied flow rates within the limits of breakthrough values are not a particular problem. As flow rate increases, the filter run time decreases and the necessary head increases. Over an entire run the average flow rate is the controlling variable. Because clogging is a storage phenomenon, the total applied volume is more important than the rate of application. Thus, for example, filters should be designed on an average flow rate basis.

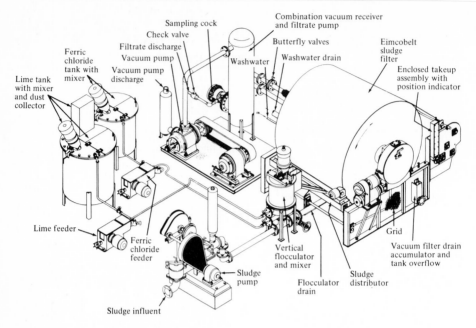

FIGURE 5-25
Typical vacuum-filter system. (*Municipal Equipment Division, Envirotech Corp.*)

Precoat Filters

Vacuum filters are the principal type of precoat filter used in water and waste-water treatment, and discussion here will center on this device. Physically, vacuum filters consist of a sectioned drum covered by a filter fabric. As the drum rotates through a slurry tank, a vacuum is applied to the submerged sections, liquid passes through the filter fabric, and the solids are retained. Solids are then removed from the belt by a scraper mechanism, as shown in Fig. 5-25. Hydraulics of vacuum filtration are similar to depth filtration in that flow is laminar and Darcy's law applies. Additional complexities exist because the filter cake, which also acts as the filter medium, varies in depth and is compressible. If we consider a section of the filter cake such as is shown in Fig. 5-26, a modified form of the Kozeny equation [Eq. (5-21)] can be written.

$$\frac{1}{A}\frac{dV_w}{dt} = \frac{\phi_z^3}{5(1-\phi_z)^2 a_v}\frac{dp_z}{\mu\,dz} \qquad (5\text{-}29)$$

where A = cake area

V_w = volume of water filtered

a_v = specific surface of particles, l^2/l^3

and ϕ, μ, and z are as previously defined.

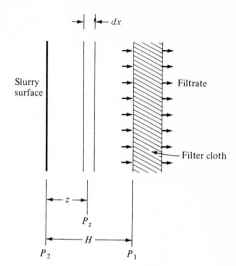

FIGURE 5-26
Section of vacuum filter.

The differential depth dz can be rewritten in terms of the filtrate volume dV_w, the mass of solids m deposited per unit of filtrate, the solid-particles density ρ_s, the porosity ϕ_z, and the filter area A.

$$dz = \frac{m\, dV_w}{(1 - \phi_z)\rho_s A} \qquad (5\text{-}30)$$

Replacing dz in Eq. (5-29) by Eq. (5-30) gives

$$\frac{dV_w}{dt} = \frac{A^2 \phi_z^3 \rho_s}{5m(1 - \phi_z)a_v^2 \mu} \frac{dp_z}{dV_w}$$

$$= \frac{A^2}{m\bar{\zeta}_z \mu} \frac{dp_t}{dV_w} \qquad (5\text{-}31)$$

where $\bar{\zeta}_z$ is the specific resistance of the filter cake and is based on the flow of filtrate through a unit mass of cake covering a unit area.

Assuming that m is a constant and that dV/dt is constant over short time periods allows integration of Eq. (5-31) between p_1 and p_2.

$$\frac{dV_w}{dt} = \frac{A^2}{V_w \mu m} \int_{p_1}^{p_2} \frac{dp_t}{\bar{\zeta}}$$

Knowledge of the relationship between $\bar{\zeta}$ and pressure is primarily empirical. A commonly used expression is

$$\bar{\zeta} = \bar{\zeta}'(p_2 - p_1)^{n'}$$

where $\bar{\zeta}'$ is not dependent upon pressure.

$$\frac{dV_w}{dt} = \frac{A^2}{V_w \mu m \bar{\zeta}'} \frac{\Delta p}{(1-n')\Delta p^{n'}} \qquad (5\text{-}32)$$

The coefficient n' has values ranging from 0.01 for highly incompressible materials to 0.9 for highly compressible materials. Often the function $\bar{\zeta}' \Delta p^{n'}$ is combined for filters using constant pressure.

$$\frac{dV_w}{dt} = \frac{A^2 \Delta p}{V_w \mu m \bar{\zeta}''} \qquad (5\text{-}33)$$

Equation (5-33) is commonly used in a vacuum-filter design. An initial resistance R_c must be included, which includes the resistance of the filter cloth and the initial layer of cake. The final forms of the equation are then

$$\frac{dV_w}{dt} = \frac{A^2 \Delta p}{V_w \mu m \bar{\zeta}'' + \mu A R_c} \qquad (5\text{-}34)$$

$$\frac{t}{V_w} = \frac{\mu m \bar{\zeta}'' V_w}{2A^2 \Delta p} + \frac{\mu R_c}{A \Delta p} \qquad (5\text{-}35)$$

The integral form [Eq. (5-35)] is useful in determining the overall specific resistance and the initial resistance from experimental data. Once these values are determined, the expression can be used in the design of continuously operating units. In applying the expression to continuous-flow units, it must be remembered that vacuum is applied for only a fraction of the cycle time.

EXAMPLE 5-3   A sludge is to be concentrated by vacuum filtration. Experimental studies at three vacuum values 15, 30, and 60 cm Hg, using a 4.7-cm-diameter filter and a 25 g/l sludge concentration, have resulted in the data presented in Table 5-3. Determine the value of the specific resistance for this sludge.

The slope of the t/V_w versus V_w curves is determined from Fig. 5-27.

$$(\text{Slope})_{15} = 3.73$$
$$(\text{Slope})_{30} = 2.20$$
$$(\text{Slope})_{60} = 1.30$$

$$\bar{\zeta}''_p = \frac{2(\text{slope})_p \Delta p A^2}{\mu m}$$

$$m = 25 \text{ g/l} = 0.025 \text{ g/cm}^3$$

$$\bar{\zeta}''_{15} = \frac{2(3.73)(15)(13.5)(982)(\pi/4)^2(4.7)^4}{0.01(0.025)}$$

$$= 1.79 \times 10^{12} \text{ cm/g}$$

$$\bar{\zeta}''_{30} = \frac{2(2.20)(3.0)(13.5)(482)(\pi/4)^2(4.7)^4}{0.01(0.025)}$$

$$= 2.13 \times 10^{12} \text{ cm/g}$$

$$\bar{\zeta}''_{60} = \frac{2(1.30)(60)(13.5)(982)(\pi/4)^2(4.7)^4}{0.01(0.025)}$$

$$= 2.50 \times 10^{12} \text{ cm/g}$$

The initial resistance can now be calculated.

$$R_{c_p} = \frac{A\,\Delta p}{\mu}\left(\frac{t}{V_w} - \frac{\mu m \bar{\zeta}'' V_w}{2A^2\,\Delta p}\right)$$

The easiest method of solving this expression is to choose the point $t = 0$ on the curves in Fig. 5-27. All three curves converge at $V_w = -3.0 \text{ cm}^3$.

$$R_{c_{15}} = \frac{(\pi/4)(4.7)^2(202.5)(082)(3.73)(3.0)}{0.01}$$

$$= 3.86 \times 10^9 \text{ cm}^{-1}$$

$$R_{c_{30}} = \frac{(\pi/4)(4.7)^2(405)(982)(2.20)(3.0)}{0.01}$$

$$= 4.55 \times 10^9 \text{ cm}^{-1}$$

$$R_{c_{60}} = \frac{(\pi/4)(4.7)^2(810)(982)(1.30)(3.0)}{0.01}$$

$$= 5.38 \times 10^9 \text{ cm}^{-1}$$

Table 5-3   VOLUME OF FILTRATE, cm³

Time, min	Δp, cmHg		
	15	30	60
10	11.7	15.4	20.0
20	16.9	22.2	28.8
40	24.3	31.9	41.3
80	34.7	45.6	59.1
120	42.7	56.1	72.7
160	49.3	64.9	84.1
200	55.3	72.7	94.2
240	60.7	79.8	103.4
280	65.6	86.3	111.8
320	70.2	96.3	119.6

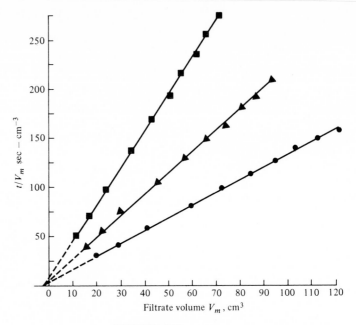

FIGURE 5-27
Vacuum-filtration resistance plot for Example 5-3.

Although three points are inadequate for determining a relationship between pressure and specific resistance, plotting these values gives a qualitative picture of the dependence (Fig. 5-28). ////

Vacuum Filtration of Wastewater Treatment Sludges

Vacuum filtration of primary wastewater sludge results in a sludge cake with about 30 percent solids content. Sludge cake from raw activated sludge usually has a solids content of about 20 percent. Well-digested sludges may result in filter cakes as high as 40 percent solids. In most cases, chemical conditioning of organic or biological sludges is necessary to promote flocculation and increase the drainage rate. Chemicals used include alum, ferric chloride, ferric sulfate, and lime. Dosage and choice of chemical conditioner are based on laboratory or pilot studies. Because the conditioning chemicals react with carbonate and bicarbonate, the quantity necessary increases with increasing alkalinity.

Vacuum applied to the submerged zones ranges from 30 to 70 cmHg. The nonsubmerged zone may be subjected to a drying vacuum of 50 to 70 cmHg.

Commercially available vacuum filters range in size from 1.5 to 4.0 m. Face lengths vary from 1 to 6 m. Rotational speeds are of the order of 0.25 m/min.

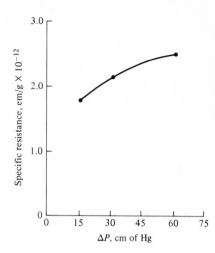

FIGURE 5-28
Specific resistance versus ΔP.

PROBLEMS

5-1 Sediment washed out of the mountains in California contains a great deal of clay. A large fraction of this material does not settle out in the Sacramento–San Joaquin Delta even though velocities are very low but does settle in San Pablo and San Francisco Bays. Explain why this might occur.

5-2 For what particle-size range is the Stokes' law assumption valid? Assume quiescent conditions and use specific gravities of 1.1, 1.5, and 2.0.

5-3 A pilot plant stake and stator flocculation-sedimentation system has been tested over a range of rotational speeds, and the data are presented in Table 5-4. Is there any relation between the Camp-Stein model and the data? How would you use this data in the design of a walking-beam flocculator? Is sedimentation-tank effluent turbidity a good way to estimate flocculator performance?

Table 5-4 FLOCCULATION DATA FOR PROB. 5-3

r/min	1	2	3	6	8	10	15
Turbidity, Jtu	5.0	1.8	1.2	1.1	1.1	1.6	5.2

5-4 Settling tests have been conducted on a wastewater containing 250 mg/l of non-floccuable suspended solids. The test system is similar to that shown in Fig. 5-7, and the sampling point is at a depth of 2 m. Using the following data, develop a curve of overflow rate as effluent suspended-solids (SS) concentration for an *ideal* process.

Time, min	SS, mg/l
0	250
1	225
2	195
3	170
4	138
5	83
6	57
8	32
10	20
14	13
20	8
30	5
40	3

5-5   A settling test is run on a domestic sewage. Flocculant settling is assumed when an apparatus similar to the system shown in Fig. 5-9 was used. Using the data presented in the following table, develop a curve of effluent suspended-solids concentration as a function overflow rate.

SUSPENDED SOLIDS, mg/l

	Time, min						
Depth, m	0	10	20	30	40	50	60
0.3	250	198	152	97	75	52	31
0.6	250	225	180	149	109	80	70
0.9	250	230	200	163	138	110	82
1.2	250	235	206	175	158	129	107
1.5	250	240	210	187	167	137	120
1.8	250	243	214	192	170	158	130

5-6   Results of sludge-thickening experiments on a biological sludge are presented in the following table. Use the data to make a plot of maximum thickened sludge concentration as a function of underflow velocity.

X_i, mg/l	1000	2000	3000	4000	5000	6000	7000
v_i, cm/min	16.7	8.8	5.0	3.1	1.9	0.6	0.17

5-7   A tracer study has been run on a rectangular sedimentation tank with dimensions of 8 m wide, 2.5 m deep, and 50 m long. The study was run using the design flow rate, 80 l/s. At time $t = 0$, 1000 g of tracer was added to the influent. Effluent tracer concentrations are shown in the following table. Determine the hydraulic efficiency of the tank. How would this data be used in process design of (a) a primary settling tank or (b) a sludge thickener?

t, min	C_e, mg/l $\times 10^2$	t, min	C_e, mg/l $\times 10^2$
10	20	140	74
20	37	160	68
30	49	180	62
40	60	200	56
50	67	220	50
60	72	240	45
70	76	260	39
80	79	280	34
90	80	300	30
100	80	350	20
110	78	450	9

5-8 For the unstratified-sand filter beds shown in Fig. 5-29, use the Bernoulli and continuity equations to determine the pressure profile through the sand bed. Assume steady-state-condition turns exist.

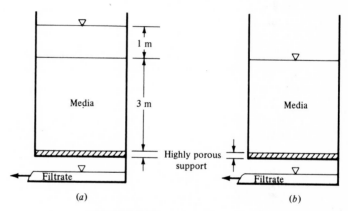

FIGURE 5-29
Flow conditions in unstratified filter beds for Prob. 5-8.

REFERENCES

1. VERWEY, E. J., and J. G. OVERBEEK: "Theory of the Stability of Lyophysic Colloids," Elsevier Publishing Company, Amsterdam, 1948.
2. STUMM, W., and C. R. O'MELIA: Stoichiometry of Coagulation, *J. Am. Water Works Assoc.*, vol. 60, p. 514, 1968.
3. STUMM, W.: Metal Ions in Aqueous Solution, in S. D. Faust and J. V. Hunter (eds.), "Principles and Applications of Water Chemistry," John Wiley & Sons, Inc., New York, 1967.
4. STUMM, W., and J. J. MORGAN: "Aquatic Chemistry," Wiley-Interscience, New York, 1970.

5. HARRIS, H. S., W. J. KAUFMAN, and R. B. KRONE: Orthokinetic Flocculation in Water Purification, *J. Sanit. Eng. Div., ASCE*, vol. 92, SA6, p. 95, 1966.

6. SWIFT, D. L., and S. K. FRIEDLANDER: The Coagulation of Hydrosols by Browian Motion and Laminar Shear Flow, *Colloid Sci.*, vol. 19, p. 621, 1964.

7. SMOLUCHOWSKI, M.: Versuch einer mathematischen Theorie der Koagulationskinetic kolloider Losungen, *Z. Phys. Chem.*, vol. 92, p. 155, 1918.

8. KRONE, R. B.: A Study of the Rheologic Properties of Estuarial Sediments, Sanit. Eng. Res. Lab., University of California, Berkeley, 1963.

9. CAMP, T. R., and P. C. STEIN: Velocity Gradients and Internal Work in Fluid Motion, *Boston Soc. Civ. Eng.*, vol. 30, p. 219, 1943.

10. ECKENFELDER, W. W., JR., and D. J. O'CONNOR: "Biological Waste Treatment," Pergamon Press, New York, 1961.

11. KYNCH, G. J.: A Theory of Sedimentation, *Trans. Faraday Soc.*, vol. 48, p. 166, 1952.

12. DICK, R. I.: Sludge Treatment, in W. Weber (ed.), "Physiochemical Processes for Water Quality Control," Wiley-Interscience, New York, 1972.

13. HOWELL, J.: Personal communication, Swansea, Wales, December 1973.

14. HANSEN, S., and G. L. CULP: Applying Shallow Depth Sedimentation Theory in Water Treatment Works, *J. Am. Water Works Assoc.*, vol. 59, p. 1134, 1967.

15. CULP, R. L., and G. L. CULP: "Advanced Wastewater Treatment," Van Nostrand Reinhold Company, New York, 1971.

16. CULP, G. L., and R. L. CULP: "Modern Concepts in Water Treatment," Van Nostrand Reinhold Company, New York, 1974.

17. COULSON, J. M., and J. F. RICHARDSON: "Chemical Engineering," vol. 2, 2d ed., Pergamon Press, New York, 1968.

18. TCHOBANOGLOUS, G., and R. ELIASSEN: Filtration of Treated Sewage Effluent, *J. Sanit. Eng. Div., ASCE*, vol. 96, p. 243, 1970.

19. TCHOBANOGLOUS, G.: A Study of the Filtration of Treated Sewage Effluent, Ph.D. thesis, Stanford University, Stanford, Calif., 1968.

20. KOZENY, J.: Ueber kapillare Leitung des Wassers im Boden, *Sitzungsber. Akad. Wiss. Wien, Abt. 3A*, vol. 136, p. 276, 1927.

21. FAIR, G. M., and L. P. HATCH: Fundamental Factors Governing the Streamline Flow of Water through Sand, *J. Am. Water Works Assoc.*, vol. 25, p. 1551, 1933.

6

MICROBIOLOGY AND WATER AND WASTEWATER TREATMENT

Microorganisms are important in both water and wastewater treatment, but for entirely different reasons. Removal or killing (disinfection) of microorganisms is an important step in water treatment because they are a threat to public health or cause tastes, odors, and undesirable growths in pipes and tanks. In wastewater treatment, microorganisms provide the functional basis for a number of treatment processes. A major concern in wastewater treatment is the development and maintenance of suitable microbial cultures.

This chapter will discuss the general characteristics of microorganisms found in water and wastewater treatment systems and methods of disinfection. The following chapters will develop applications of microbial systems to wastewater treatment.

Microorganisms found in water and wastewater fall into four general groups: viruses, procaryotic organisms, eucaryotic organisms, and simple invertebrates. The procaryotic and eucaryotic organisms are single celled, while the invertebrates are multiple celled. Viruses are essentially nonliving particles that interact with living organisms. Bacteria, a procaryotic group, are the organisms of principal interest in both water and wastewater treatment and will be discussed first.

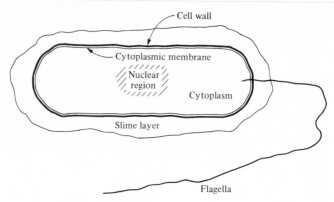

FIGURE 6-1
Schematic diagram of a bacterium illustrating common structures.

6.1 BACTERIA

Bacteria are important in water and wastewater treatment because a number of species are pathogenic (disease causing) and because bacterial cultures can be used to remove organic material and some undesirable minerals from wastewaters. They include a wide-ranging class of organisms that have as a common characteristic the procaryotic cell.[1] This type of cell is also characteristic of one other group, the blue-green algae, and is typified by a lack of internal membranes (Fig. 6-1). Bacterial cells are not internally uniform but have distinct areas of specialized metabolic activity or storage. Differentiation is due to the concentration of specific chemical species, such as deoxyribonucleic acid in the nuclear region.

Beyond this similarity of having a basic cell structure, few absolutes stand. Most bacteria are *chemoheterotrophic*; i.e., they utilize organic materials as a source of energy and carbon. A few species oxidize reduced inorganic compounds (such as NH_3) for energy and use CO_2 as their carbon source. These bacteria are called *chemoautotrophs*. Finally, some bacteria are *photosynthetic* and use light as their energy source and carbon dioxide for their carbon source. Chemoheterotrophic bacteria are the most important in wastewater treatment because they break down organic material. Chemoautrotrophic bacteria play an important role in wastewater treatment also, particularly the nitrifying bacteria that oxidize ammonia (N^{3-}) nitrogen to nitrate nitrogen (N^{5+}).

Although all bacteria have the same basic cell structure, there are several distinct shape groupings. Spirochetes have a spiral shape around an axial filament, are motile, moving in a corkscrew fashion, and are best known through the work on one member, *Treponema pallidum*, which causes syphilis. Cocci have a general spherical shape and include a large number of bacteria species occurring in biological waste treatment processes. Bacilli are rod shaped and are also extremely common in biological wastewater treatment processes.

Because the terms cocci and bacilli are somewhat qualitative and because their shape is somewhat dependent on environmental conditions, a more satisfactory method of classification has been by method of motility and certain structural characteristics. The myxobacteria, for example, all exhibit what is called "gliding" movement and characteristically have thin, flexible cell walls. All myxobacteria are chemoheterotrophic rods. A second grouping, the spirochetes, have been described previously. Because their shape and method of movement are unique, they are easily set apart. They also have thin flexible walls and are all chemoheterotrophs. All other bacteria fall into a group called the *eubacteria*. This group includes chemoheterotrophs, chemoautotrophs, and photosynthetic bacteria. A relatively thick, rigid cell wall is a general characteristic, and if they are motile, motility is due to the action of long, whiplike flagella. Bacteria flagella are approximately 100 A diameter and may be as long as 10 μm (Ref. 1).

Characteristic dimensions of bacteria are of the order of micrometers. Typical diameters of rod-shaped cells range from 0.5 to 2.0 μm, although some species fall outside this range. Cell lengths generally range from 1.0 to 5.0 μm. Diameters of spherical cells fall into the same general range, 0.5 to 3.0 μm. Density of wet bacteria measured by centrifugation is nearly always close to 1.1 g/cm^3. Thus the mass of one cell is very small. For a spherical cell 1.0 μm in diameter, the volume is 5.2×10^{-13} cm^3, and the mass is 5.7×10^{-13} g. Because bacteria are small, the surface area per unit mass is relatively high. For the example cell, the ratio is 5.5 cm^2/g. Nutrient uptake by bacteria, the essential or key process in biological wastewater treatment, involves transfer of nutrients through the cell boundary; thus the available surface area is related to rate of uptake. Examples of the different types of bacteria are shown in Fig. 6-2.

The Bacterial Cell

The procaryotic cell characteristic of all the lower protists (organisms without cell specialization) is differentiated from the eucaryotic cell [common to *all* higher organisms—including higher protists (protozoa, algae, fungi), plants and animals] by the simplicity of structure and particularly the absence of intracellular membranes. The basic structure of the procaryotic cell is shown in Fig. 6-1, and the chemical makeup of the cell is given in Table 6-1. A definable boundary for the "living" or reactive portion of the cell is provided by the cytoplasmic membrane. Although some enzyme-catalyzed reactions occur outside the membrane, none are reactions that release energy for use by the cell. Additionally, the cell can survive without a cell wall but cannot survive without an intact cytoplasmic membrane. All material, organic or inorganic, metabolized by the cell must pass through the membrane. Transport mechanisms are only beginning to be understood, but current (1976) belief is that most molecules pass through

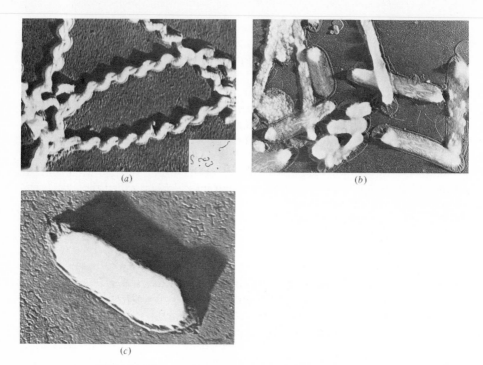

(a)

(b)

(c)

FIGURE 6-2
Common types of bacteria: (a) Spirillum, (b) cocci, and (c) bacillus. (*Courtesy of Professor F. Pocchiari, Instituto Superiore di Sanità.*)

Table 6-1   APPROXIMATE ELEMENTARY
COMPOSITION OF
BACTERIAL CELLS[2]

Element	Percentage of dry weight
Carbon	50
Oxygen	20
Nitrogen	14
Hydrogen	8
Phosphorous	3
Sulfur	1
Potassium	1
Calcium	0.5
Magnesium	0.5
Chlorine	0.5
Iron	0.2
All other	~0.3

the membrane due to reactions with specific enzyme systems called *permeases.*[3-7] Molecules for which permease systems do not exist do not enter the cell and, therefore, are not metabolized. This is a partial explanation for selective utilization of nutrients by bacteria and a reason why the use of mixed cultures is necessary in wastewater treatment. Each bacterial species has a different group of compounds it can metabolize. Thus mixed cultures have the capability of metabolizing a wide range of compounds. A bacterial species may have the ability to synthesize a permease system for a particular organic material even though the system is not normally present. When the organic species is present, the system is synthesized, and the material is metabolized. This "acclimation" process is particularly important in wastewater treatment plant start-up and is of concern where wastewaters change qualitatively on short notice.

Inside the bounding cytoplasmic membrane, most of the volume is taken up by the cytoplasm. This material is relatively uniform throughout the cell and has a highly granular appearance.[8] Granule size is approximately 100 to 200 Å, and the ribonucleic acid (RNA) content is high throughout the region. Density of the granule packing decreases and granule size increases with cell age.[9] Most metabolic activities of the cell are carried out in the cytoplasm. Organic materials are broken down here for the purpose of releasing energy and for providing structural elements necessary for growth. Molecules necessary to the cell, such as enzymes, are synthesized in the cytoplasm as well, and these processes serve as a partial sink for the energy released by organic oxidation. (*Note:* chemoautotrophic and photosynthetic organisms do not oxidize organics for energy but do have analogous processes.)

The cytoplasm also serves as a storage area for cellular reserve materials. Starch, glycogen (a polysaccharide very similar to starch), and poly-β-hydroxybutyric acid (an ester of β-hydroxybutyric acid) are the most common reserve materials (Fig. 6-3). Most bacterial species synthesize only one type of reserve material, but a few synthesize more. Under most conditions, the quantity of reserve materials in a cell is small, but in some cases (such as stoichiometric nitrogen deficiency) reserve materials may accumulate to the extent of 50 percent of the cell's dry weight. When an external (exogenous) carbon source is not available, the reserve materials are utilized by the cell.

Bacteria also store inorganic materials in the cytoplasm. For example, when nutrient starvation (other than phosphate) occurs, polyphosphates, linear polymers of orthophosphate, are stored in bodies called *volutin granules.* Sulfate-poor cultures are particularly notable for their production of volutin granules.[1] When sulfate is added to the culture, the polyphosphates are metabolized and the granules disappear.

Near the center of the cell is the nuclear region containing the genetic material of the bacteria, which consists of a single strand of deoxyribonucleic acid (DNA) that is believed to be circular.[10] Bacterial chromosomes are approximately 1 mm in length, which in consideration of the overall size of the cell

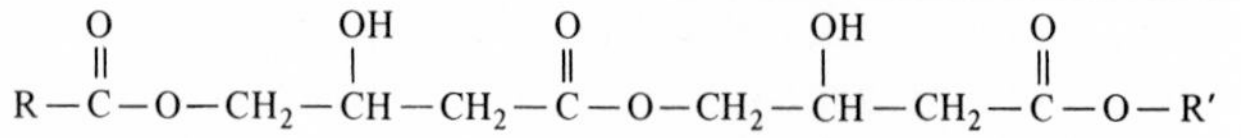

$$O \qquad\qquad OH \qquad\qquad O \qquad\qquad OH \qquad\qquad O$$
$$\parallel \qquad\qquad | \qquad\qquad\quad \parallel \qquad\qquad | \qquad\qquad\quad \parallel$$
$$R-C-O-CH_2-CH-CH_2-C-O-CH_2-CH-CH_2-C-O-R'$$

FIGURE 6-3
Poly-β-hydroxybutyric acid.

indicates an exceedingly tight packaging arrangement. For this reason the nuclear region, although not separated from the cytoplasm by a membrane (unlike the eucaryotic cell of higher organisms), is relatively easy to identify chemically. Integrity of the nuclear region is maintained by the intramolecular linkages of the DNA molecule.

Returning to the outer portion of the cell (Fig. 6-1), we can now consider the structures outside the cytoplasmic membrane. Nearly all procaryotic organisms (including the blue-green algae) have a relatively rigid cell wall. Presence of this structure allows the cell to survive under changing osmotic pressure or in the presence of high local shear forces, as well as giving the cell a defined shape. Although a few bacterial species do not have cell walls, they can only survive in isotonic environments (equal ionic concentration inside and outside the cell) and, consequently, are of no significance in water and wastewater treatment. Thickness of the cell wall varies from 100 to 800 Å depending on the species of bacteria and environmental factors. Chemical makeup of the wall varies with species also, but complex polymers made up of carbohydrates and amino acids called *muriens* are common to all bacterial cell walls in amounts varying from 10 to 50 percent by weight. Muriens are responsible for the strength of the cell wall, and destruction of the murien content results in lysis or breakup of the cell. An example of this is the action of penicillin, which inhibits the synthesis of murien and thus is active against most bacterial species.

Passage of nutrients through the cell wall is not well understood. Enzymatic reactions, analogous to those in the cytoplasmic membrane, do not exist, and the wall is not involved in osmotic pressure control. This would leave transport methods such as diffusion through pores and the formation of a cell wall–nutrient solution through which specific molecules would diffuse. Comparison of nutrient uptake rates of whole cells and cells of which the walls have been removed might provide some insight into the mechanism, but such work has not been done at this time.

While nearly all bacteria have cell walls, the presence of a capsule or slime layer is more of a variable and is often a function of the cell's environment. These layers vary in thickness from being only detectable by chemical means to being visibly larger than the cell. Chemical content consists of various organic polymers, with the particular type dependent upon species or in some cases subspecies. Typical materials are polysaccharides, complex polysaccharides (containing a number of different subunits), and polypeptides. Cellulose, a glucose polysaccharide constituent of plant but not of bacterial cell walls, is found in slime layers and capsules of some species of bacteria.

Removal of the slime layer does not affect the metabolism of the cell or the rate of growth.[1] However, certain antigenic properties are changed. An example is the loss of virulence of *Streptococcus pneumoniae* if the capsule layer is absent.

Slime layers are believed to function as a "binder" for bacterial floc particles. These particles, consisting of large numbers of individual cells, form in biological wastewater treatment processes and make possible the removal of the bacteria by gravity sedimentation. Slime layers are far more significant and extensive in "old" or resting cultures of the type maintained in the activated-sludge process. Microscopic examination of floc particles demonstrates the presence of large quantities of extracellular material, which would seem to support the theory. Additionally, rapidly growing cultures characteristically have small slime layers, and flocculation is correspondingly low.

Some species of bacteria are motile, and the most general form of motility is due to the whiplike action of the flagella (Fig. 6-1). The base of the flagella is believed to be in the cytoplasmic membrane, and the structure may be 10 times as long as the cell. Width of the flagella is usually between 120 and 185 Å. Flagella are constructed entirely of protein molecules, which typically have a single kind of molecular subunit. The particular subunit varies with species and strain, however.

Although motile bacteria are not particularly desirable from a wastewater treatment point of view, they provide insight into more general properties of cultures through their "tactic response." Motile bacteria respond to physical and chemical gradients by moving from less favorable to more favorable conditions. Because the rate of tactic response is relatively fast, an observer can note favorable and unfavorable conditions by watching their movement under a microscope. Growth rates would be expected to be higher in the favorable regions; this information can be extrapolated to studies of nonmotile bacteria.

Bacterial Metabolism

Bacterial metabolism may be divided into two parts: catabolism, the breakdown of organic materials, and anabolism, the synthesis of molecules needed by the cell. Catabolism and anabolism are closely interrelated, and when the overall process of cellular metabolism is considered, separation of the two is impossible. Two types of interrelationships between catabolism and anabolism occur. One is turnover in which certain cell molecules are synthesized, utilized for a period of time, and then for some reason catabolized. The other is the use of catabolites or partially oxidized catabolites as skeletons for synthesis. Because of these interrelationships, separate stoichiometric equations cannot be written for organic

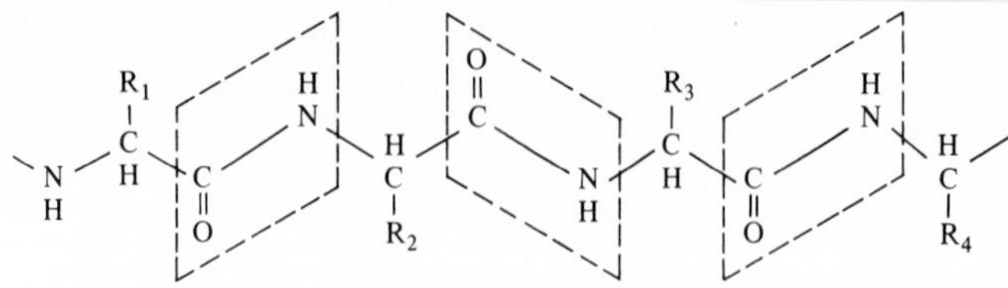

FIGURE 6-4
Peptide bonds, the primary structure of a protein.

breakdown and for growth, but instead equations must be of the following form:

$$v_1[A] + v_2[B] \rightarrow v_3[CO_2] + v_4[H_2O] + v_5[C] + v_6[\text{cells}] \qquad (6\text{-}1)$$

where CO_2, H_2O, and cells are common end products and C would represent any end products peculiar to a particular culture system. The reactants A and B might represent the organic material and a limiting nutrient concentration. An empirical cell formulation may be used for the cells produced. A commonly used formulation, $C_5H_7O_2N$, was developed by Porges et al.[11] Phosphorus is also an important nutrient, and a formulation can be derived from data in the literature that takes this element into account, $C_{42}H_{100}N_{11}O_{13}P$.[13] Both formulations have virtually the same mass fractions for each of the elements. Mass fractions may change due to changes in stoichiometry [of Eq. (6-1)], but such changes would probably be small. Burkhead and Ward[12] have also reported changes in constituent mass fractions with changes in organic species metabolized.

Metabolic reactions within the cell are virtually all enzyme catalyzed. In addition, certain reactions exterior to the cell, such as those which break down large molecules to soluble fractions, are catalyzed by extracellular enzymes. All enzymes are proteins and thus are made up of large numbers of amino acids bound together by peptide bonds (Fig. 6-4) in what is termed the *primary structure*. Size of a protein molecule ranges from molecular weights of several thousand up to several million. Thus the number of amino acid groups

$$(R_i\text{--}\overset{\overset{\displaystyle NH_2}{|}}{C}\text{--}COOH)$$

making up a given protein is large. Only about 20 amino acids occur commonly in nature, and nearly all of them are α amino acids having both the carboxyl group and the amino group attached to the α carbon. Each amino acid has a side chain which gives the molecule specific characteristics.

Many reactive groups exist along the length of a protein molecule, and because of their close proximity to each other, there is a strong tendency for interaction. One such interaction is the formation of a hydrogen bond between carboxyl oxygen groups and amide nitrogen groups as shown in Fig. 6-5. Hydrogen bonding results in two classes of "secondary" structure: helical and sheet. The most important type of structure with respect to enzymes and, therefore, wastewater treatment is helical. Pauling and Corey[13] found the most

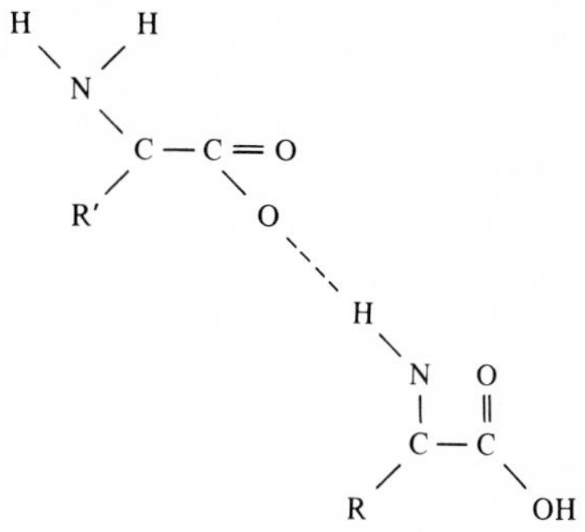

FIGURE 6-5
Hydrogen bonding between two amino acids.

stable form was the α helix. Side-chain groups in this helix often interact, and the result is a complicated bending of the helix into a "globular" form called the *tertiary* structure.

Enzymes are characterized by the manner in which side-chain groups are formed into a geometric configuration by the tertiary structure to produce a catalytically active site. Often activity is dependent upon the attachment of cofactors of which there are three general types: (1) Prosthetic groups that are tightly bound to the protein. Flavin adenine dinucleotide (FAD) is an example that will appear in later discussion. (2) Coenzymes that readily disassociate from the enzyme. Nicotinamide-adenine dinucleotide (NAD^+) and its phosphate ester, $NADP^+$, are important examples. (3) Metal activators, ions that are bound by chelation to some side-group moiety. Examples are K^+, Mn^{2+}, Mg^{2+}, Ca^{2+}, and Zn^{2+}.

Reactions involving enzymes are understandably complex. A reaction model discussed in Chap. 3, the Michaelis-Menten model, has commonly been used to describe enzymatic reactions. Here the intermediate formed is a *substrate-enzyme complex*

$$A + Enz \underset{k_2}{\overset{k_1}{\rightleftharpoons}} [AEnz] \overset{k_3}{\longrightarrow} B + Enz \qquad (6\text{-}2)$$

The model assumes the quasi-steady-state hypothesis applies and that the substrate-enzyme complex concentration is constant. Equation (2-11) can be rewritten, assuming total enzyme concentration is constant, to give

$$\frac{dC_A}{dt} = -\frac{\mu_{max} C_A}{K + C_A} \qquad (6\text{-}3)$$

Figures 2-5 and 2-6 remain useful in describing the kinetics of simple enzymatic reactions.

Many obvious modifications of the Michaelis-Menten model are possible, and several have been found useful in enzymology. Among these are models with reversible complex formation, bisubstrate reactions, and reactions in which product concentration controls the reaction rate. Equations for each can be constructed in a manner similar to that for Eq. (6-3). A further complication is introduced by the fact that reactions occur in long sequences and the rates are interrelated, much like a wave running down a rope or a dam backing up a stream. Thus each reaction must be characterized both individually and in light of its position in a system of reactions. There is no general reaction equation.

A further complexity that must be considered is the inhibition of enzyme-catalyzed reactions. A number of types of inhibition exist. Direct competition for the active site by molecules similar to the reactant can occur, and this is called *competitive inhibition*. *Uncompetitive inhibition* occurs when the inhibitor affects the quantity of enzymes available but not the rate at which individual enzymes react with the desired reactant. An example of uncompetitive inhibition would be inhibition of enzyme synthesis. Noncompetitive inhibition occurs when the inhibitor molecule attaches to the enzyme close enough to the active site to block attachment by the reactant molecule. A fourth form is *allosteric inhibition* in which the inhibitor attaches to the enzyme at some remote point but changes the bond structure to the point that reactivity is decreased. For additional information on enzyme kinetics and inhibitions, see Mahler and Cordes[14] or Dixon and Webb.[15]

The actual role of enzyme kinetics in wastewater treatment is not well established. If a system is truly reaction limited, enzyme kinetics are extremely important. There are several transport steps that would appear to have distinct possibilities of limiting the overall rate, however. Thus a pragmatic viewpoint must be taken in most cases with the hope that someone will elucidate what is going on at some time in the future.

Complete discussion of the reaction systems involved in oxidation and synthesis of molecules within the cell is beyond the scope of this text. Certain central pathways for organic breakdown can be discussed with profit, however. The reader should remember that these pathways, while common, are not constituents of all bacteria and that alternate schemes are often available also. We will not discuss individual reactions or the enzymes connected with each reaction. Reactants and products will be given at important reference points and will be treated much like mass points in a lumped-parameter system, even though this is not a precise analogy. Carbohydrate breakdown is a central process and will be discussed first, along with the associated energy metabolism. Protein and lipid (fat) metabolism will then be discussed in relation to carbohydrate metabolism. Although a number of carbohydrate metabolism pathways are found, only one, glycolysis coupled to the tricarboxylic acid cycle, will be presented for the sake of simplicity.

Carbohydrate Metabolism

Glycolysis is sometimes referred to as *anaerobic carbohydrate metabolism* because energy production occurs without the use of free oxygen from outside the cell.

Energy released in oxidation reactions is captured in the form of the high-energy phosphate ester bonds of adenosine triphosphate (ATP). This energy is available for use in endothermic reactions within the cell. A second form of energy storage also occurs through the production of the reduced form of the coenzyme NAD^+, $NADH + H^+$. Release of the stored energy for the production of ATP does not occur under anaerobic conditions but is important as will be discussed.

A complete picture of carbohydrate breakdown must begin with the macro-molecules, which cannot penetrate the cell membrane. These molecules may have molecular weights of the order of several hundred thousand and must be broken down into subunits outside the cell by extracellular hydrolytic enzymes, then transported through the cytoplasmic membrane for metabolism. Discussion here will deal only with glucose metabolism. Other materials are normally converted into glucose or one of the degradation products of glucose and enter the pathway at that point. A schematic diagram of the glycolytic pathway is shown in Fig. 6-6.

Energy in the form of ATP must be added to initiate the glycolysis sequence. Two moles of ATP are produced per mole of glucose entering glycolysis, and thus there is a net energy production. The phosphate ester bond of ATP stores approximately 8000 cal/mol. Production of ATP in glycolysis is called *substrate-level phosphorylation.*

When pyruvic acid has been produced, two general pathways are available. If the organism contains a tricarboxylic acid (TCA) cycle, the preferred pathway involves this terminal oxidation scheme. Large amounts of $NADH + H^+$ are produced in the TCA, however, and a method of oxidation to NAD^+ must be available. Operation of the TCA cycle as a terminal oxidation scheme depends on the presence of an exogenous electron acceptor such as oxygen being available in the cell's environment to oxidize the $NADH + H^+$. Thus, under anaerobic (oxygen-free) conditions, the TCA cycle is used only for production of particular carbon skeletons for synthesis and interaction with protein and lipid metabolism.

Under anaerobic conditions, pyruvic acid is converted to an end product that is excreted into the surrounding media. The reaction or reactions have the purpose of oxidizing $NADH + H^+$ to NAD^+. Most microorganisms capable of anaerobic fermentation produce a number of end products, although one compound may be the primary product. Table 6-2 provides an example of the range of anaerobic fermentation products produced by a common enteric bacteria, *Escherichia coli.*

While every organism is different and produces different fermentation products for a given set of conditions, there is a general situation that the end products will be toxic to the bacteria if allowed to accumulate. Of particular importance are the organic acids that if allowed to accumulate result in lowering of the system pH. Anaerobic treatment processes are, therefore, dependent on some technique for pH control, preferably further oxidation of the organic acids. The conventional process used is methane fermentation, and this subject will be discussed in Chap. 10.

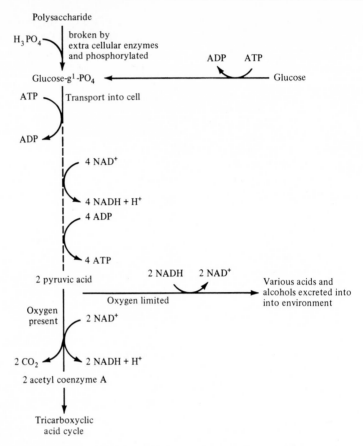

FIGURE 6-6
Schematic of glycolytic pathway (stoichiometric coefficients are for 1 mol of glucose entering the pathway).

Many bacterial species are facultative; i.e., they function under anaerobic and aerobic conditions. Most facultative species utilize glycolysis or similar pathways under both anaerobic and aerobic conditions. When free oxygen is available, pyruvic acid is oxidized to acetyl coenzyme A, which in turn enters the TCA cycle (often called the *Krebs cycle* after H. A. Krebs who developed much of the original theory). The TCA cycle (Fig. 6-7) is a cyclic system for complete oxidation of pyruvic acid to carbon dioxide. Carboxyl units are removed at two points in the cycle. Thus, for every mole of acetyl coenzyme A entering the cycle, a corresponding number of carbon atoms (two) leave as carbon dioxide. Energy released through oxidation reactions is stored in three

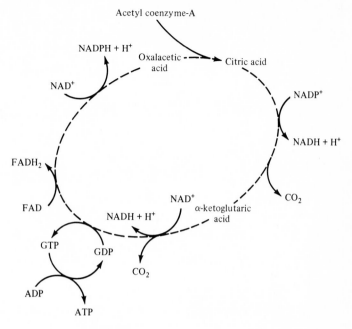

FIGURE 6-7
Simplified schematic of the tricarboxylic acid cycle (TCA).

Table 6-2   **FERMENTATION PRODUCTS OF**
ESCHERICHIA COLI[16, 17]
MOL/100 MOL GLUCOSE
FERMENTED

	pH 6.2	pH 7.8
2,3-Butanediol	0.3	0.26
Acetoin	0.059	0.190
Glycerol	1.42	0.32
Ethanol	49.8	50.5
Formic acid	2.43	86.0
Acetic acid	36.5	38.7
Lactic acid	79.5	70.0
Succinic acid	10.7	14.8
Carbon dioxide	88.0	1.75
Hydrogen	75.0	0.26
Carbon recovered, %	91.2	94.7
O/R balance	1.06	0.91

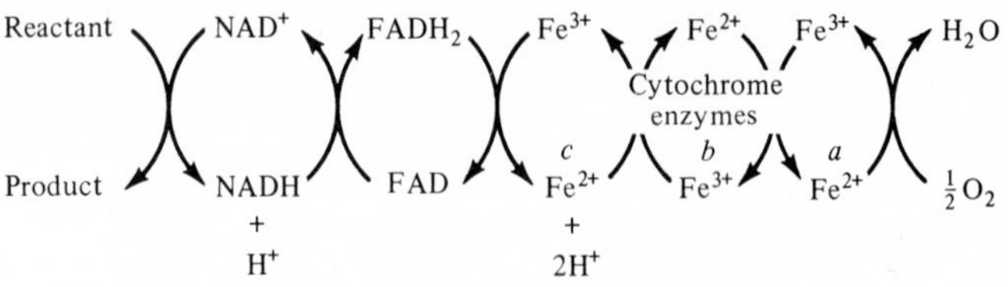

Reactant — NAD^+ — $FADH_2$ — Fe^{3+} — Fe^{2+} — Fe^{3+} — H_2O

Cytochrome enzymes

c b a

Product — $NADH + H^+$ — FAD — $Fe^{2+} + 2H^+$ — Fe^{3+} — Fe^{2+} — $\frac{1}{2}O_2$

FIGURE 6-8

Schematic of the electron transport system.

forms: the reduced coenzyme $NADH + H^+$; the reduced prosthetic group $FADH_2$; and in the phosphate ester bonds of guanosine triphosphate (GTP), which can be directly converted to ATP. Utility of the production of the other two compounds, $FADH_2$ and $NADH + H^+$, is dependent upon some scheme being available for reoxidizing them, preferably with concurrent production of ATP.

The reaction system that oxidizes $FADH_2$ and $NADH + H^+$ is called *respiration*, and in the most common form involves oxygen as a terminal exogenous electron acceptor. A system of reactions that involve enzymes with iron-containing prosthetic groups called *cytochromes* serves to oxidize the cofactors through coupled oxidation-reduction reactions. Although variations may exist among species as is the case with glycolysis and the TCA, the reaction sequence is of the form shown in Fig. 6-8. Three cytochrome enzymes are shown, but in many cases at least four are believed to be involved.

Under anaerobic conditions many facultative bacteria have the ability to utilize nitrate reductase, an enzyme that catalyzes the reduction of nitrate (NO_3^-) to nitrite (NO_2^-) and thus provides an alternate pathway for the electron transport chain. Because most bacteria are able to use nitrate as a source of nitrogen for synthesis, the ability to reduce nitrate nitrogen to ammonia- (NH_3) valence nitrogen is necessary. Only ammonia-valence nitrogen is incorporated into cell tissue. Thus the connection to the respiratory chain is not surprising. A few bacteria are able to reduce nitrite to molecular nitrogen (N_2). Connection of this process to electron transport is uncertain, but the necessity of removing nitrite, which is quite toxic to bacteria, from solution makes the process important. A second and possibly more important result of reducing nitrate nitrogen to molecular nitrogen is the resulting ability of the engineer to design biological nitrogen-removal processes. A detailed discussion of nitrate reduction will be given in Chap. 8.

In addition to nitrate, sulfate serves as an electron acceptor for some bacteria under anaerobic conditions. Details of the process are less understood than those of nitrate reduction, but the process is quite common in nature. An important end product is H_2S, a result familiar to most persons dealing with wastewater treatment.

Location of the bacterial electron transport chain is in the cytoplasmic membrane. Eucaryotic cells have a special organelle, the mitochondrian, which is itself bounded by a membrane that houses the respirometric mechanism. In either case, when the terminal electron acceptor (usually oxygen) is not available, the enzyme cofactors are locked into their reduced form, and the system effectively shuts down. A second way in which respiratory activity can be stopped is by poisoning the enzymes. Common poisons are carbon monoxide and cyanide, both of which irreversibly combine with the ferrous ion of the cytochromes.

Energy available as the result of the reactions of the TCA cycle and the electron transport chain is 215 kcal/mol acetyl-SCoA entering. An overall reaction equation would be

$$\text{Acetyl-CoA} + 2O_2 \rightarrow 2CO_2 + 2H_2O + \text{coenzyme A} + 215 \text{ kcal} \qquad (6\text{-}4)$$

While much of this energy is lost as heat, a significant portion is trapped by the cell in the form of ATP through a process known as *oxidative phosphorylation.* Eucaryotic cell respiration consistently produces 3 mol ATP/mol NADH + H^+ or 2 mol ATP/mol $FADH_2$ entering the electron transport chain. This would correspond to a ratio of inorganic phosphate uptake to oxygen utilization (P/O ratio) of 3. Bacterial P/O ratios are usually less than 1 with a reported range of 0.4 to 1.0.[18] Free energy change between the first step of electron transport, NADH + H^+ oxidation, and the terminal step, oxygen reduction, is 52 kcal, while the high-energy phosphate ester bond of ATP yields approximately 8 kcal. Thus, for a P/O ratio of 1, the energy-transfer efficiency is only 15.4 percent, and the efficiency for a P/O ratio of 3 is 46.2 percent.

Metabolism of Fats and Proteins

We have now discussed the metabolism of carbohydrates and the associated energy metabolism. Fats and proteins fit into the scheme discussed very nicely and can in effect be hung on the side of Figs. 6-7 and 6-8 much like attachments are hung onto a vacuum cleaner.

Fatty acids have the general formula $R-CH_2-CH_2-COOH$. Breakdown of fatty acids is by a process of removing two carbon units at a time (Fig. 6-9). The two carbon units are removed in the form of acetyl-CoA and enter the TCA cycle. As noted in Fig. 6-9, each cycle of the β oxidation scheme produces one $FADH_2$ and one NADH + H^+. One ATP is needed to start the first cycle, but none is needed thereafter. Additional reduced cofactors are produced during oxidation of acetyl-CoA in the TCA cycle, making fatty acids an exceedingly rich source of energy. Because ATP is not produced directly, there is considerable difficulty in anaerobic metabolism of fats, however.

Protein metabolism also is interlocked with the TCA cycle. Hydrolysis of proteins into their amino acid subunits is similar to reactions for carbohydrates. Deamination of the amino acids occurs directly in some cases or through a process of transamination (Fig. 6-10). Because deamination and transamination

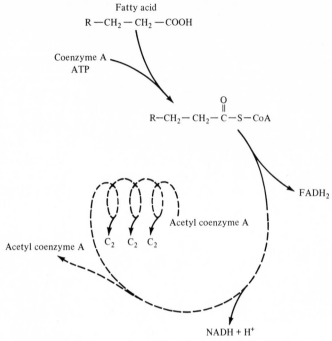

FIGURE 6-9
Simplified schematic of β oxidation scheme for fatty acid breakdown.

are reversible, there is a direct relationship between amino acid catabolism and protein synthesis. Amino acids enter the glycolytic and TCA pathways at a number of points. The most important are the interrelationship between glutamic acid and α-ketoglutaric acid, and between aspartic acid and oxalacetic acid.

Summarizing the metabolism of bacteria can best be done by use of a diagram such as Fig. 6-11. The reader must remember that the diagrams and

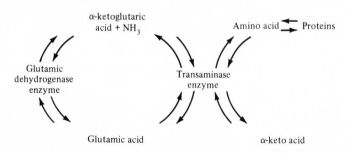

FIGURE 6-10
Schematic of reversible deamination and transamination reactions.

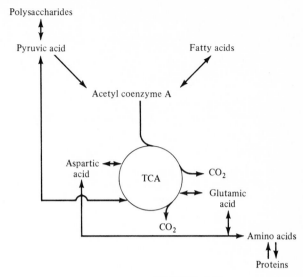

FIGURE 6-11
Interconversion in metabolism.

statements on metabolism provide an overview only. Few, if any, organisms completely fit the descriptions, but the overall pattern shown satisfies the generalizations.

Nutrient Limitation

Not shown in Fig. 6-11, but extremely important, are the essential nutrients. Many, if not most, microorganisms are unable to synthesize all the carbon skeletons necessary for growth. Different species have different requirements, but common "growth factors" are one or more amino acids or one or more of the vitamins. Additionally, certain ions must be present in trace quantities, usually because of their place in enzyme cofactors. The most important trace ions are potassium, magnesium, manganese, calcium, iron, cobalt, copper, zinc, and molybdenum. Normally, natural waters or wastewaters contain satisfactory amounts of trace elements, and mixed cultures provide organic growth factors. In laboratory studies, those ions appear as constituents on contaminants of the major inorganic nutrients (SO_4^{2-}, PO_4^{3-}, NH_4^+) added to experimental systems. Recent studies by Wood and Tchobanoglous[19, 20] and McKinney[21] have led to the conclusion that iron deficiencies may be fairly common in wastewaters, and one result of this situation is the development of filamentous cultures.

Bacterial Growth

Bacterial growth covers a wide range of subject material. This section will concentrate on cell scale and larger phenomena rather than on a discussion on the biochemistry of biosynthesis. We will begin by describing the process of cell reproduction and then move on to a discussion of batch and continuous cultures. When cultures are considered, individual cells quickly lose significance much as individual molecules lose significance when considering the paper on this page.

Individual cells go through a range of sizes as they metabolize nutrients. The cell is relatively homogeneous throughout the cytoplasm, and therefore the only structure that must be duplicated is the chromosome. When this has been accomplished, the two chromosomes migrate to the poles of the cell. Formation of the cell wall and cytoplasmic membrane through the lateral axis then begins. Completion of the membrane and wall produces two new individuals. This process is called *binary fission*. Rate of cell division varies with bacterial species, organic material being metabolized, and environmental conditions. Some species have minimum (optimal) division times of less than 30 min, while others require many hours.

Beginning with one cell in a carefully controlled environment, an investigator can synchronize cell division such that each generation of cells divide together. Synchronous growth stops after a few generations because of the number of variables that must be controlled, however. At this point the system becomes a bacterial culture, and the individuals lose significance. Wastewater treatment processes have the further complication that the cultures are mixed; i.e., they are made up of a number of different bacterial species and normally include protozoa and other predators as well.

Advantages of Mixed Cultures

System complexity provides a distinct advantage in that the culture is able to adapt to changing conditions. A change in wastewater makeup, temperature, or pH would possibly change the growth rate of species making up the culture, increasing some and decreasing others. Species with increased growth rates would become more competitive and thus more predominant in the culture. The strength of the mixed culture is that metabolism of the waste materials continues, and the process responds in a positive or stable fashion. A point should be made, however, that positive response to waste assimilation may not include continued stability with respect to other parts of a waste treatment process (e.g., sedimentation).

A disadvantage of the complex system is that reproducible data are difficult to produce. A process change may wipe out a portion of the culture that is difficult to reintroduce. This would occur for pure cultures also, of course, but

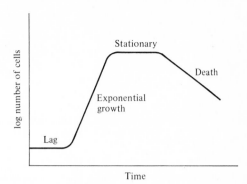

FIGURE 6-12
Typical growth curve for bacterial culture.

the problem with mixed cultures is determining effects for a given cause. The literature abounds with reported phenomena that have been observed in the field for which there are dozens of plausible explanations.

Measurement of Growth

When one moves from the observations of the growth of individual organisms to the growth of a population or culture, the method of measurement changes also. Methods that have become standard among bacteriologists are *cell counts* and *turbidity measurements*. Cell counts, usually made by plating out sample dilutions on nutrient media, are usually considered to be the primary standard with turbidity serving as an easily correlated method of following growth. Bacteriologists are particularly interested in the *viable* (cells that can reproduce are viable) cell concentration, and only viable cells produce colonies on nutrient media. A typical curve resulting from pure culture experiments is shown in Fig. 6-12.

Curves such as Fig. 6-12 result from systems that initially are limited only in the number of bacteria present, thus allowing growth to occur at the optimal or maximum rate. Normally, the initial inoculum of cells is taken from a "resting" or stationary culture. These cells do not begin metabolizing nutrients immediately on contact but need time to adjust to the new environment. The reason for this lag is not understood, but quite probably a number of enzyme systems (e.g., the transferase system which moves material across the cytoplasmic membrane) must be synthesized. Growth occurs at the maximum rate until some nutrient becomes limiting. At that time the curve begins to become less steep. Stored nutrients (glycogen, starch, β-hydroxy butyrate) may be metabolized, with resulting growth, after the limiting nutrient has been removed. Maximum growth (in terms of total mass) should occur shortly after nutrient disappearance, however. Following a period in which the number of viable cells does not increase, cells begin to die faster than they can be replaced. Death is difficult to describe, and while most models assume an exponential relationship, reproducible data are difficult to obtain.

Stoichiometry of Growth

Stoichiometric considerations would predict that for a given system linear relationships between nutrient uptake, cell growth, and for aerobic systems oxygen uptake exist as per Eq. (6-1). Because of the complex nature of cell metabolism, the stoichiometry is dependent upon the condition of the bacterial culture.[22-25] When the culture moves from the exponential growth phase toward the stationary phase (Fig. 6-12), energy required for the turnover of molecules within the cell becomes a significant fraction of the total energy expended. Under these conditions, the quantity or mass of cells produced per unit mass of nutrient consumed decreases. Batch-process stoichiometry can be determined directly from curves such as Fig. 6-12. Continuous-flow-process (which includes most wastewater treatment processes) stoichiometry can be best determined as a function of growth rate (or the inverse of growth rate mean cell residence time or cell age).

Because wastewater treatment processes utilize mixed cultures and because the nature of the engineering problems deal with quantities of materials, cell counts and turbidity do not provide satisfactory design and operation information. Engineering measurement of cell population is nearly always on the basis of mass concentration (milligrams per liter), therefore. This method has the advantage of fitting directly into mass balances and the disadvantage of providing no indication of the viable or "active" mass present. Material retained on a specified filter dried at 105°C is reported as suspended solids or mixed-liquor suspended solids (MLSS). This material may include organic constituents of the waste, inorganic particles, and dead cells as well as viable cells. Many workers believe that the volatile suspended solids (VSS), that is, the solids which volatilize at 600°, are a more satisfactory measurement of cell mass. The author feels that in most cases the extra work of determining VSS is unnecessary if one understands the system being studied. In either case, coefficients will have to be determined that account for a viable fraction. Further discussion of this subject will be included in Chap. 8.

Mass Yield

Measurement of growth on a mass-increase basis is useful for rate determinations as well as for establishing system stoichiometry. Thus we speak of mass of cells produced per unit time or, more precisely, mass increase of MLSS per unit time. Stoichiometric information would allow relating the growth rate to the mass removal rate of necessary nutrients (organic compounds, inorganic nutrients, oxygen, etc.). Cell yield, a stoichiometric parameter, can now be defined as the negative ratio of the growth rate R_g to the removal rate of the limiting nutrient R_N.

$$Y_{obs} = -\frac{R_g}{R_N} \qquad (6\text{-}5)$$

where R_g has units of mass of cells produced per unit volume per unit time and R_N has units of mass of nutrient removed per unit volume per unit time.

Observed cell yield (Y_{obs}) varies with temperature, limiting nutrient, and culture environment. Bacterial species have optimal temperatures for growth, for example. Three loose classifications are used: psychrophilic, which includes species that compete best (but do not necessarily grow best in pure culture) below 20°C; mesophyllic, which includes species with maximum growth rates between 20 and 40°C; and thermophilic, which includes species with maximal growth rates above 40°C. Not surprisingly, maximum yield is related to maximum growth rate, and one would expect psychrophilic organisms to have maximum yields at lower temperatures. A second aspect of temperature must be considered also, however. As temperature increases, the breakdown rate of reactive molecules (e.g., enzymes) within the cell increases, thus increasing the "maintenance energy" requirement.[22] The net result in mixed cultures is a general decrease in yield with increasing temperature.

Limiting-nutrient effects on yield vary with the type of nutrient. When carbon, i.e., the organic material, is limiting, yield decreases as the limitation becomes more severe. Batch cultures show this in a rather obvious fashion (Fig. 6-12), but continuous cultures provide a better demonstration of the principle because there is always some of the limiting material present. Sherrard's[25] 1971 studies on activated sludge increased the nutrient (carbon) limitation by increasing the cell age (or mean cell residence time). The effective result of his control procedure was to increase the MLSS concentration and decrease the amount of incoming carbon available to each cell. Cell yield decreased with mean cell residence time Θ_c, as shown in Fig. 6-13.

Nitrogen limitation results in another type of response. Because an energy and carbon source is still available (i.e., nitrogen is the only limiting variable), metabolism continues. Limited amounts of nitrogen are available for synthesis of proteins and nucleic acids, and instead, large amounts of storage materials and slimes are produced. Yield actually increases as nitrogen limitation increases, but the increase is not indicative of the reactive cell mass. As the limitation becomes greater, the maintenance-energy requirements of the cell become greater also (as in the carbon-limited case), and eventually a maximum yield value is reached. Beyond this value, net mass yield decreases. Moore[26] studied nitrogen-limited denitrifying cultures and reported yield data. His work is discussed in Chap. 8. Similar yield results might be expected for phosphate-limited cultures, but data are not available on the subject at this time.

Cell yield is quite clearly related to the amount of energy available for growth. A number of workers[27–33] have used mass of cells produced per mole of ATP

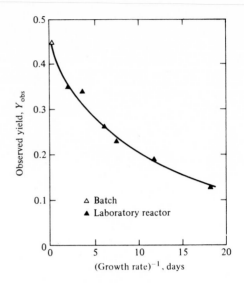

FIGURE 6-13
Observed cell yield vs cell residence time.[25]

generated in metabolism. This procedure is useful in comparing the effects of one organic substrate against another or in trying to elucidate certain aspects or pathways of metabolism. Few municipal wastewaters are consistent enough in their constituents to make the approach feasible for wastewater treatment, however. A more significant problem with utilizing such a method for process design or control is that maintenance requirements are not taken into account when Y_{ATP} is determined. Thus moving from a maximum-growth–minimum-maintenance situation destroys the model validity.

An extension of this approach uses energy released by the overall reactions of reactant to end product to predict yield. This technique utilizes an empirically obtained conversion efficiency. The conversion efficiency would be expected to vary with environmental conditions just as the yield does, and therefore predictive capability of the approach is limited to regions in which experimental data and substantial information about the energy source are available.

A point to remember in considering the energy-available approach is that the problem is treated as if a molecule either is completely oxidized or goes directly into synthesis. Clearly, a glucose molecule might be oxidized to a TCA cycle intermediate such as α-ketoglutaric acid with the expected release in energy and conversion efficiency and then enter the pathway to synthesis. This means that energy-conversion efficiencies are purely empirical when overall processes are used for their determination.

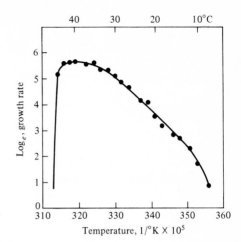

FIGURE 6-14
Growth rate of *E. coli* as a function of temperature.[34]

Temperature Effects

Rate of growth R_g is dependent upon temperature, limiting-nutrient concentration, cell concentration, and environmental conditions. Temperature-rate relationships have been a matter of continual study for both bacteriologists and engineers. The Arrhenius relationship [Eq. (6-6)] is often used to describe temperature effects.

$$R_g = Ae^{-E'/RT} \qquad (6\text{-}6)$$

Activation energies E' can be established for a given nutrient-bacterial species concentration. An example of data from temperature-growth studies is given in Fig. 6-14. Death of the culture occurs rapidly at higher temperatures, and the relationship clearly does not provide a perfect fit in any region.

Convenience led to the use of a modified form of Eq. (6-6) in wastewater treatment studies.

$$R_g = R_{g(20)} \psi^{T-20} \qquad (6\text{-}7)$$

where $R_{g(20)}$ = rate of growth at 20°C
 ψ = constant.

Values of ψ reported vary widely, but most are between 1.0 and 1.15, depending upon the process studied and the temperature range used. A widely used rule of thumb (the van't Hoff rule) is that biological reaction rates double every 10°. This would correspond to a ψ of 1.072. The great majority of the studies on biological wastewater treatment processes have resulted in temperature coefficients below 1.05, however. Ingraham's[34] data (Fig. 6-14) also places the value of ψ at considerably less than 1.07.

Growth Rate Model

Nutrient and cell concentration would be expected to affect growth rate in a manner similar to chemical reactions. Monod,[35] working with pure cultures under both conditions, found that the initial portions of the growth curve (Fig. 6-12) could be described by a Michaelis-Menten type of relationship.

$$R_g = \frac{r_{max} C}{K_m + C} X \qquad (6\text{-}8)$$

where X is the cell concentration or the MLSS concentration, and the other terms are as before.

A number of studies have utilized the Monod model to describe continuous-flow biological processes. Most of the successful studies have utilized pure cultures and high growth rates, notably the works of Herbert and his co-workers,[36, 37] Fencl,[38] and Powell.[39] While the model has been used for waste treatment process studies, the results have usually been ambiguous. A significant part of the problem seems to be in the relatively crude methods of measuring the variables. Measurement of cell mass concentration has been described already. Organic material is usually measured on a gross or overall basis also, partly because of the difficulty in measuring individual waste components. A second reason is the need to determine effluent quality on an overall basis, however. One would suspect that the limiting material might well be a small fraction of the total quantity present, thus making the measurement techniques quite insensitive.

Moore and Schroeder[40] and Requa and Schroeder[41] have reported data for denitrification processes that fit the Monod model extremely well. Their experimental systems were quite simple, and only a single organic nutrient (methanol) was used. Nitrogen was the growth-limiting material in both studies.

Powell,[42] Van Uden,[43] Lee et al.,[44] and Grady and Roper[45] have proposed extensions of the Monod model that include consideration of maintenance-energy requirements, transport limitations, and multiple-product formation. Application of these extensions to wastewater treatment processes is difficult for the same reasons that make application of the Monod model itself difficult. Kinetic approaches to process modeling will be considered as individual processes are discussed.

6-2   VIRUSES

Viruses fall into the crack between living and nonliving things. They are not complete organisms, being made up of a protein-protective coating surrounding a strand of nucleic acid (Fig. 6-15). All viruses are obligate parasites in that the way they reproduce involves infecting living cells and redirecting the synthesis reactions toward producing new viral particles.[46, 47]

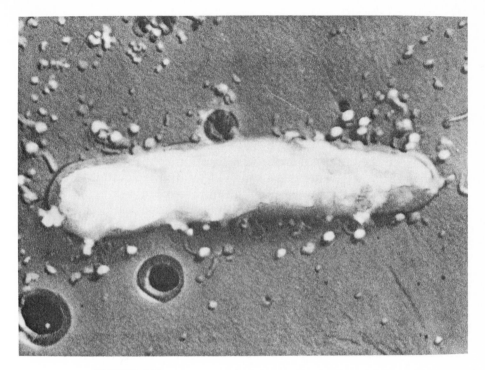

FIGURE 6-15
Bacterial viruses (bacteriophage) together with bacterial hosts. (*JEOL Ltd., Tokyo.*)

Size of viruses varies between 1×10^{-8} and 30×10^{-8} m, with size being an important and reproducible characteristic of the particular type. Polio virus has a characteristic dimension of 1.2×10^{-8} m, bacteriophage (viruses that attack bacteria) usually have dimensions of the order of 6.5 to 9.5×10^{-8} m, and smallpox virus has a characteristic dimension of 25×10^{-8} m. Viral particle shape is also characteristic of a particular type. Shape is usually regular and is often made up of a number of plane surfaces.

Viral Reproduction

Replication of viral particles begins with adsorption of the virus to a susceptible cell. Viruses are very host specific, and thus the type of cell is important. Animal viruses penetrate the cell in their entirety, while only the nucleic acid of bacteriophage enters the cell. The viral nucleic acid then migrates to the nucleus of the cell and, in some manner, redirects the synthetic mechanisms toward producing new viral particles. After a number of new particles have been assembled, the cell membrane breaks, and the particles are released into the environment. Time lapse from penetration to release may be as short as 20 min.

Importance of Viruses in Water and Wastewater Treatment

The primary area of concern with respect to viruses in water is public health. A number of viral diseases have been classified as waterborne, including polio, hepatitis, and Coxsackie virus infections. Concentrations of viruses in sewage are typically of the order of 1 to 2 ml^{-1}, and it is presumed that most water supplies have considerably lower concentrations.[48, 49] A single viron can presumably result in an infection, and therefore, the low concentrations are not entirely satisfying.

Viral activity does not decrease with time outside the host. Prevention of infections by viruses in water thus requires either their removal or disinfection. A number of studies have been made on removal of viruses by coagulation, sedimentation, and filtration.[50-55] A general conclusion has been that these processes can remove up to 99 percent of the viral particles in water. In most of the studies, the tests were run on water samples that had been inoculated with a high concentration of viruses ($> 10^5$/l). Whether these high removals occur for lower initial concentrations is not completely known at this time. A number of studies have shown that viruses which have been adsorbed or caught by coagulant floc retain their activity.

Chlorine and ozone have both been shown to be viral disinfectants.[49] Ozone is more effective than chlorine, but in either case contact times necessary are relatively long. Further discussion of this subject will be given in the section on disinfection later in this chapter.

6-3 PROTOZOA

The protozoa are a group of generally motile, single-celled organisms that do not have a cell wall. In most cases protozoa are predatory, often feeding on bacteria. Protozoa are also characterized by the eucaryotic cell, which has internal membranes and is considerably more complex than the procaryotic cell of the bacteria.

Protozoa are important in wastewater treatment because they feed on bacteria. A protozoan commonly found in water treatment processes is vorticella.

The major public health problem associated with the protozoa is amebic dysentery. This disease is caused by an organism called *Entamoeba histolytica*.

6-4 ALGAE

The algae are a highly varied group of photoautotrophic organisms. All algae use CO_2 as a carbon source and light as an energy source. A number of subgroups exist, with the most important in water and wastewater treatment being the blue-greens and greens.

Blue-green Algae

This group is quite different from the other algae types because of its procaryotic cell. In many ways the blue-green algae might be considered the most versatile of all living organisms. Although photosynthesis is the primary energy-conversion process, some blue-green algae have been observed to metabolize complex organics. Nitrogen fixation has been observed to occur with many blue-green species. The blue-green algae always have unspecialized cells; i.e., each cell is independent of others with respect to metabolism. Certain species typically grow in filamentous forms and often cause problems in streams and lakes. Notable among this group are *Nostoc* and *Oscillatoria*.

Two groups of blue-green algae, *Microcystis* and *Anabaena*, produce toxins that can result in sickness or death to birds and mammals drinking the water.[56] Excretions from blue-green algae often cause taste and odor problems in water. Oxidation of the odor-causing compounds with chlorine or ozone often improves the situation, but additional treatment with activated carbon is usually necessary to produce a palatable product. Taste and odor problems usually occur during the warm months of late summer and are widespread among communities using surface-water sources in the Midwestern United States.

Green Algae

The green algae, like all other groups except the blue-greens, are characterized by the eucaryotic cell. They have a rigid cell wall similar to plants and are strict photopautotrophs. Two unicellular green algae, *Chlorella* and *Senodesmous*, are commonly found to be the predominant algae species in wastewater oxidation ponds (Chap. 8). Both groups would serve as satisfactory feed additives for cattle if inexpensive harvesting techniques were available.[57]

Other Algae of Interest

A number of algae groups are of general interest because of particular properties. The diatoms have a hard, silica shell, which in fossil form gives us diatomataceous earth. Other algae groups grow in colonial forms and may achieve lengths as great as 30 m. Kelp are the principal example of these groups. Colored waters, such as the red tides that occur in some areas of the world, are also often due to algae.

Photosynthesis

Photosynthesis is the process by which organisms utilize light energy for synthesis of sugars from carbon dioxide and water. The sugars are then used in metabolism in much the same manner as was described for bacterial metabolism. Oxygen is a by-product of the photosynthetic process as shown in Eq. (6-9).

$$CO_2 + H_2O \xrightarrow{\text{sunlight}} CH_2O + O_2 \qquad (6\text{-}9)$$

Eucaryotic algae have photosynthetic pigments called *chlorophylls*, which are similar to those of green plants. Blue-green and green algae chlorophylls are similar, but the photosynthetic structure of the green algae is more complex.

A number of bacterial species also carry out photosynthesis, but oxygen is not a product. The purple bacteria and the green bacteria are both obligate anaerobes, and their photosynthetic reactions are quite different from those of algae and plants. Purple bacteria sometimes predominate near the surface of organically overloaded oxidation ponds (see Chap. 8). These ponds are very obvious from the air, being a dark but bright purple.

Photosynthetic pigments differ in their principal adsorption wavelengths. Oxygen-generating photosynthesis usually involves wavelengths between 3500 and 7500 Å[1], which is approximately the same as the visible region. Bacterial photosynthesis extends into the near-infrared region to about 9000 Å.

Algae and Eutrophication of Natural Waters

Eutrophication is a phenomenon that includes many factors such as increases in turbidity, sediment, productivity, and average temperature. Algae are nearly always a significant factor in the eutrophication process because they add organic material to the system. When nutrients necessary for algal growth are present, "blooms" (large increases) in the algae population occur in warm weather. Large differences in oxygen content of the water may occur between day and night during blooms. In some cases the high rate of photosynthesis occurring during the day may result in supersaturation of the water. At night algal respiration continues, and oxygen depletion occurs. When algae die, they sink to the bottom of the lake or stream and are oxidized by bacteria, resulting in sludge deposits and oxygen depletion.

6-5 DISINFECTION

Disinfection interpreted in a very broad sense would include the removal and killing (or rendering nonviable) of viruses and organisms found in water and wastewater. Coagulation, flocculation, sedimentation, and filtration often remove more than 99 percent of the organisms in water. This result is an artifact of normal process design methods, and the objective of removing organisms is not included in the quantitative design process. High removals in the physical treatment process steps are helpful, but more complete removals are necessary. United States Public Health Service Drinking Water Standards[58] require that potable water be free of coliform organisms. Discharge requirements placed on many wastewater treatment plants require effluents to meet bacterial quality for contact sports. Chemical disinfection is necessary to meet these requirements.

Desirable Properties of Disinfectants

High reaction rate at low concentrations is the most desirable property of a disinfectant. Rate of disinfection is related to the detention time required for satisfactory results, and the concentration necessary affects chemical costs. Persistence of the disinfectant is desirable in water treatment where prevention of regrowth of organisms is wanted but undesirable in wastewater treatment because of possible toxicity to fish in the receiving water.

Disinfection rates often follow a logarithmic decay pattern as described by Chick's law.[48]

$$\frac{dN}{dt} = -kN \qquad (6\text{-}10)$$

where N is the number concentration or organisms and k is a rate coefficient that varies with pH, temperature, and disinfectant concentration.

Chlorine

Chlorine has been used as a disinfectant in water since 1896. Two forms of chlorine are in common use, chlorine gas and hypochlorite compounds. In water, chlorine reacts very rapidly to form hypochlorous and hydrochloric acids.

$$Cl_2 + H_2O \rightarrow HOCl + HCl$$

$$HOCl \rightleftharpoons H^+ OCl^-$$

$$K_i = \frac{(H^+)(OCl^-)}{(HOCl)}$$

Values of the ionization constant K_i at different temperatures are given in Table 6-3. Hypochlorous acid and hypochlorite ion are the principal disinfectants, and their total concentration is referred to as *free residual chlorine*.

Hypochlorite compounds such as $Ca(OCl)_2$ or $NaOCl$ also form hypochlorite ion and hypochlorous acid upon dissolving in water.

$$Ca(OCl)_2 + 2H_2O \rightarrow 2HOCl + Ca(OH)_2$$

$$NaOCl + H_2O \rightarrow HOCl + NaOH$$

Table 6-3   VALUES OF K_i FOR HYPOCHLOROUS ACID

Temperature, °C	0	5	10	15	20	25
$K_i \times 10^8$ mol/l	1.5	1.7	2.0	2.2	2.5	2.7

Cost of chlorine in the form of hypochlorite is usually about 1.5 to 2.5 times as great as in the form of gas. In some cases, the convenience and relative safety of hypochlorite are more important than cost.

Chlorine reacts with a wide range of reducing agents including Fe^{2+}, Mn^{2+}, $NO_2{}^-$, H_2S, and most of the organic compounds found in water. The major effect of these reactions is to consume chlorine without providing disinfection. Sung[59] found that humic acid interfered with the disinfecting action of chlorine even at normally used free residual chlorine concentrations.

Ammonia reacts with hypochlorous acid to form compounds called *chloramines*.

$$NH_3 + HOCl \rightarrow NH_2Cl + H_2O \quad \text{(monochloramine)}$$

$$NH_2Cl + HOCl \rightarrow NHCl_2 + H_2O \quad \text{(dichloramine)}$$

$$NHCl_2 + HOCl \rightarrow NCl_3 + H_2O \quad \text{(trichloramine)}$$

The chloramines are slow-acting disinfectants.[49] They are rarely used in primary disinfection operations but occasionally are used in distribution systems because of their persistence.

Rate of Disinfection with Chlorine

Destruction rate is dependent upon temperature, pH, and chlorine concentration. The type of organism being destroyed is also a major factor. Viruses are commonly more difficult to kill than bacteria. Cysts (such as those of *Endoameba hystolytica*) are more difficult to kill than either bacteria or viruses. Temperature and pH effects are twofold. Both factors affect viability directly, and both factors affect the hypochlorous acid equilibrium. The latter effect is generally the greatest because temperature and pH ranges encountered in practice are normally in the acceptable range for bacterial growth. Temperature effects can be described by an Arrhenius-type relationship.[48]

$$\ln \frac{t_1}{t_2} = \frac{E'(T_2 - T_1)}{RT_1T_2} \quad (6\text{-}11)$$

where t_1, t_2 = time to obtain a specified percentage kill at temperatures T_1 and T_2, K

E' = activation energy, cal

R = gas constant, 1.00 cal/K-mol

Typical values for the activation energy of chlorine and the chloramines taken as a group are given in Table 6-4.

Concentration effects on disinfection can, within limits, be described by Eq. (6-12):

$$C^n t_p = \text{constant} \quad (6\text{-}12)$$

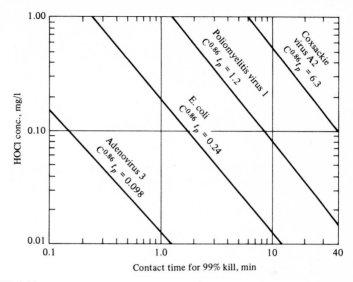

FIGURE 6-16
Concentration of free residual chlorine and contact time necessary for 99 percent kill at 0 and 6°C.

where C = concentration of disinfectant
t_p = time to achieve a given percentage kill
n = experimentally derived constant

Data presented by Berg[61] on contact time necessary for 99 percent kill as a function of HOCl concentration are shown in Fig. 6-16.

EXAMPLE 6-1 Determine the concentration of chlorine that must be added to a water to ensure 99 percent kill of *E. coli* at 20°C in 10 min at pH 7.0.

Table 6-4 ACTIVATION ENERGIES FOR AQUEOUS CHLORINE AND CHLORAMINES AT NORMAL TEMPERATURES[60]

Compound	pH	E, cal
Aqueous chlorine	7.0	8,200
	8.5	6,400
	9.8	12,000
	10.7	15,000
Chloramines	7.0	12,000
	8.5	14,000
	9.5	20,000

Assume that Fig. 6-16 is precise at 6°C from Eq. (6-11).

$$t_{6°} = 10 \exp \frac{E'(T_2 - T_1)}{RT_1 T_2}$$

$$= 10 \exp \frac{8200(293 - 279)}{199(293)(279)}$$

$$= 20.3 \text{ min}$$

From Fig. 6-16,

$$C^{0.86} t_{6°} = 0.24$$

$$C = 0.006 \text{ mg/l HOCl}$$

Because the pH and HOCl concentration is known, the hypochlorite-ion concentration can be determined.

$$K_i = 2.5 \times 10^{-8} = \frac{10^{-7}(OCl^-)}{1.15 \times 10^{-7}}$$

$$(OCl^-) = 2.88 \times 10^{-8} \text{ mol/l} = 0.0015 \text{ mg/l}$$

Remembering that 1 mol chlorine gives 1 mol hypochlorous acid in solution,

$$(Cl_2) \text{ added} = 1.15 \times 10^{-7} + 2.88 \times 10^{-8} + 1.44 \times 10^{-7} \text{ mol/l}$$

$$= 0.01 \text{ mg/l}$$

This would be the residual concentration to be maintained after other demands had been met. In most cases, several milligrams per liter are necessary, and residual values of the order of 0.5 to 1.0 mg/l are required by code.   ////

Advantages and Disadvantages of Chlorine as a Disinfectant

Chlorine is a very effective bactericide at low concentrations. In addition, the rate of chlorine disinfection is satisfactory, particularly in water treatment where product water is stored for times ranging from hours to days. A third advantage is that chlorine residuals can be maintained in treated water. This provides additional safety by preventing regrowth of bacteria.

A number of disadvantages of disinfection with chlorine also exist. Because of the toxicity of the material, the public health hazard associated with its use is significant. Safety procedures at treatment plants are, in general, adequate for plant personnel. Accidents during shipment are a potential problem, particularly where very large quantities are involved.

Chlorinated wastewater treatment plant effluents are often very toxic to fish. The cause of this toxicity may be free residual chlorine or may be reaction products of chlorine and wastewater constituents. Chlorinated hydrocarbons are a product of wastewater chlorination.

Chlorinated hydrocarbons are also produced when organic-containing water is treated for domestic use. Many of these products are known or suspected carcinogens. Concentrations of these materials are very low, and information on relationships between the presence of these materials and incidence of cancer is still under study.

Ozone

Disinfection with ozone (O_3) has been practiced since 1906, although its oxidizing and disinfecting properties were recognized much earlier.[62] Ozone disinfection rates are higher than those for chlorine at the same concentration,[63-65] and ozone is considerably more effective in inactivation of viruses.

Ozone oxidizes many materials found in water, and therefore the quantity needed varies from situation to situation. The only residual is dissolved oxygen, and thus there is no protection against regrowth of organisms. A major advantage of ozone is the lack of toxic by-products.

Because of ozone's instability, it must be generated on site. Costs of ozonation of domestic waters are high, being approximately three times those for chlorine disinfection.

Bromine and Iodine

The halogens, bromine and iodine, have often been suggested as suitable disinfectants. Iodine does not combine with ammonia, and bromine disinfection is not significantly affected. Molecular iodine (I_2) and hypoiodous acid are equally good bactericides, but hypoiodous acid is far more effective in inactivating viruses.

Bromine chemistry is similar to that of chlorine. The effective bactericidal form is HOBr. Because ammonia-bromine combination does not greatly affect the disinfection process, bromine may have a place as a water disinfectant. Like iodine, bromine's relatively high cost will probably prevent its wide application.

PROBLEMS

6-1 Consider two reaction systems (Fig. 6-17), one an ideal plug-flow reactor, the other an ideal CFSTR. A nutrient solution with a limiting nutrient A at a concentration C_{Ai} is fed into the two systems at equal flow rates. Assuming steady-state conditions and the kinetics of Eq. (6-8), determine the ratio of the effluent concentrations, C_{Ae}, of the two systems. Use hydraulic residence times of 1, 12, and 24 h; α values of 0.1, 0.5, and 1.0; and $r_{g,\,max} = 24$ d^{-1}, $r_{0,\,max} = 48$ d^{-1}, and $K_m = 1$ mg/l.

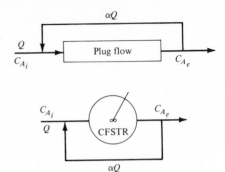

FIGURE 6-17
Reaction systems for Prob. 6-1.

6-2 The diagrams for Prob. 6-1 show recycle lines. Are these lines necessary? Why?

6-3 Determine the value of ψ, the temperature coefficient of Eq. (6-7), that might be expected in a strongly diffusion-limited system. Diffusion coefficients can be found in Sherwood and Reid.[66] Compare this value with that derived from Fig. 6-14.

6-4 Catabolism and synthesis are in some ways similar or analogous to the competitive reactions discussed in Chap. 2. Construct an attainable-region diagram using Eqs. (6-5) and (6-8) for biological reaction systems.

6-5 A conservative estimate of the ratios of rates for different reaction schemes can be obtained from the ratios of energy available. Determine these ratios for systems using oxygen, nitrate, nitrite, and sulfate as electron acceptors. Why are the values conservative?

6-6 A characteristic of many anaerobic fermentation processes is the production of large amounts of heat. Examples include silage and composting operations. For an alcohol fermentation designed to have a final product concentration of 12 percent ethyl alcohol by weight with a cell yield of 0.2 g cells/g glucose added, determine the amount of cooling water initially at 10°C necessary to maintain the fermentation at 22°C. Assume that the mixture has a heat capacity of 1.09 cal/°C. Assume cells produced are at the same energy level as glucose and that the alcohol production is represented by $C_6H_{12}O_6 \rightarrow 2C_2H_6O + 2CO_2$.

6-7 The ratio of mixed-liquor suspended-solids concentration (MLSS) to mixed-liquor volatile solids concentration (MLVSS) does not remain constant with changing operating conditions in wastewater treatment processes. Explain why this might be so and determine the primary variables. Can the statement in the text that in most cases MLSS gives a satisfactory estimate of cell mass be justified? What does this really mean with respect to Eqs. (6-5) and (6-8)?

6-8 Sketch the shape of the effluent cell mass and limiting-nutrient concentrations for an ideal stirred-tank reactor (CFSTR) without recycle as a function of residence time. Define the washout rate. Assume the cell yield is constant. How would the curves be altered if a more realistic cell-yield function were used?

REFERENCES

1 STANIER, R. Y., M. DOUDOROFF, and E. A. ADELBERG: "The Microbial World," 3d ed., Prentice-Hall, Inc., Englewood Cliffs, N.J., 1970.

2 LURIA, S. E.: The Bacterial Protoplasm: Composition and Organization, in I. C. Gunsalas and R. Y. Stanier (eds.), "The Bacteria," vol. I, Academic Press, Inc., New York, 1960.

3 HENDLER, R. W.: "Protein Biosynthesis and Membrane Biochemistry," John Wiley & Sons, Inc., New York, 1968.

4 COLHEN, G. N., and J. MONOD: Bacterial Permeases, *Bacteriol. Rev.* vol. 21, p. 169, 1957.

5 SALTON, M. R. J.: Structure and Function of Bacterial Cell Membranes, *Ann. Rev. Microbiol.*, vol. 21, p. 417, 1967.

6 ROSEMAN, S.: Carbohydrate Transport in Bacterial Cells, in L. E. Hokin (ed.), "Metabolic Transport," Academic Press, Inc., New York, 1972.

7 POUYSSEGUR, J. M., P. FAIK, and H. L. KORNBERG: Utilization of Gluconate by Eschericia Coli, *Biochemistry*, vol. 140, p. 193, 1974.

8 MURRAY, R. G. E.: The Internal Structure of the Cell, in I. C. Gunsalas and R. Y. Stanier (eds.), "The Bacteria," vol. 1, Academic Press, Inc., New York, 1960.

9 BRADFIELD, J. R. G.: in E. T. C. Spooner and B. A. D. Stecker (eds.), "Bacterial Anatomy," p. 296, Cambridge University Press, New York, 1956.

10 CAIRNS, J.: The Chromosome of Eschericia Coli, *Coldspring Harbor Symp. Quant. Biol.*, vol. 28, p. 43, 1963.

11 PORGES, N., L. JASEWICZ, and S. R. HOOVER: Biological Oxidation of Dairywaste VII, *Proc. 10th Ind. Waste Conf.*, 1953.

12 BURKHEAD, C. E., and S. L. WARD: Composition Studies on Activated Sludges, *Proc. 24th Ind. Waste Conf.*, 1969.

13 PAULING, L., and R. B. COREY, *Proc. Int. Wool Textile Res. Conf.*, 1975.

14 MAHLER, H. R., and E. H. CORDES: "Biological Chemistry," Harper & Row, Publishers, Incorporated, New York, 1966.

15 DIXON, M. and E. C. WEBB: "Enzymes," 2d ed., Academic Press, Inc., New York, 1965.

16 STOKES, J. L.: Fermentation of Glucose by Suspensions of Eschericia Coli, *Bacteriology*, vol. 57, p. 147, 1949.

17 BLACKWOOD, A. C., A. C. NEISH, and G. A. LEDINGHAM: Dissimilation of Glucose at Controlled pH Values by Pigmented and Non-Pigmented Strains of Eschericia Coli, *Bacteriology*, vol. 72, p. 497, 1956.

18 SMITH, L.: Cytochrome Systems in Electron Transport, in I. C. Gunsalas and R. Y. Stanier (eds.), "The Bacteria," vol. 2, Academic Press, New York, 1961.

19 WOOD, D. K., and G. TCHOBANOGLOUS: Trace Elements in Biological Waste Treatment with Specific Reference to the Activated Sludge Process, *Proc. 29th Ind. Waste Conf.*, 1974.

20 WOOD, D. K., and G. TCHOBANOGLOUS: Trace Elements in Biological Waste Treatment, *Water Pollut. Control Fed.*, vol. 47, p. 1933, 1975.

21 CARTER, J. L., and R. G. MCKINNEY: Effects of Iron on Activated Sludge Treatment, *J. Environ. Eng. Div., ASCE*, vol. 99, p. 135, 1973.

22 NILSON, E. H., A. G. MARR, and D. J. CLARK: The Maintenance Requirement of Eschericia Coli, *Ann. NY Acad. Sci.*, vol. 102, p. 536, 1963.

23 SHERRARD, J. H., and E. D. SCHROEDER: Cell Yield and Growth Rate in Activated Sludge, *Water Pollut. Control Fed.*, vol. 45, p. 1889, 1973.

24 FRIEDMAN, A. A., and E. D. SCHROEDER: The Effect of Temperature on Growth Rate and Yield in Activated Sludge, *Water Pollut. Control Fed.*, vol. 44, p. 1433, 1972.

25 SHERRARD, J. H.: Control of Cell Yield and Growth Rate in the Completely Mixed Activated Sludge Process, Ph.D. dissertation, University of California, Davis, 1971.

26 MOORE, S. F., and E. D. SCHROEDER: An Investigation of the Effect of Residence Time on Anaerobic Bacterial Denitrification, *Water Res.*, vol. 4, p. 685, 1970.

27 SERVIZI, J. A., and R. H. BOGAN: Free Energy as a Parameter in Biological Treatment, *J. Sanit. Eng. Div., ASCE*, vol. 89, SA3, p. 17, 1963.

28 SERVIZI, J. A., and R. H. BOGAN: Thermodynamic Aspects of Biological Oxidation and Synthesis, *Water Pollut. Control Fed.*, vol. 36, p. 535, 1964.

29 BAUCHOP, T., and S. R. ELSDEN: The Growth of Micro-organisms in Relation to Their Energy Supply, *Gen. Microbiol.*, vol. 23, p. 457, 1960.

30 HADJIPETROU, L. P., J. G. GERRITS, F. A. G. TEULINGS, and A. H. STOUTHAMER: Relation between Energy Production and Growth in Aerobacter Aerogenes, *Gen. Microbiol.*, vol. 36, p. 159, 1964.

31 GUNSALAS, I. C., and C. W. SCHUSTER: Energy Yielding Metabolism in Bacteria, in I. C. Gunsalas and R. Y. Stanier (eds.), "The Bacteria," vol. 2, Academic Press, Inc., New York, 1961.

32 MCCARTY, P. L.: Thermodynamics of Biological Synthesis and Growth, *Proc. Int. Conf. Water Pollut. Res.*, 1964.

33 MCCARTY, P. L.: Energetics and Bacterial Growth, *Proc. 5th Rudolf Res. Conf.*, 1969.

34 INGRAHAM, JOHN L.: Growth of Psychrophilic Bacteria, *Bacteriology*, vol. 76, p. 75, 1958.

35 MONOD, J.: The Growth of Bacterial Cultures, *Annu. Rev. Microbiol.*, vol. 3, 1949.

36 HERBERT, D., R. ELLSWORTH, and R. C. TELLING: The Continuous Culture of Bacteria— A Theoretical and Experimental Study, *Gen. Microbiol.*, vol. 14, p. 601, 1956.

37 HERBERT, D.: A Theoretical Analysis of Continuous Culture, *Symp. Continuous Culture of Micro-Organisms*, SCI Monograph no. 12, 1961.

38 FENCL, Z.: A Theoretical Analysis of Continuous Culture Systems, in I. Malek and Z. Fencl (eds.), "Theoretical and Methodological Basis of Continuous Culture of Micro-Organisms," Academic Press, Inc., New York, 1966.

39 POWELL, G. O.: The Growth of Micro-Organisms as a Function of Substrate Concentration, *Proc. 3d Int. Symp. Continuous Culture of Micro-Organisms*, 1967.

40 MOORE, S. F., and E. D. SCHROEDER: Effect of Nitrate Feed Rate on Dentrification, *Water Res.*, vol. 5, p. 445, 1971.

41 REQUA, D. A., and E. D. SCHROEDER: Kinetics of Packed Bed Dentrification, *Water Pollut. Control Fed.*, vol. 45, p. 1696, 1973.

42 POWELL, E. O.: Theory of the Chemostat, *Lab. Pract.*, vol. 14, p. 1145, 1965.

43 VAN UDEN, N.: Transport Limited Growth in the Chemostat and Its Competitive Inhibition, a Theoretical Treatment, *Arch. Mikrobiol.*, vol. 58, p. 145, 1967.

44 LEE, S. S., A. P. JACKMAN, and E. D. SCHROEDER: A Two State Microbial Growth Kinetics Model, *Water Res.*, vol. 9, p. 491, 1975.

45 GRADY, C. P. L., and R. E. ROPER: A Model for the Bio-oxidation Process, *Water Res.*, vol. 8, p. 471, 1974.

46 LURIA, S., and J. DARNELL: "General Virology," 2d ed., John Wiley & Sons, Inc., New York, 1967.

47 STENT, A. (ed.): "Papers on Bacterial Viruses," 2d ed., Little Brown and Company, Boston, 1965.

48 METCALF AND EDDY, INC.: "Wastewater Engineering," McGraw-Hill Book Company, New York, 1972.

49 CULP, G. L., and R. L. CULP: "New Concepts in Water Purification," Van Nostrand Reinhold Company, New York, 1974.

50 BERG, G., R. B. DEAN, and D. R. DAHLING: Removal of Poliovirus 1 from Secondary Effluent by Lime Flocculation and San Filtration, *J. Am. Water Works Assoc.*, vol. 59, p. 193, 1968.

51 COOKSON, J. T.: Mechanism of Virus Adsorption on Activated Carbon, *J. Am. Water Works Assoc.*, vol. 60, p. 52, 1969.

52 THORUP, R. T., et al.: Virus Removal by Coagulation with Polyelectrolytes, *J. Am. Water Works Assoc.*, vol. 61, p. 97, 1970.

53 CHAUDHURI, M., and R. S. ENGLEBRECT: Removal of Viruses from Water by Chemical Coagulation and Flocculation, *J. Am. Water Works Assoc.*, vol. 61, p. 563, 1970.

54 MANWARING, J. F., M. CHAUDHURI, and R. S. ENGLEBRECT: Removal of Viruses by Coagulation and Flocculation, *J. Am. Water Works Assoc.*, vol. 62, p. 298, 1971.

55 CHAUDHURI, M. et al.: Virus Removal by Diatamaceous Earth Filtration, *J. Environ. Eng. Div.*, ASCE, vol. 100, p. 937, 1974.

56 PELCZAR, M. J., and R. D. REID: "Microbiology," 3d ed., McGraw-Hill Book Company, New York, 1972.

57 DODD, J. C.: Harvesting Algae with a Paper Precoated Belt-type Filter with Integral Dewatering and Drying, Ph.D. Dissertation, University of California, Davis, 1972.

58 U.S. Public Health Service Drinking Water Standards, *U.S. Public Health Serv. Publ. 32*, 1962.

59 SUNG, R.: Ph.D. thesis, Department of Civil Engineering, University of California, Davis, 1974.

60 FAIR, G. M., et al.: The Behavior of Chlorine as a Water Disinfectant, *J. Am. Water Works Assoc.*, vol. 40, p. 1051, 1948.

61 BERG, G.: The Virus Hazard in Water Supplies, *N. Engl. Water Works Assoc.*, vol. 78, p. 79, 1964.

62 VENOSA, A. D.: in F. L. Evans, III (ed.), "Ozone in Water and Wastewater Treatment," Ann Arbor Science Publishers, Inc., Ann Arbor, Mich., 19

63 BRINGMAN, G.: Determination of Lethal Activity of Chlorine and Ozone on E. Coli, *Hyg. Infektionskr.*, vol. 139, pp. 130, 333, 1954: *Water Pollut. Abstr.*, vol. 28.

64 WURHMANN, K., and J. MEYRATH: The Bactericidal Action of Ozone Solution, *Schweiz. Z. Allg. Pathol. Bakteriol*, vol. 18, p. 1060, 1955: *Water Pollut. Abstr.*, vol. 29, p. 223, 1956.

65 KESSEL, J. F., et al.: Comparison of Chlorine and Ozone as Virucidal Agents of Poliomyelitis Virus, *Proc. Soc. Exp. Biol. Med.*, vol. 53, p. 71, 1943: *Water Pollut. Abstr.*, vol. 17, p. 239, 1944.

66 SHERWOOD, T., and R. C. REID: "The Properties of Gases and Liquids," 2d ed., McGraw-Hill Book Company, New York, 1966.

7

WASTEWATER CHARACTERISTICS IMPORTANT IN BIOLOGICAL TREATMENT

Biological wastewater treatment processes are similar in most respects to the reaction systems discussed in Chap. 2. Two additional factors must be taken into consideration: Cells are both reactants and products, and reaction stoichiometry is not constant. These factors make the characteristics of the wastewater being treated extremely important. All reaction systems are affected by pH and temperature. Biological processes are also affected by the nutrient or nutrients that limit growth, the growth rate of the culture, and type of waste being treated.

Process performance and operation is related to these factors in a number of ways. For example, filamentous bacteria tend to be more competitive under conditions where inorganic ions are the limiting nutrients.[1, 2] Many industrial wastes have very low nitrogen and phosphorus concentrations, and these nutrients must be added if the activated-sludge process in which filamentous bacteria are highly undesirable is to be used. Extremely low nutrient concentrations can also result in incomplete conversion of organic material present in a wastewater.

Wastewater treatment processes can be designed and operated to take advantage of wastewater characteristics. Because the stoichiometry of the conversion reactions varies with process operating conditions, a low nitrogen, phosphorus, or other inorganic nutrient concentration can be accommodated by designing the system to have a low cell yield. Recent work by Sherrard[3-5] and Wood and Tchobanoglous[1, 6] has improved understanding of these relationships.

The most important wastewater characteristic with respect to biological treatment is organic concentration. In nearly all cases, the waste being treated is a mixture of organic materials, and monitoring individual constituents is virtually impossible. For this reason, overall measurements of the amount of organic material present are nearly always used. Important inorganic nutrients such as nitrogen and phosphorus are more easily monitored individually. These materials are usually reported both as the total amount present and in the various forms present (e.g., organic, ammonia, nitrite, and nitrate nitrogen). In the case of nitrogen, reporting the various forms is particularly important because of differences in biological availability.

7-1 ORGANIC CONCENTRATION AND OXYGEN DEMAND

Most biological treatment processes have the primary purpose of removing organic material from a wastewater. Some organic materials are more difficult to degrade than others. Many industrial wastewaters contain high concentrations of organic compounds that can be oxidized chemically (i.e., register as chemical oxygen demand) but that cannot be oxidized by bacteria to any great extent. Examples of this type of material are lignins, cellulose, and slightly soluble oils and greases. Thus a biological treatment process can be expected to remove only the biodegradable fraction of the organic material present, and at least two approaches to measuring organic concentration must be considered. The first type measures the total amount of organics present. Examples of this type of test include the chemical oxygen demand (COD) determination,[7] which measures the oxygen necessary for conversion of the organic material to carbon dioxide and water, and the total organic carbon (TOC) determination. This latter test is an instrumental method that measures the carbon dioxide produced on combustion. Both tests have limitations in terms of being approximations of the true total concentration and in sensitivity and precision. Thus the applicability of the techniques and the exact meaning of the results must be judged on a case-by-case basis.

The second type of test measures the biodegradable organic concentration present. Two approaches to measuring biodegradable organic concentration are in current use: measurement of oxygen uptake resulting from bacterial conversion of organics present and measurement of the change in organic concentration (as COD or TOC) due to biological activity. Measurement of oxygen uptake is an indirect method but has two significant advantages: Oxygen uptake is a process operation parameter in aerobic biological processes, and oxygen uptake is an important characteristic in determining effluent quality. For these reasons, techniques based on oxygen demand have become the fundamental methods of measuring organic concentration.

7-2 BIOCHEMICAL OXYGEN DEMAND

The most widely used and accepted measure of biodegradable organic content of a wastewater is the 5-d, 20°C biochemical oxygen demand (BOD_5) value. A brief description of the test is given below:

Brief BOD_5 Procedure

1 Sample of wastewater is obtained and transported as quickly as possible to the laboratory.

2 Dilutions of the sample are made with an aerated nutrient solution[7] that will provide new samples with a maximum BOD_5 of less than 6 mg/l.

3 A bacterial "seed," usually settled domestic sewage, is added in amounts of the order of 5 ml/l if the wastewater is sterile (as many industrial effluents are).

4 Standard 300-ml BOD bottles are filled with the diluted wastewater and sealed. Blanks containing only seeded dilution water are also made up.

5 Oxygen content of at least two samples and two blanks is determined immediately.

6 Samples are incubated at 20°C for 5 d (120 h), and the oxygen content of remaining samples and blanks is determined.

7 BOD_5 is calculated from the following equation:

$$BOD_5 = D^*[(DO_{t=0} - DO_{t=5})_{sample} - (DO_{t=0} - DO_{t=5})_{blank}] \qquad (7\text{-}1)$$

where D^* = dilution factor

Rate of BOD exertion and ultimate or total BOD values are of interest in both design and operation of wastewater treatment processes. A first-order rate model is usually fitted to BOD data, and the rate coefficient and ultimate value are then obtained by calculation.

$$\frac{dC_B}{dt} = -kC_B \qquad (7\text{-}2)$$

$$BOD_t = C_{B_0}(1 - e^{-kt}) \qquad (7\text{-}3)$$

where C_B = BOD remaining (i.e., actual concentration present at time t)

C_{B_0} = initial concentration ($t = 0$) (usually referred to as BOD_L)

BOD_t = cumulative BOD exerted up to time t

The first-order BOD model, Eq. (7-2), was originally used as a matter of convenience, and the choice of the 5-d period for evaluation resulted from problems with reproducibility at shorter periods.[8]

Two major defects of the traditional BOD model [Eq. (7-2)] exist. First, the first-order relationship does not include cell concentration as a parameter and does not correspond to established microbial growth relationships. Second, the 5-d evaluation period is arbitrary and has no relationship to events occurring in the reaction system.

Defects of BOD_5

The second defect is particularly important because of the nature of bacterial growth. As shown in Fig. 6-15, a lag phase of unpredictable length occurs before active growth begins. Length of lag will affect the value of the 5-d BOD by moving the curve along the time axis. The stationary phase shown in Fig. 6-15 is due to exhaustion of the limiting nutrient. This corresponds to the stoichiometric end point of nonbiological reaction systems and should be a reproducible value of oxygen uptake. Oxygen demand exerted beyond this point is due to breakdown on the experimental system, and this process is not a reproducible phenomenon.

Reaction models such as Eq. (6-18) or Eq. (7-2) only apply as long as the reaction is taking place. In the case of the BOD reaction system, once the organic material is converted into new cellular mass and carbon dioxide the reaction rate relationship must change. Oxygen demand exerted beyond this point is due to process breakdown, usually referred to as *endogenous respiration*, and to predators (protozoa and higher forms) feeding on the bacteria. Endogenous respiration can generally be thought of as including both use of stored materials and of consumption of material released from dead cells. Figure 7-1 presents a summary of these relationships.

Nitrification in the BOD Test

At some point in time, ammonia-nitrogen oxidation begins. Because oxygen is used in this biochemical reaction process, a BOD is recorded as shown in Fig. 7-1. An estimate of the amount of the nitrogenous BOD can be obtained from Eq. (7-4):

$$\text{Nitrogenous BOD} \approx 4.6[\text{ammonia N} + \text{organic N}] \qquad (7\text{-}4)$$

Nitrifying bacteria grow slowly, and nitrification generally does not begin in the BOD bottle at an incubation time of less than 9 or 10 d. The oxygen demand exerted is important in many receiving waters and must be considered in the overall design process.

Ordinarily, nitrogenous BOD is reported separately from carbonaceous BOD. Thus, when a BOD figure is given in the literature either as BOD_5 or BOD_L, it should be assumed that nitrogenous BOD is excluded.

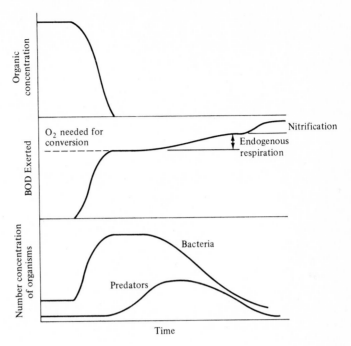

FIGURE 7-1
Relationships between organic-removal BOD exertion and microorganism
populations.

BOD Rate

The problem of the rate expression being independent of cell concentration
extends beyond the BOD system. A general rate expression, valid for biological
wastewater systems, in general, is needed. Monod's[9] expression for growth rate
[Eq. (6-8)] has been used successfully in pure and mixed cultures.[10-16] Because
a stoichiometric relationship between growth and organic removal rate should
exist, the removal rate should be of the same form.

$$R_0 = - \frac{r_{0,\,max}\,CX}{K_m + C} \qquad (7\text{-}5)$$

where $r_{0,\,max}$ = maximum specific removal rate, t^{-1}, corresponding to $r_{g,\,max}$
C = organic concentration, m/l^3
X = cell mass concentration, m/l^{-3}
K_m = same saturation coefficient as in Eq. (6-18)

Both the growth and removal rates are often written in their unit or specific rate forms.

$$r_g = \frac{r_{g,\,max}\,C}{K_m + C} \qquad (7\text{-}6)$$

$$r_0 = -\frac{r_{0,\,max}\,C}{K_m + C} \qquad (7\text{-}7)$$

Stoichiometry of the BOD System

The organic concentration in Eqs. (7-4), (7-6), and (7-7) is the concentration that the bacteria "see" in the immediate vicinity of the cells. When the BOD value is used as a measure of the biodegradable organic concentration, the ultimate BOD (BOD_L) must be used because the bacteria are affected by the total amount of biodegradable organic material present, not the fraction which will be oxidized. Thus a method of determining the ultimate BOD must be developed.

Busch[17-19] recognized this problem in 1952, and over a period of 10 years, worked out a useful system of stoichiometric evaluation of the oxygen-demand content of wastewaters. He first considered the need to establish an end point for the organic conversion reaction. By carefully following the BOD exertion curve for pure, soluble substrates, he found that a sharp decrease in oxygen uptake rate occurred at a reproducible value. His index of reproducibility was grams of oxygen uptake per gram of substrate originally present, and he found that when he used filtered or sonicated sewage seed, which he assumed was nearly free of predator organisms, the BOD value at the point where the sharp decrease in reaction rate, or "plateau," occurred was reproducible within 5 percent. Typical BOD progression curves are shown in Fig. 7-2, and plateau BOD values of Busch and other workers for pure, soluble organic compounds are given in Table 7-1. Busch assumed that the plateau in the oxygen-demand curve was due to depletion of available substrate, and this assumption was later confirmed by Schroeder[20] and Parisod.[21] If the plateau is due to substrate depletion, then the cell mass produced up to the plateau is the net yield for the reaction. All

Table 7-1 PLATEAU BOD VALUES FOR VARIOUS SUBSTRATES, mg O_2/mg substrate

Substrate	Plateau BOD	Refs.
Glucose	0.42	18, 20, 21
Glutamic acid	0.38	18, 21, 24
Sodium acetate	0.58	20
Fructose	0.42	20
α-Ketoglutaric acid	0.39	20
Sorbitol	0.39	21

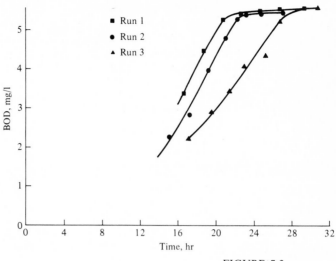

FIGURE 7-2
BOD progression curves.

the organic material originally present can be accounted for in terms of oxygen used or as cell material produced. The total or ultimate biochemical oxygen demand is then the plateau BOD value plus the oxygen demand of the cells produced (i.e., if the cells produced were biologically oxidized). Cell BOD is a bit difficult to determine, because allowing the cells to degrade by natural processes takes a very long period of time. Oxidizing them by chemical processes is faster but not very precise at low concentrations. Busch and his coworkers[19, 22, 23] concluded that the most practical method of estimating the total BOD was to measure the plateau value, determine cell production gravimetrically, and utilize an empirical cell formulation based on cell-constituent ratios such as that of Porges et al.[25] or from data such as Lüria's.[26] Using Porges' et al.[25] formulation, $C_5H_7O_2N$, and data for glucose as the sole carbon source, Busch[18] reported the following relationship for the BOD bottle system:

$$24C_6H_{12}O_6 + 59O_2 + 17NH_3 \rightarrow 17C_5H_7O_2N + 59CO_2 + 110H_2O$$

For this relationship, there is 1.41 g O_2 demand per gram of cells produced. An assumption is made that all cell material produced will eventually be oxidized. This procedure gave satisfactory results, and the results were independent of the experimental system; i.e., the same total oxygen-demand value was obtained with BOD bottles and with respirometers, which use higher cell concentrations or "mass culture" apparatus. This latter point is particularly interesting and important because respirometric BOD determination methods had been introduced in the 11th edition of "Standard Methods"[7] and dropped from the 12th edition because of poor correlations with the BOD_5. Plateau BOD values are

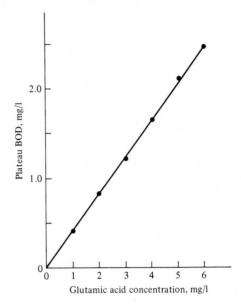

FIGURE 7-3
Plateau BOD as a function of initial substrate concentration.[19]

linear with initial substrate concentration (Fig. 7-3) for a given substrate, while BOD_5 values tend to increase with dilution because of effects of the blanks on the calculation. Recent studies on respirometric BOD determination by Flegal[24] and Parisod[21] have further validated the plateau theory and the use of respirometric methods. Their work has provided insight into the effect of temperature on stoichiometry and given further evidence that the use of an arbitrary test time (for example, 5 d) is incorrect. Flegal[24] determined that the plateau stoichiometry is constant between 10 and 37°C. This result makes careful incubation at constant temperature unnecessary.

Introduction of the concept of stoichiometry into the BOD determination led to the development of a relatively fast method of determining the ultimate oxygen demand (BOD_L), the approximate cell yield, and the oxygen necessary to treat any given waste.[27] This method, the total biological oxygen-demand test ($T_b OD$), uses the difference in chemical oxygen demand (COD) resulting from biological conversion as the basic parameter. The COD test chemically oxidizes *most* organic compounds found in wastewaters to carbon dioxide, water, and nonorganic compounds such as NH_3. Certain compounds only partially oxidize or do not oxidize at all. Therefore, the method must be modified in a few cases, but the concept does not change.

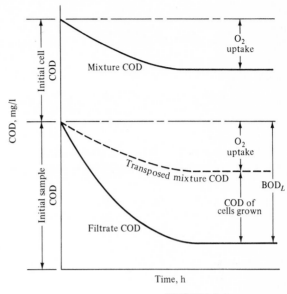

FIGURE 7-4
Typical T_bOD test results.

Simplified Procedure for the T_bOD Test

1 An acclimated cell suspension, preferably from a waste treatment plant treating the wastewater of interest, is obtained and washed.
2 COD and mass concentration determinations are made on the cell suspension.
3 COD of the wastewater is determined.
4 The cell suspension and the wastewater are mixed in predetermined proportions in a 1- or 2-l graduated cylinder.
5 The mixture is aerated by a diffuser stone.
6 Samples are taken and analyzed for total (mixture) COD, filtrate (0.45 μm) COD, and suspended solids at convenient time intervals.
7 Makeup water is added as needed to replace evaporation.

Plotting the results of a T_bOD determination for a soluble (passing a 0.45-μm filter) wastewater gives curves such as those shown in Fig. 7-4.

The conversion oxygen demand is the amount of oxygen necessary to convert the organic material in the wastewater to cells and CO_2. This value corresponds conceptually to the plateau BOD but will have a higher value because with the higher cell concentrations used in the T_bOD test, cell yields will be lower and stoichiometry will be somewhat different. The change in filtrate COD is the total oxygen demand removed as a result of biological action, and by definition is the ultimate BOD (BOD_L). A particular point should be made that there is usually a residual COD value remaining which does not enter into the T_bOD calculation. This residual is a basic problem with simple use of the COD test alone. Some fraction of the COD of a wastewater is not biologically degradable or will be converted to nonbiodegradable products. The T_bOD test provides an indication of the biologically degradable and nondegradable organic concentrations.

Subtracting the conversion oxygen demand from the T_bOD gives the oxygen demand of the cells produced. This value can be used or the one determined gravimetrically, depending on the need. These two values should correlate, but this correlation is dependent upon the accuracy of the empirical cell formulation and the precision of the gravimetric test.

When the wastewater contains a significant amount of nonsoluble organic material, the relationships shown in Fig. 7-5 become clouded because cell solids are indistinguishable from wastewater solids. Mullis and Schroeder[28] proposed extending the T_bOD test to these systems by assuming that all incoming solids are converted. They reported on studies with domestic sewage which gave linear COD/T_bOD ratios for seasonal periods and suggested that the T_bOD test be run intermittently throughout the year to maintain the validity of the relationship for plant operational purposes. Because their correlation was empirical, it cannot be extended beyond a given wastewater for a given period of time, but it does provide a straightforward test for operational control.

Three important parameters are provided by the T_bOD test. One is an estimate of the ultimate *biological* oxygen demand. This value can be determined by the plateau test also. Although the plateau method is more precise, it is also more difficult. A second parameter is an estimate of the oxygen demand that must be satisfied in the actual treatment plant. Validity of this estimate depends on the relationship of the culture used in the test to the one that is or will be in the treatment plant. The third parameter, cell yield, has the same restrictions as the oxygen-demand estimate. Stoichiometry (i.e., cell yield and oxygen consumption) is highly dependent upon the manner of process operation,[29-31] but initial estimates obtained from simple batch experiments are inexpensive and will be in the general design range.

EXAMPLE 7-1   Data from the analysis of a dairy wastewater are given below. Determine the BOD_L and the residual COD to be expected after treatment. Estimate the quantity of oxygen and air which must be supplied for aerobic biological treatment.

Sample volume	1 l
Sample COD	768 mg/l
Culture volume	1 l
Culture COD	395 mg/l

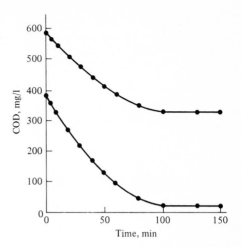

FIGURE 7-5
Plot of data for Example 7-1.

Time, min	0	5	10	20	30	40	50	60	80	100	130	150
Mix. COD	583	565	548	506	469	436	403	377	340	320	319	317
Filt. COD	381	346	317	259	206	169	117	90	51	22	21	20

SOLUTION

1 Plot data as COD versus time as shown in Fig. 7-5.
2 Note that a dilution factor of 2 is incorporated into the procedure when the culture and sample are mixed.
3 Ultimate BOD of the diluted sample is the initial filtrate COD minus the final filtrate COD. The wastewater BOD_L is therefore

$$BOD_L = 2(381 - 20) = 722 \text{ mg/l}$$

4 The residual COD after treatment is

$$\text{Residual COD} = 768 - 722 = 46 \text{ mg/l}$$

5 The estimated quantity of oxygen necessary for treatment is found by subtracting the final mixture COD from the initial value and multiplying by the dilution factor.

$$O_2 \text{ necessary} = 2(583 - 317) = 532 \text{ mg/l}$$

Because air is only 20 percent oxygen, and oxygen-transfer efficiencies are of the order of 5 percent, the amount of air necessary will be approximately 53.2 g/l (41 1/1).   ////

Use of Total Organic Carbon

In the 1960s instrumental methods of determining organic carbon concentrations of wastewaters were introduced.[32, 33] These techniques have the same problem as the COD test in that biodegradable organic carbon is not determined. A modified $T_b OD$ procedure can, of course, be used, and this is quite useful in some cases. The exact meaning of organic carbon concentration has not been determined with respect to waste treatment or water pollution. Clearly, high concentrations are bad, but just what a high concentration is has not been established. The significance of carbon concentration is highly dependent upon the chemical species present. In this sense the parameter is not as useful as oxygen demand because oxygen demand, while being a general parameter, gives direct stoichiometric information.

7-3 INORGANIC NUTRIENTS

Inorganic nutrient concentrations in wastewaters are important because these materials are essential cellular constituents. With the exception of nitrogen, inorganic nutrients are generally required in low amounts.[34] Many are listed as trace elements because a measurable presence is all that is necessary to support maximum growth. Absence of an essential nutrient may result in complete lack of growth. An insufficient quantity to allow for stoichiometric conversion of the organic material to cells and normal end products may result in either incomplete conversion or predominance of undesirable bacterial species (e.g., the filamentous bacteria *Sphaerotilus natans*) in the culture.[6] Estimation of the stoichiometric requirements can be made from the composition of bacterial cells and a knowledge of cell yield for a given set of operating conditions. Typical composition of bacterial cells was given in Table 6-1, and the relationship between cell yield and operating conditions will be discussed in Chap. 8.

For a given wastewater, the maximum nutrient requirements can be estimated from batch studies in which the maximum growth rate is attained. This growth rate constraint can be satisfied by having an initially high concentration of the limiting nutrient and a relatively low concentration of cells. A definition of high nutrient concentration is a condition where $C \gg K$ in Eqs. (6-8) and (7-4). An excellent estimate of the maximum yield (Y) of cells per unit mass of the limiting nutrient converted can be obtained under these conditions. Low nutrient and high cell mass concentrations result in lower net yields because maintenance-energy requirements become significant. This subject will be discussed in the following chapter.

EXAMPLE 7-2 An industrial wastewater analysis has provided the following information:

Plateau BOD	1640 mg/l
BOD_L	2600 mg/l
Total nitrogen	27 mg/l
Organic N	19 mg/l
NH_3-N	8 mg/l
NO_2^--N	0 mg/l
NO_3^--N	0 mg/l
Total P	7 mg/l
Fe	1.5 mg/l
Mg	5.2 mg/l
Ca	31.8 mg/l
S	5.0 mg/l

Determine the quantity of each nutrient that must be added to provide for complete BOD_L removal.

$$\text{BOD converted to cells} = 2600 - 1640 = 960 \text{ mg/l}$$

$$\text{Cell mass produced} = \frac{960 \text{ mg BOD}_L/\text{l}}{1.41 \text{ mg BOD}_L/\text{mg cells}}$$

$$= 681 \text{ mg/l cells}$$

Nitrogen required = 0.14(681) = 95.3 mg/l
  Additional N needed = 95.3 − 27 = 68.3 mg/l
  (assuming all organic N is available)

Phosphorus required = 0.03(681) = 20.4 mg/l
  Additional P necessary = 13.4 mg/l

Iron required = 0.002(681) = 1.4 mg/l
  No additional Fe required

Magnesium required = 0.005(681) = 3.41 mg/l
  No additional magnesium required

Calcium required = 0.005(681) = 3.41 mg/l
  No additional calcium required

Sulfur required = 0.01(681) = 6.8 mg/l
  Additional S required = 1.8 mg/l   ////

7-4 SOLIDS

Suspended solids are important characteristics in biological wastewater treatment because they are often biodegradable, and their oxidation may be the rate-controlling variable in many situations. Ordinarily the BOD of the suspended solids would be measured in the BOD test used, and the rate of oxidation would be included in any rate studies conducted.

7-5 TEMPERATURE AND pH

Effects of temperature and pH on bacterial cultures and biochemical reactions were discussed in Chap. 6. Problems related to temperature and pH characteristics of wastewaters occur in anaerobic processes that are extremely temperature and pH sensitive with respect to process stability. Many industrial wastewaters are either highly acidic (e.g., wine stillage, which typically has a pH of 3.5 and an acidity greater than 1000 mg/l $CaCO_3$) or alkaline (e.g., laundry wastes, which typically have a pH greater than 9.0 and alkalinities of 250 mg/l $CaCO_3$). Conventional biological treatment processes do not operate well outside the 6.5 to 8.5 pH region, and acidic or alkaline characteristics must be modified in some manner. Possible methods of pH modification include neutralization, dilution with another effluent, and control of the biological reaction process. The latter method can be used where the cause of high or low pH condition is an organic material. Wastewaters containing significant concentrations of organic acids often have low pH values, for example, and can be effectively handled by matching the process removal rate to the mass input rate of the acids.

Domestic wastes usually have a near-neutral pH and a temperature in the range of 15 to 25°C, depending on the season and climate. These temperatures are well below optimum values for bacterial growth, but this is not a major design or operating constraint in most cases. Some industrial wastewaters, such as those from distilleries, may have temperatures of 65°C or greater. Anaerobic processes may be well adapted to treating these wastewaters.

7-6 TOXICITY

Many wastewater constituents are toxic to living organisms. Unfortunately, most elements in nature are toxic to one organism or another in some concentration, and thus the question of toxicity must be answered, at least partially, in terms of public values and attitudes. The simplest case is where a wastewater constituent is toxic to the bacteria in the treatment process.

Typical problem materials are copper, chromate, chlorine, cyanide, and phenol. Mixed cultures of microorganisms can be acclimated to phenol concentrations of 100 mg/l or greater if the quantities are slowly increased over a period of weeks. Acclimation to heavy metals or cyanide does not occur, however. These materials must be removed or diluted before a biological treatment process can operate satisfactorily. Clearly, these conditions must be considered in the wastewater characterization process. A BOD will not be exerted if the bacteria are unable to grow.

Public values and attitudes are important where the final discharge may be toxic to living organisms. Even well-treated wastewaters will alter the ecology of a receiving water to some extent. In many cases, toxic materials in a discharge cause severe damage to natural aquatic populations. Fish populations are the most common cause of public concern. In coastal areas, protection of shellfish, such as oysters, crabs, and shrimp, has also been a major issue. Large fish kills often result in a strong public response, while discharge of toxicants which inhibit reproduction or destroy food-chain links receives less publicity. While both situations are serious, the recovery time of the population from a fish kill is generally less than from damage to the reproductive cycle or the food chain.

7-7 BIOASSAYS

Estimation of the effects of a discharge on the natural population is usually made through use of a bioassay. These studies are most often conducted on fish, but the concept can be used with any organism. Test organisms are selected and placed in a range of wastewater dilutions from 0 (control) to full strength (undiluted). The dilution water used would normally be the receiving water for the discharge, and the organism selected should be the most sensitive or important organism in the receiving water. Until recently, values were reported as mean toxicity limits, the concentration at which half the test organisms died within a specified time interval. An increasingly used criteria is to require that a specified percentage of the test organisms survive for a given time period. For example, a paper mill located in Anderson, Calif., was required to produce a wastewater in which 90 percent of the salmon fry placed in the undiluted effluent survived 96 h.

Bioassays based on death of the test organism have obvious defects. Chronic effects and damage of food-chain components are not determined. Specification of the test organism may not always be straightforward either. For example, sensitivity of eggs, fry, and adults to a given toxicant is often quite different. Fisheries biologists should have control over the bioassay procedures wherever possible.

7-8 WASTEWATER CHARACTERIZATION STUDIES

All wastewaters should be characterized before process design is begun. Where new industrial installations are being constructed, special problems exist. If the installation is similar to others in existence (e.g., an oil refinery), a reasonable characterization may be obtained from their wastewaters. No two manufacturing installations are identical, and consideration of the differences should be included in the conceptual design. Wherever possible, wastewater streams should be eliminated. More important, however, is the determination of which streams should be mixed. Segregation of highly acid or alkaline wastes from streams with high organic contents will often greatly reduce treatment problems.

When new products are to be manufactured, it is extremely important to determine what wastewater treatment problems will result before the installation goes into production. Wastewater samples should be available from process pilot plants, and these can be used for both wastewater characterization and for treatment studies.

Volume of Flow

Flow rate and flow rate variation are important factors in process design. A number of units in most treatment systems must be designed on the basis of peak flow rate. This requires flow rate studies and also provides a reason to minimize flow rate variation whenever possible.

Organic Concentration

In the beginning of this chapter it was stated that organic concentration is the single most important characteristic of a wastewater with respect to biological treatment. Currently, BOD_5 is the most commonly used measure of organic content. Ultimate BOD is a much better parameter for use in treatment process design. In many cases, the COD provides an adequate estimate of the BOD_L, but this fact should always be determined experimentally. Many wastewaters have high CODs and very low BODs due to the presence of nonbiodegradable organics or toxic materials.

Residual organic concentrations to be expected after treatment should also be estimated in characterization studies. A satisfactory estimate of this value is given by the difference between the COD and the BOD_L.

Physical Characteristics

Temperature, pH, and suspended-solids concentrations are all variables having direct impact on biooxidation processes. Values of these variables would be expected to change with flow rate and season. This information should be available to the treatment process designer.

Toxicity

With respect to biological wastewater treatment, BOD analysis will provide an initial indication of the presence of toxic materials. If a wastewater has a high COD or TOC and little BOD, a toxicity study is probably necessary. In most cases, toxicity problems can be solved by eliminating one or a few wastewater streams.

PROBLEMS

7-1 A mass-culture $T_b OD$ test has been run on an industrial waste, and the data are presented below. The culture used in the test was taken from an activated-sludge process with a mean cell residence time of 5 d which is used to treat the waste. Determine the ultimate BOD of the wastewater, and estimate the oxygen demand and cell yield expected in the treatment plant.

Sample volume	1 l
Sample COD	880 mg/l
Sludge volume	1 l
Sludge COD	640 mg/l

Time, min	Mixed-liquor COD, mg/l	Filtrate COD, mg/l
0	760	440
15	682	306
30	614	187
45	572	112
60	552	80
75	540	60
90	558	55
105	538	55
120	538	55

7-2 The wastewater in Prob. 7-1 is from a food-processing plant and contains very little nitrogen. (*a*) Determine the amount of nitrogen that should be added to the waste. (*b*) Estimate the cost of nitrogen addition for a plant producing 8×10^6 l (2.1 mgd) of wastewater if ammonia costs \$0.25/kg. What would be the result of using nitrate rather than ammonia nitrogen?

7-3 What filter porosity should be used in separating the mixed liquor from the residual effluent in systems such as Prob. 7-1.

7-4 Maximum cell-division rates at 20°C are of the order of $1 \ h^{-1}$, and maximum cell yields are of the order of 0.4 g cells/g BOD_L removed. Using the relationships developed in Chap. 7, estimate the maximum organic concentration (as BOD_L) that can be treated aerobically as a function of cell concentration. Assume $K_m = 20$ mg/l. Plot the results as C_{max} versus X and the fraction of the maximum removal rate r_0/r_{0max} versus X.

7-5 Discuss sample preservation problems that must be solved where laboratory facilities are not available at the sampling site. What precision is needed in the BOD test with respect to design, monitoring, and reporting to regulatory agencies?

7-6 Composite samples are often used in characterization and monitoring studies. These samples are weighted according to volumetric flow. Compare the usefulness, strengths, and defects of composite sampling for treatment plant design and effluent monitoring.

REFERENCES

1.   WOOD, D. K., and G. TCHOBANOGLOUS: Trace Elements in Biological Waste Treatment with Specific Reference to the Activated Sludge Process, *Proc. 29th Ind. Waste Conf.*, 1974.

2.   CARTER, J. L., and R. E. MCKINNEY: Effects of Iron on Activated Sludge Treatment, *J. Environ. Eng. Div., ASCE*, vol. 99, p. 135, 1973.

3.   SHERRARD, J. H.: Mathematical and Operational Relationships for the Completely Mixed Activated Sludge Process, *Water Sewage Works*, vol. 121, p. 84, 1974.

4.   STALL, T. R., and J. H. SHERRARD: Effect of Wastewater Stoichiometry and Mean Cell Residence Time on Phosphorus Removal in the Activated Sludge Process, *Water Pollut. Control Fed.*, vol. 47, 1975.

5.   SHERRARD, J. H., and E. D. SCHROEDER: Stoichiometry of Industrial Biological Wastewater Treatment, *Proc. 30th Ind. Waste Conf.*, 1975.

6.   WOOD, D. K.: Iron Limitation and Growth of Filamentous Cultures, M.S. thesis, University of California, Davis, 1974.

7.   "Standard Methods for the Examination of Water and Waste Water," 13th ed., American Public Health Association, New York, 1971.

8.   METCALF, L., and H. P. EDDY: "Sewerage and Sewage Disposal," McGraw-Hill Book Company, New York, 1930.

9.   MONOD, J.: The Growth of Bacterial Cultures, *Annu. Rev. Microbiol.*, vol. 3, 1949.

10. HERBERT, D.: Continuous Culture of Micro-Organisms, *Symp. Continuous Culture of Micro-Organisms*, SCI Monograph no. 12, 1961.

11. HERBERT, D., R. ELLSWORTH, and R. C. TELLING: The Continuous Culture of Micro-Organisms, *Gen. Microbiol.*, vol. 14, p. 601, 1956.

12. MARR, A. G., E. H. NILSON, and D. J. CLARK: The Maintenance Requirement of Escherichia Coli, *Ann. NY Acad. Sci.*, vol. 134, 536, 1963.

13. MALEK, H., and S. FENCL (eds.): "Continuous Culture of Micro-Organisms," Academic Press, Inc., New York, 1963.

14. GRAM, A. L.: Reaction Kinetics of Aerobic Biological Processes, Rept. no. 2, IER Series 90, *Sanit. Eng. Res. Lab.*, University of California, Berkeley, 1956.

15. STUM-ZOLLINGER, E., et al.: Discussion of Kinetics of Aerobic Removal of Organic Wastes, *J. Sanit. Eng. Div., ASCE*, vol. 90, SA4, p. 107, 1964.

16. NOVICK, A.: Growth of Bacteria, *Annu. Rev. Microbiol.*, vol. 9, 1955.

17. BUSCH, A. W.: M.S. thesis, Massachusetts Institute of Technology, Cambridge, 1952.

18. BUSCH, A. W.: BOD Progression in Soluble Substrates, *Sewage Ind. Wastes*, vol. 30, p. 1336, 1958.

19. BUSCH, A. W., and H. N. MYRICK: BOD Progression in Soluble Substrates III—Short Term BOD and Bio-oxidation Solids Production, *Water Pollut. Control Fed.*, vol. 33, p. 897, 1961.

20. SCHROEDER, E. D.: Importance of the BOD Plateau, *Water Res.*, vol. 2, p. 803, 1968.

21. PARISOD, J. P.: BOD Progression in Mixed Substrates, M.S. thesis, University of California, Davis, 1974.

22. LEWIS, J. W., and A. W. BUSCH: BOD Progression in Soluble Substrates VII—The Quantitative Error due to Nitrate as a Nitrogen Source, *Proc. 19th Ind. Waste Conf.*, p. 847, 1964.

23. GRADY, C. P. L., JR., et al.: BOD Progression in Soluble Substrates VI—Cell Recovery Techniques in the T_bOD Test, *Proc. 18th Ind. Waste Conf.*, 1963.

24. FLEGAL, T. M., and E. D. SCHROEDER: Temperature Effects on BOD Stoichiometry and Oxygen Uptake Rate, *Water Pollut. Control Fed.*, vol. 49, 1976.

25. PORGES, N., et al.: Principles of Biological Oxidation, in B. J. McCabe and W. W. Eckenfelder, Jr. (eds.), "Biological Treatment of Sewage and Industrial Wastes," Reinhold Publishing Corporation, New York, 1956.

26. LURIA, S. E.: The Bacterial Protoplasm: Composition and Organization, in I. C. Gunsalas and R. Y. Stanier (eds.), "The Bacteria," vol. 1, Academic Press, Inc., New York, 1960.

27. HISER, L. L., and A. W. BUSCH: An 8-hour Biological Oxygen Demand Test Using Mass Culture Aeration, *Water Pollut. Control Fed.*, vol. 36, p. 505, 1964.

28. MULLIS, M. K., and E. D. SCHROEDER: A Rapid Biochemical Oxygen Demand Test Suitable for Operational Control, *Water Pollut. Control Fed.*, vol. 43, p. 209, 1971.

29. SHERRARD, J. H., and E. D. SCHROEDER: Cell Yield and Growth Rate in Activated Sludge, *Water Pollut. Control Fed.*, vol. 45, p. 1889, 1973.

30. SHERRARD, J. H., and E. D. SCHROEDER: Importance of Cell Growth Rate and Stoichiometry to the Removal of Phosphorus from the Activated Sludge Process, *Water Res.*, vol. 6, p. 1951, 1972.

31. WUHRMANN, K.: Factors Affecting Efficiency and Solids Production in the Activated Sludge Process, in B. J. McCabe and W. W. Eckenfelder, Jr. (eds.), "Biological Treatment of Sewage and Industrial Wastes," Reinhold Publishing Corporation, New York, 1955.

32. SCHAEFFER, R. B. et al.: Application of a Carbon Analyzer in Waste Treatment, *Water Pollut. Control Fed.*, vol. 37, p. 1545, 1965.

33. VAN HALL, C. E., and V. A. STENGER: An Instrument for Rapid Determination of Carbonate and Total Carbon in Solution, *Anal. Chem.*, vol. 39, p. 503, 1967.

34. STANIER, R. Y., M. DOUDOROFF, and E. ADELBERG: "Microbial World," ed., Prentice-Hall, Inc., Englewood Cliffs, N.J., 1972.

8

ACTIVATED-SLUDGE AND OTHER SUSPENDED-CULTURE PROCESSES

Suspended-culture wastewater treatment processes are, in a conceptual sense, straightforward extensions of the homogeneous reactors discussed in Chap. 2. The conversion processes are far more complex than we have discussed previously, often involving both soluble and nonsoluble reaction-substrates predation of the bacteria by higher organisms, and in some cases photosynthesis. Because the organisms are distributed through the medium, all these reactions or systems of reactions can be included in homogeneous or pseudohomogeneous rate terms, however.

8-1 ACTIVATED SLUDGE

Activated sludge was briefly discussed in Chap. 1. As was noted at that point the process consists of an aeration tank, a sedimentation tank, and a recycle system for the settled culture (sludge) (Fig. 8-1). Because the culture is made up of organisms grown at the expense of the incoming organic material, some method of wasting must be provided. Some cells are lost in the process effluent, but this method of wasting is undesirable as the objective of the process is to

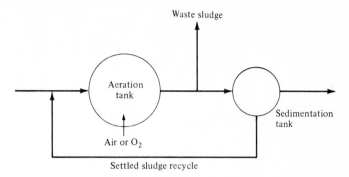

FIGURE 8-1
Schematic diagram of typical activated-sludge process.

produce a low-carbon, low-suspended-solids effluent. The best method of wasting is directly from the aeration tank, as indicated in Fig. 8-1, primarily because of the difficulty in controlling the mass rate of wasting at any other point. Wasting from the mixed liquor is easiest in large plants but can be incorporated into most processes. The amount of wasting necessary depends on the characteristics of the incoming wastewater and the mode of process operation.

Until the 1950s, most activated-sludge processes were designed as nominal plug-flow systems. The aeration tank in these processes was usually of the order of 3.5 to 4.5 m deep, 6.0 to 9.5 m wide, and 30 to 100 m long. While there is a considerable amount of backmixing in long narrow tanks such as these, significant concentration gradients exist. Dye studies have consistently demonstrated that conventional activated-sludge aeration tanks are not good approximations of stirred tanks or plug-flow processes.[1]

During the 1950s, a number of modifications of the conventional design were introduced. The most important of these modifications were the completely mixed process and the extended-aeration process. Completely mixed activated-sludge processes were introduced in Europe and America at about the same time. Busch[2] and McKinney[3] contributed a great deal to the early development of the concept, but the greatest factor in its acceptance has been the stability of the process relative to conventional configurations and the fact that contrary to most reaction systems, CFSTR and nominal plug-flow units of the same volume produce effluents of similar quality.

Several physical advantages of CFSTR activated-sludge processes also exist.[4] Because of the mixing, a great deal of dilution of incoming material is provided. When toxic materials or slugs of degradable organics are introduced into the influent stream, this dilution capacity prevents process upset. Further discussion of this point will be given later. In addition, the stirred tank provides a more uniform biological environment for the culture. While there is still some question about the effects of this uniformity, it is generally felt the CFSTR processes are more stable than nominal plug-flow processes.

According to the reaction models discussed in Chap. 2, the nominal plug-flow process should be preferable to a CFSTR process on the basis of necessary reactor volume (Fig. 2-19), but as McKinney[5] noted in 1963, this is in general not the case. This point is quite important, and a detailed discussion of it will be given later in the chapter.

Extended-aeration processes are simply systems in which the cell wasting rate is considerably lower than in conventional processes. High cell concentrations and oxygen demand rates and low cell production rates result from this mode of operation. The importance of extended aeration's introduction was that "cell age" began to be recognized as a useful parameter in process control. This recognition was not immediate but gradually entered the literature over a 15-year period.[6–9] Cell age (sometimes called solids age, sludge age, or mean cell residence time) is not a completely new parameter, a similar term having been developed in the 1930s (Ref. 10). The earlier version of the parameter involved suspended solids in the influent as well as cells produced in the reactor. An understanding of the relationship between growth rate, cell yield, and cell age did not exist, and consequently, the term was used only as a gross design parameter.

8-2   PROCESS PARAMETERS

Because of the manner in which biological treatment processes evolved, a number of quasi-empirical design and operating parameters have come into common usage. Among the most important of these parameters are the organic loading rate and food-to-microorganisms ratio. Reaction rate and stoichiometric concepts have been introduced only relatively recently. Consequently, the bulk of process experience has been related to the empirical parameters.

Organic Loading Rate

Once the organic content of a wastewater is known, attention can be turned to the organic loading rate. This term is defined in a number of ways in the literature: mass per time, mass per unit organism mass per time, and mass per unit tank volume per unit time. The latter rate is derived from experience with operating plants and has no real relation to actual process parameters. For this reason, we will not consider mass per unit tank volume per unit time here. The other two loading rates will be termed the *mass organic loading rate* M_0 and the *unit mass loading rate* m_0, respectively. In the literature, the unit mass loading rate m_0 is often called the *food-to-microorganisms ratio*.[3]

$$M_0 = QC_i \qquad (8\text{-}1)$$

$$m_0 = \frac{QC_i}{VX} \qquad (8\text{-}2)$$

where V = tank volume

X = cell concentration

Q = volumetric flow rate

C_i = influent organic concentration

In this text organic concentration will be normally given in units of total (ultimate) biological oxygen demand (BOD_L) because this value reflects the actual biodegradable organic concentration in the wastewater. Cell mass concentration will be referred to as *mixed-liquor suspended-solids* (MLSS) or *mixed-liquor volatile suspended solids* (MLVSS) concentration. An assumption will be made that these values are proportional to the actual active microbial mass concentrations. In most wastewaters, incoming suspended solids are not high enough to invalidate this assumption, but in some cases, more definitive terms must be developed and used. Measures such as RNA, DNA, and enzyme concentration are quite useful. These parameters are, in general, not proportional to each other or to MLSS or MLVSS; thus, a direct relationship between them is impossible. For a given treatment system a relationship can be established for a modified unit loading rate (and as we shall see, unit reaction rate) that can be used for operation and design, however:

$$m_0^* = \frac{QC_i}{VX_0^*} \qquad (8\text{-}3)$$

where X_i^* is the concentration of some cell parameter such as RNA, DNA, or a specific enzyme.

Rate of Organic Removal and Cell Growth

Organic and cell mass balances on the aeration tank of Fig. 8-1 result in Eqs. (8-4) and (8-5) for CFSTR processes and Eqs. (8-6) and (8-7) for ideal plug-flow processes.

CFSTR

$$QC_i + \alpha QC_u + VR_o = (1 + \alpha)QC + V\frac{dC}{dt} \qquad (8\text{-}4)$$

$$QX_i + \alpha QX_u + VR_g = (1 + \alpha)QX = V\frac{dX}{dt} \qquad (8\text{-}5)$$

Plug flow

$$-(1 + \alpha)\frac{\partial C}{\partial \Theta} + R_o = \frac{\partial C}{\partial t} \qquad (8\text{-}6)$$

$$-(1 + \alpha)\frac{\partial X}{\partial \Theta} + R_g = \frac{\partial X}{\partial t} \qquad (8\text{-}7)$$

where Θ is the hydraulic residence time V/Q. At steady state the first two expressions reduce to algebraic equations, and the second two reduce to ordinary differential equations. Boundary conditions for the steady-state forms of Eqs. (8-6) and (8-7) are

BC 1: $$\Theta = 0 \qquad C^* = \frac{C_i + \alpha C_u}{1 + \alpha} \qquad (8\text{-}8)$$

BC 2: $$\Theta = 0 \qquad X^* = \frac{X_i + \alpha X_u}{1 + \alpha} \qquad (8\text{-}9)$$

We have discussed the functional form of the rate terms to some extent in Chap. 7 and also in Chap. 6. Monod's[12] expression for rate of growth [Eq. (7-4)] and its asymptotic forms for $C \rightarrow 0$ and $C \rightarrow \infty$ are generally accepted. We must remember that the Monod expression is empirical and was developed from studies of batch cultures under high growth rate conditions and with pure (single-species strain) cultures. In wastewater treatment, we deal with mixed cultures at very low growth rates and continuous-flow systems. For example, Selna[13] found that activated-sludge processes responding to square-wave-type transients in organic concentration behaved in a "Monod" fashion to the increase in substrate concentration (and loading rate) but were distinctly non-Monod in their return to steady-state conditions. He also found that the response characteristics varied in magnitude with "sludge age." Selna used a peptone (yeast extract) as a substrate for his studies and determined a saturation constant of the order of 20 mg/l as COD for that portion of the curves responding in a Monod fashion.

Activated-sludge processes make use of flocculant microbial cultures (Fig. 8-2), i.e., cultures in which large numbers of cells agglomerate together. Maintaining the culture in this form is necessary because cell separation in the clarifier-thickener can not be obtained for a dispersed cell culture. Although the process of agglomeration is not completely understood, it is believed that the binding agent is the slime layer produced by many bacterial species under growth-limited conditions. Why the slime layer is produced is also not clearly understood, but it appears that the layer is involved in protection of the cell during the resting stage (Fig. 6-15). In batch cultures, increased slime-layer depth and flocculation are noticed as substrate concentrations approach zero and the culture's growth rate begins to decline.

8-3 MASS-TRANSFER LIMITATIONS ON REMOVAL RATE

Consideration of flocs and the fact that organic material and nutrients must be transported into the floc and through the individual organisms' slime layers forces the consideration of mass transport rate limitation of the activated-sludge process. The most obvious method of attacking this problem is to develop an effectiveness-factor model similar to the one in Chap. 2. Using the Monod expression, the

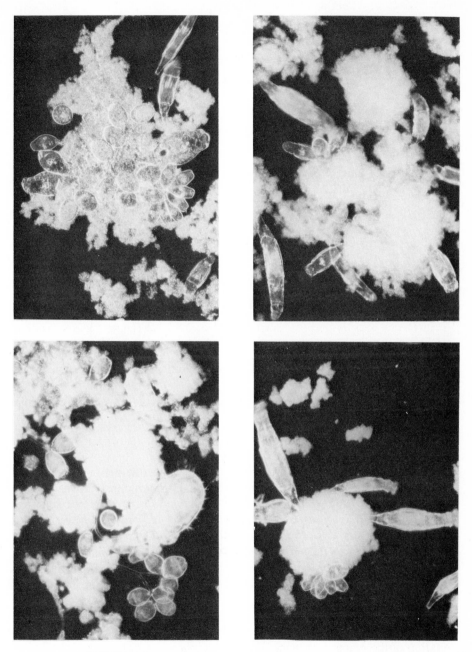

FIGURE 8-2
Dark-field photomicrographs of activated-sludge floc and associated microfauna.
(*Courtesy of Dr. J. W. Nunley.*)

overall rates of growth and substrate uptake, R_g and R_0, become

$$R_g = E \frac{r_{max} C_b X}{K_m + C_b}$$

$$R_0 = -E \frac{r_{0,max} C_b X}{K_m + C_b} \qquad (8\text{-}10)$$

where C_b = liquid-phase organic concentration adjacent to floc particles
  E = effectiveness factor

Atkinson et al.[14] have derived an expression for the effectiveness factor E

$$E = \begin{cases} 1 - \dfrac{\tanh\,(k_2 V_p/A_p)}{k_2 V_p/A_p}\left(\dfrac{\Phi_p}{\tanh \Phi_p} - 1\right) & \Phi_p \le 1 \qquad (8\text{-}11) \\[3mm] \dfrac{1}{\Phi_p} - \dfrac{\tanh\,(k_2 V_p/A_p)}{k_2 V_p/A_p}\left(\dfrac{1}{\tanh \Phi_p} - 1\right) & \Phi_p \ge 1 \qquad (8\text{-}12) \end{cases}$$

where

$$\Phi_p = \frac{(k_2 V_p/A_p)(C_b/K_m)}{\sqrt{2}(1 + C_b/K_m)}\left[\frac{C_b}{K_m} - \ln\left(1 + \frac{C_b}{K_m}\right)\right]^{-1/2} \qquad (8\text{-}13)$$

and

$$k_2 = \left(\frac{r^*_{0,\,max}\,a}{K_m \mathscr{D}}\right)^{1/2}$$

where a = external area of microorganisms per unit volume of microbial mass
  V_p/A_p = floc volume-to-area ratio
  $r^*_{0,\,max}$ = maximum rate of substrate uptake per unit floc area per unit time, $m/l^2 t$

If we consider values for the coefficients in Eqs. (8-11) and (8-12) in a reasonable range, a set of curves similar to Fig. 2-11 can be constructed. The saturation coefficient can be expected to be in the range 0.1 to 10 mg/l, and thus $K_m^{1/2}$ should vary between approximately 0.3 and 3. Maximum removal rate per unit area $r_{0,\,max}$ is of the order of 1.7×10^{-5} mg/cm²-s. If we assume spherical particles, V_p/A_p reduces to $d_p/6$ and a reduces to $6/d_{cell}$. Cells can be considered to be of the order of 2×10^{-4} cm in diameter, and thus a is of the order of 3×10^4 cm^{-1}. The diffusion coefficient is of the order of 10^{-5} cm²/s. Using these values, C_b/K_m ratios between 0.01 and 100 and particle diameters between 0.01 and 1 cm result in the values given in Table 8-1 and the curves shown in Fig. 8-3.

As can be observed from Fig. 8-3, when $k_2 V_p/A_p$ is less than 0.5 the effectiveness factor is approximately 1. From Table 8-1, we see that high K_m values and reasonably small particles (0.5 mm) are needed to prevent diffusion limitation of the reaction system. Thus, for a given particle size and K_m value, we would expect rate curves such as those shown in Fig. 8-4. Two factors are immediately apparent. One factor is that direct inspection of experimental data may not lead to the conclusion that a given system is diffusion limited. All the curves in Fig. 8-4 have the same shape, and obtaining more than one curve

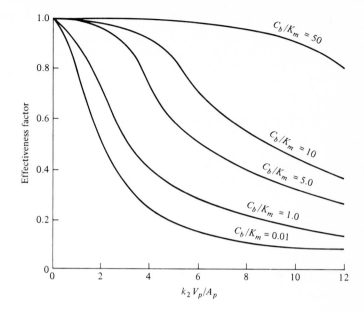

FIGURE 8-3
Effectiveness factor as a function of particle characteristics and substrate concentration.

Table 8-1 EFFECTIVENESS FACTORS FOR IDEAL SPHERICAL FLOCS

	Particle diameter, cm									
	0.01		0.05		0.1		0.5		1.0	
	K_m, mg/l									
$\dfrac{C_b}{K_m}$	0.1	10	0.1	10	0.1	10	0.1	10	0.1	10
0.01	0.702	0.995	0.170	0.898	0.085	0.702	0.017	0.170	0.008	0.085
0.10	0.728	0.996	0.170	0.909	0.090	0.728	0.018	0.179	0.009	0.090
0.50	0.739	0.997	0.219	0.938	0.019	0.016	0.022	0.219	0.011	0.109
1.0	0.878	0.998	0.271	0.960	0.136	0.878	0.027	0.271	0.014	0.136
5.0	0.966	1.00	0.503	0.989	0.259	0.966	0.051	0.503	0.023	0.259
10.0	0.982	1.00	0.698	0.994	0.360	0.982	0.072	0.698	0.036	0.360
50.0	0.997	1.00	0.980	0.999	0.805	0.997	0.165	0.980	0.082	0.805
100.0	0.998	1.00	0.990	0.999	0.981	0.998	0.234	0.990	0.118	0.981
$k_2 \dfrac{V_p}{A_p}$	1.19	0.119	5.95	0.595	11.9	1.19	59.5	5.95	119	11.9

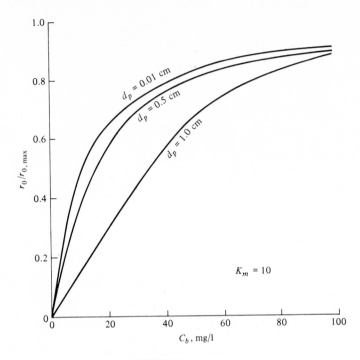

FIGURE 8-4
Effect of substrate concentration on removal rate.

would be quite difficult. Second, we need to be concerned with diffusion limitations in the operating region of activated-sludge processes (floc diameters of 0.05 to 0.5 cm and C_b of the order of 20 mg/l). In practice, this dependence will depend on the shape of the floc. If the flocs are relatively compact (quasi-spherical or cylindrical), diffusion limitations can definitely be expected. If flocs are planar and have "arms" or are highly porous, diffusion may not be limiting.

A qualitative indication of the extent of diffusion limitation can be obtained by comparing the overall uptake rates at different temperatures. Dispersed, rapidly growing bacterial cultures have Q_{10} values (the ratio of the rates at $10°C$ intervals) of between 1.7 and 2.2, in most cases. Values of Q_{10} reported in the literature vary between 1.0 and 2.3, and values of Q_{10} for the diffusivity are of the order of 1.3. Thus diffusion limitation would seem to be probable in some cases.

Table 8-1 and Figs. 8-3 and 8-4 are based upon assumed values of $r_{0,\,max}$, a, and $\mathscr{D}$. Values used were representative and would not be expected to vary by more than one order of magnitude from the values used. The maximum substrate uptake rate would be expected to vary inversely with the area per unit volume parameter a, and thus changes in these parameters would tend to damp each other out. Diffusion-coefficient values can be estimated by the equation of Wilke and Chang.[15]

$$\mathscr{D}_{12} = 7.4 \times 10^{-8} \frac{(\phi M_2)^{1/2} T}{\mu_2 \tilde{V}_1^{0.6}} \qquad (8\text{-}14)$$

where $\mathscr{D}_{12}$ = diffusivity of solute 1 in solvent 2
ϕ = association parameter of solvent (2.6 for water)
M_2 = molecular weight of solvent
T = temperature, K
μ_2 = solvent viscosity, cP
$\tilde{V}_1$ = partial molal volumes, cm^3/g, of solute at its normal boiling point

An estimation of $\tilde{V}_1$ for larger molecules, such as glucose, can be made by adding up the number of carbon, hydrogen, oxygen, and nitrogen atoms, plus any double bonds, and multiplying by 7 (Ref. 16). Glucose thus has a partial molal volume of 175 cm^3/g. This method does not work for small molecules such as oxygen, which has a partial molal volume of 25.6 cm^3/g.

Diffusivities experienced in flocs may be considerably less than the 10^{-5} cm^2/s assumed, but it seems unlikely they are less than 10^{-6} cm^2/s. Because variation is as the square root of the diffusivity, large differences from the estimated values for effectiveness factor are not expected under most circumstances.

Our concern with the limiting or controlling parameters in the rate expression is due to the need to predict behavior of a biological treatment process when environmental changes occur. For example, if substrate removal rate is reaction controlled ($E \approx 1$), oxygen uptake rate will increase much faster with temperature than if the system were mass-transport controlled ($E \approx 0$).

Real cultures are made up of a distribution of particle sizes, and for this reason, Eq. (8-10) is not of direct utility. We could develop modified forms of Eq. (8-10) from various possible particle-size distributions, but this would greatly multiply the complexity of the expression without adding to its usefulness. Our purpose in discussing the effectiveness factor was to develop a qualitative "feel" for probable system behavior. Having gone through this exercise, a simplified version of Eq. (8-10), which combines mass transport and reaction rate limitations into a coefficient k_0, can be utilized:

$$R_0 = -\frac{k_0 CX}{K_m + C} \qquad (8\text{-}15)$$

where the subscript b has been dropped from the organic concentration term.

8-4 CELL YIELD

A basic stoichiometric parameter in any biological process is the cell yield. In Chap. 6 the yield was defined as the negative ratio of the growth rate to the removal rate [Eq. (6-5)]. This method of defining yield gives the actual or observed mass of cells produced per mass of limiting nutrient converted. Observed yield (Y_{obs}) decreases with decreasing growth rate due to increasing maintenance-energy requirements of the microbial culture. An expression for this relationship can be derived by relating the growth rate R_g to the removal rate R_0, as suggested by Pirt,[17] Van Uden,[18] and others.[19-21]

$$R_g = -YR_0 - bX \qquad (8\text{-}16)$$

where Y = maximum yield
$\qquad b$ = maintenance-energy coefficient, t^{-1}

Dividing both sides of Eq. (8-16) by the MLSS concentration X gives the expression in terms of the specific growth and removal rates.

$$r_g = -Yr_0 - b \qquad (8\text{-}17)$$

The observed yield is then defined as the negative ratio of the unit rates.

$$Y_{obs} = Y + \frac{b}{r_0} \qquad (8\text{-}18)$$

In considering Eq. (8-18), it must be remembered that the removal rate $(R_0$ or $r_0)$ is a negative term, and thus the observed yield decreases with decreasing removal rate. As was noted in Eq. (8-16), the removal rate is proportional to the growth rate, and thus the statement that the observed yield decreases with decreasing growth rate is also true. A more useful relationship for prediction of process operation, performance, and control will be developed in the next section.

More complex models for cell yield have been proposed. Grady and Roper[22] have suggested a modified yield model which takes into account the viability of cells in a culture, the death rate of viable cells, and the decay rate of nonviable cells. Their model is basically the same as Eq. (8-16) but includes additional terms to account for the additional factors considered. Lee et al.[23] have proposed a model in which bacteria in floc particles have different growth rates and yields than bacteria dispersed through the culture. They demonstrated that this model could simulate observed culture response to transient organic loadings. Both Roper and Grady's and Lee et al.'s models may become useful as the amount of data available for determination of coefficient values increases. At the present time, these models are useful as a conceptual basis for research.

8-5   PROCESS OPERATION, PERFORMANCE, AND CONTROL

A convenient way of discussing the activated-sludge processes is to begin with CFSTR systems at steady state. After identifying some basic relationships, other situations can be discussed. Mass balances around the aeration tank of a CFSTR process were presented in Eqs. (8-4) and (8-5).

Growth Rate

Rearranging Eq. (8-5) and assuming that the mass rate of cells into the system, QX_i, is negligible results in Eq. (8-19):

$$R_g = \frac{Q}{V}[(1 + \alpha)X - \alpha X_u] \qquad (8\text{-}19)$$

where V/Q can be replaced by Θ_H, the hydraulic residence time. A cell mass balance around the clarifier (Fig. 8-1) gives

$$[(1 + \alpha)Q - Q_w]X = \alpha Q X_u + (Q - Q_w)X_e \qquad (8\text{-}20)$$

In properly operating processes, the effluent suspended-solids concentration X_e is small, and the last term in Eq. (8-20) can be neglected. Substituting for αX_u in Eq. (8-19) results in two very simple expressions for the growth rate.

$$R_g = \frac{Q_w X}{V} = \frac{X}{\Theta_C} \qquad (8\text{-}21)$$

$$r_g = \frac{Q_w}{V} = \frac{1}{\Theta_C} \qquad (8\text{-}22)$$

where Θ_C is termed the *mean cell residence time* (MCRT) or the *solids retention time* (SRT).

Equation (8-22) can be used to develop an extremely useful expression for the observed yield. Remembering that the specific removal rate can be written as

$$r_0 = \frac{-1}{Y}(r_g + b) = \frac{-1}{Y}\left(\frac{1}{\Theta_C} + b\right)$$

the observed yield can be written as

$$Y_{obs} = \frac{Y}{1 + b\Theta_C} \qquad (8\text{-}23)$$

Organic Removal Rate

Returning to Eq. (8-4), the mass balance on organic material, and solving for the removal rate R_0 gives

$$R_0 = -\frac{1}{\Theta_H}[C_i + \alpha C_u - (1 + \alpha)C] \qquad (8\text{-}24)$$

where the subscript u indicates that the underflow or recycle stream concentration and the organic concentration values should reflect the biodegradable material present. Normally, ultimate BOD will be used, although biodegradable organic carbon would also be satisfactory. The underflow organic concentration C_u varies with the method of measurement. If specific influent and effluent organic species are being measured, C_u will be close to zero. If an overall measure is used (BOD_L), C_u will be approximately the same as C. Because the latter situation is nearly always the case in wastewater treatment, Eq. (8-24) can be rewritten

$$R_0 = -\frac{(C_i - C)}{\Theta_H} \qquad (8\text{-}25)$$

Reactor Solids and Organic Concentrations

An expression for the organic concentration in the reactor can be derived by combining Eqs. (8-15), (8-17), (8-22) and (8-23):

$$\frac{1}{\Theta_C} = Y \frac{k_o C}{K_m + C} - b$$

$$C = \frac{K_m(1 + b\Theta_C)}{Yk_0 \Theta_C - b\Theta_C - 1} \qquad (8\text{-}26a)$$

$$C = \frac{K_m}{Y_{obs} k_0 \Theta_C - 1} \qquad (8\text{-}26b)$$

The solids concentration in the reactor can be predicted by combining Eqs. (8-21), (8-23), and (8-25).

$$\frac{X}{\Theta_C} = Y_{obs} \frac{C_i - C}{\Theta_C}$$

$$X = Y \frac{C_i - C}{1 + b\Theta_C} \frac{\Theta_C}{\Theta_H} \qquad (8\text{-}27)$$

Inserting typical observed coefficient values into Eqs. (8-23), (8-26), and (8-27) allows the development of curves for organic and solids concentrations and for the observed yield. Figures 8-5 to 8-7 were developed using the following values:

$$b = 0.06 \ \mathrm{d}^{-1}$$
$$K_m = 20 \ \mathrm{mg/l}$$
$$Y = 0.4$$
$$k_0 = 2.0 \ \mathrm{d}^{-1}$$
$$C_i = 500 \ \mathrm{mg/l} \ BOD_L$$

The washout line shown on each of the three curves is the minimum MCRT value for the system. At lower values of Θ_C, the cells are washed out faster than they can reproduce. True washout would occur at a lower Θ_C value. The value predicted in Figs. 8-5 to 8-7 results from the low value of k_0 used. This value is representative of flocculant cultures and includes effects of mass-transfer

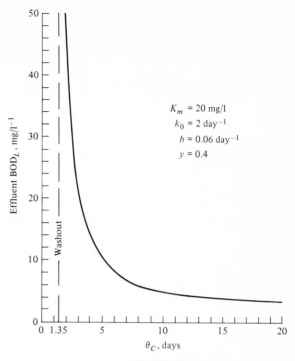

FIGURE 8-5
Effluent BOD_L as a function of θ_C.

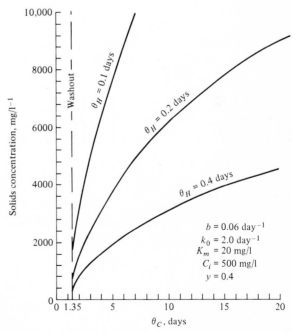

FIGURE 8-6
Mixed-liquor suspended-solids concentration as a function of θ_C and θ_H.

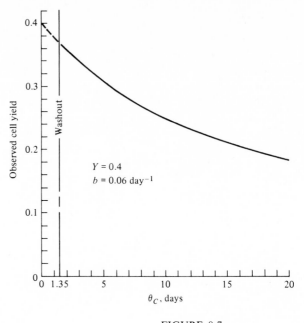

FIGURE 8-7
Cell yield as a function of θ_c.

limitations and a nonviable fraction of the solids mass. Actual minimum Θ_C values are in the neighborhood of 3 d because of the necessity of maintaining the culture in a flocculant state.

Significance of Rate Models

Our interest in the rate models is associated with predicting operating conditions and effluent quality. A particularly important feature of the models is that the effluent organic concentration is not influenced by the influent concentration. This result occurs because a change in the available mass rate of organic input results in a change in cell mass that balances out the system. This would seem to reflect the real world in most cases, but a few studies have demonstrated an effect of influent concentration. Grady and his coworkers[24] studied pure and mixed bacterial cultures and pure and complex substrates and found that influent organic concentration affected the effluent concentration. They used low residence times and reactors without recycle in their studies. Moore and Schroeder[25] and Kaufman[26] have reported similar results with denitrification processes.

Experience with activated sludge has consistently demonstrated that a wide operating region (with respect to Θ_C) exists in which the effluent organic-concentration variation is small. Most activated-sludge processes are designed to operate within the 3- to 10-d MCRT range, and one reason is that these values are in a region that gives satisfactory effluent organic concentrations and insensitivity to moderate changes in operating parameters.

At values of Θ_C below 3 d, the organic concentration C begins to rise sharply, even though the actual values may still be acceptable. In practice, as Θ_C decreases, activated-sludge processes begin to deflocculate; i.e., slime-layer production decreases and floc particles become smaller or mixed with dispersed cells, and predators such as rotifers are washed out. The result is that the culture does not settle well. Because the secondary clarifier is designed for separation at definite rates, the slower-settling deflocculated cells are washed out of the tank in the effluent and the cell concentration drops more rapidly then predicted by the model. Thus a minimum value of Θ_C in the neighborhood of 3 d should be used in design and operation. Process stability should be greater at longer residence times, also.

Limitations are also imposed by long cell residence times, and the model [Eq. (8-26)] does not satisfactorily describe this situation. As residence time increases, a larger and larger fraction of the cell mass becomes nonrespiring organic material, e.g., dead cells or cell fragments.[27, 28] Eventually, the flocs begin to break up, organic solids are lost in the effluent, and the process becomes unstable. The model does not predict this process, but as the effects of extremely long cell residence times are undesirable, our primary goal is to define a boundary for the effective use of the models. The maximum value of Θ_C (or the minimum value of growth rate) that should be used is not clearly defined by the literature. Extended-aeration activated-sludge processes ($\Theta_C > 10$ d) often have tendencies to discharge solids intermittently, but this result is due to overloading of the clarifier (see Chap. 5) in many cases. Bisogni and Lawrence[29] and Selna[13] are virtually the only workers who have considered settling rate as a function of cell age. Selna's studies were with nonsteady-state processes and do not directly apply to this discussion. The work of Bisogni and Lawrence was over a Θ_C range of 0.28 to 12 d. Their cultures were essentially nonflocculated at Θ_C less than 2 d. Between 2 and 4 d the settling rate was approximately 0.017 cm/s (360 gal/ft²-d), and between 4 and 12 d the settling rate linearly increased to approximately 0.161 cm/s (3400 gal/ft²-d). A maximum value was not determined, and consequently there is no information available on maximum Θ_C values from this standpoint. In real processes, MLSS concentration increases with increasing cell residence time, at least in the operating region. Because settling rate goes through a maximum with increasing MLSS concentrations (see Chap. 5), the beneficial effects of long cell residence times may be negated.

The value of the rate coefficient k_0 used strongly affects the organic concentration for a given Θ_C value. In the operating region ($\Theta_C > 3$ d) C_b varies nearly inversely with k_0, and thus some knowledge of the rate coefficient's value is desirable. Figure 8-5 was generated using $k_0 = 2$ d^{-1}. This value is of the order of the values obtained in conventional activated-sludge processes. In cultures with flocs consisting of 100 percent respiring cells and without mass-transfer limitation, the value of k_0 would be about 24 d^{-1}. This would decrease the residual BOD value somewhat. Actual activated-sludge plants operating on either municipal sewage or industrial wastes generally produce an effluent with an ultimate BOD of 15 to 40 mg/l. This value includes suspended solids washed over the clarifier weir, and the soluble BOD fraction is usually of the order of 5 to 10 mg/l.

Hydraulic Residence Time

Up to this point, little has been said about the hydraulic residence time Θ_H. In practice, hydraulic residence time is an important parameter because tank volume is fixed by the value chosen. In Eq. (8-27) and Fig. 8-6, the hydraulic residence time is seen to strongly affect the cell concentration of a CFSTR activated-sludge process, but not the organic concentration. Thus practical limits on hydraulic residence time in the aeration-basin values are related to settling rate and the related separation problems. For example, if Θ_H is very low, the MLSS concentration may be so great that the secondary clarifier's capacity is exceeded. Municipal waste treatment hydraulic residence time values traditionally used are of the order of 0.2 d. These values produce acceptable MLSS concentrations of 1500 to 5000 mg/l (assuming good operation). Some consideration should be given to these alternatives, but considerable experimental work needs to be done before design practice will be generally altered.

Plug-flow Processes

Until the late 1950s nearly all activated-sludge aeration tanks were designed as long, narrow (length-to-width ratio > 5) tanks that were considered to be reasonably good approximations of plug-flow systems. Tracer studies on typical aeration tanks have resulted in hydraulic efficiencies between 60 and 80 percent,[1] and there has been considerable discussion in the literature as to whether or not the plug-flow model adequately describes these systems. One suggestion has been to use the Wehrner and Wilhelm model[30] presented in Chap. 2, but this model is based on the first-order kinetics, and biological treatment systems are at best pseudo first order in very limited regions. A second approach has been to assume that the long narrow tanks are ideally mixed. One support for this approach is that the same design criteria (air requirements, hydraulic detention time, and cell residence time) have been found satisfactory for CFSTR processes as for nominal plug-flow systems. Milbury et al.[31] commented on this indirectly in 1964 in suggesting that many conventionally designed systems were in fact CFSTRs. Tracer studies on conventionally designed aeration basins rarely support this argument, however, and there is little question about the existence of longitudinal organic concentration gradients in nominal plug-flow tanks. The observations of Milbury and others cannot be ignored, however, just as 60 years' experience with other phases of activated sludge cannot be ignored. An explanation can be developed from a mass balance on a plug-flow process using the rate models developed in this chapter and the same steady-state form of Eqs. (8-6) and (8-7).

$$-\frac{k_0 CX}{K_m + C} = (1 + \alpha)\frac{dC}{d\Theta} \qquad (8\text{-}28)$$

$$\frac{k_g CX}{K_m + C} = (1 + \alpha)\frac{dX}{d\Theta} \qquad (8\text{-}29)$$

where both k_0 and k_g may be considered to include any necessary effectiveness factors. The initial ($\Theta = 0$) value of the variables C^* and X^* must include the recycle term.

$$\text{At } \Theta = 0 \qquad C = C^* = \frac{C_i + \alpha C_u}{1 + \alpha}$$

$$X = X^* = \frac{X_i + \alpha X_u}{1 + \alpha}$$

Assuming that the recycle-stream organic contribution is small allows us to write

$$C^* = \frac{C_i}{1 + \alpha} \qquad (8\text{-}30)$$

We can also safely assume, in most cases, that X_i is small. The growth ΔX through one pass of the tank is then

$$\Delta X = X - X^* = Y_{obs}(C^* - C) \approx Y_{obs} C^*$$

Substituting this value of ΔX into a mass balance around the secondary clarifier gives

$$(1 + \alpha)QX^* = \alpha QX_u = [(1 + \alpha)Q - Q_w](X^* + Y_{obs} C^*)$$

which reduces to

$$X^* = \frac{(1 + \alpha)\Theta_C - \Theta}{\Theta} Y_{obs} C^* \qquad (8\text{-}31)$$

Here an assumption is made that Y_{obs} is constant through the tank and depends only on Θ_C. In the plug-flow case, Θ_C must be considered an average over the length of the tank, and thus MCRT has less physical meaning. Substituting for X and integrating between 0 and Θ_H and between C^* and C results in Eq. (8-32).

$$(1 + \alpha)\frac{dC}{d\Theta} = -\frac{k_0 C}{K_m + C}[X^* + Y_{\text{obs}}(C^* - C)]$$

$$\frac{K_m + C}{C(X^* + Y_{\text{obs}})(C^* - C)} dC = -\frac{k_0}{1 + \alpha} d\Theta_H$$

$$\left[\frac{K_m}{C(X^* + Y_{\text{obs}})(C^* - C)} + \frac{1}{X^* + Y_{\text{obs}}(C^* - C)}\right] dC = -\frac{k_0}{1 + \alpha} d\Theta_H$$

$$\frac{k_0}{1 + \alpha}\Theta_H = \frac{K_m}{X^* + Y_{\text{obs}} C^*} \ln\left[\frac{C^*}{X^*C}(X^* + Y_{\text{obs}} C^* - Y_{\text{obs}} C)\right]$$

$$+ \frac{1}{Y_{\text{obs}}} \ln \frac{X^* + Y_{\text{obs}} C^* - Y_{\text{obs}} C}{X^*} \qquad (8\text{-}32)$$

Process performance with respect to organic removal can be predicted by inserting appropriate parameter and coefficient values into Eq. (8-32). Because the MLSS concentration is a function of Θ_H, the solution will always be trial and error. The following values were used in developing the curves in Fig. 8-8.

$$C_i = 500 \text{ mg/l}$$
$$\Theta_H = 0.25 \text{ d}$$
$$K_m = 20 \text{ mg/l}$$
$$Y = 0.4$$
$$b = 0.06 \text{ d}^{-1}$$

Curves are presented for three values of the rate coefficient k_0: 24 d^{-1}, 1 d^{-1}, and 0.5 d^{-1}. Dashed lines in Fig. 8-8 are based on the solution to Eq. (8-26) for CFSTR processes and are shown for comparison.

Equation (8-32) is constrained to small values of effluent organic concentration because of the approximation made in Eq. (8-30). Figure 8-8 was limited to values of C below 30 mg/l for this reason.

Actual values of k_0 occurring in practice are evidently of the order of 2 to 10 d^{-1}. In this region, effluent qualities of the order of 10 to 20 mg/l BOD_L can be achieved, and as noted earlier, the difference between an ideal plug-flow reactor and an ideal CFSTR would be difficult to measure by conventional tests. Thus, the choice of process configuration cannot be made on the basis of effluent quality obtained at a given residence time.

Design values of the cell age Θ_C are fixed by the need to produce a flocculant, easily settleable culture. In practice, this means using Θ_C values greater than 3 d. In terms of removing soluble organic material, mixing thus becomes irrelevant; i.e., the CFSTR will produce an effluent that is not measurably different from an ideal plug-flow process. Nonideal systems (i.e., those with

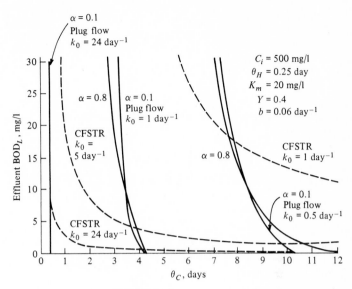

FIGURE 8-8
Comparison of plug-flow and CFSTR performance.

backmixing) have characteristics between ideal plug-flow reactors and ideal CFSTRs, and thus we do not need to be concerned with the problem. Short circuiting the passage of a portion of the incoming wastewater through the aeration vessel *without significant contact* with the culture is not likely to happen but in any case will not greatly affect the results. This interesting result is substantiated by a vast amount of field data.

We must remember that the result illustrated in Fig. 8-8 does not mean that CFSTR and plug-flow systems are identical, only that with the imposed design constraints on Θ_C, they will produce an effluent of the same quality. A major difference between the processes is the organic concentration gradient existing in plug-flow or nominal-flow processes. This gradient must have an effect on process operation and stability. Oxygen uptake rates will decrease along the length of a plug-flow tank, and this factor must be considered in design. As yet the effects of concentration gradients on the culture have not been determined. Another difference is the dilution factor provided for toxic materials by CFSTR processes. This is particularly important where treatment plants receive industrial wastes.

8-6 EXTENDED-AERATION PROCESSES

Extended-aeration processes are, in simplest terms, activated-sludge processes with high values of Θ_C. The process concept grew out of the "total oxidation" scheme proposed in the late 1950s (Refs. 6, 32). As pointed out previously, when Θ_C increases, the net cell yield decreases. Thus if cells are not wasted at all, the net yield should decrease to 0, and a secondary waste (cells) will not be produced. In practice, one of two events occurs: The MLSS concentration increases to the point that the secondary clarifier is overloaded, or the "old" sludge tends to deflocculate. In either case, large amounts of solids are lost. Gaudy and his coworkers[33, 34] have recently reported on experiments in which a centrifuge was used and all the solids were returned to the aeration chamber. Their conclusion was that total oxidation was theoretically possible. Practical application of total oxidation is questionable on an economic basis, however.

In practice, extended aeration is most often encountered in "package" wastewater treatment plants. These plants are sold as complete units that include aeration tank, the necessary pumps, compressors, and other appurtenances. Package plants are normally designed for flows of less than 75,000 l/d (20,000 gpd). In most cases, sludge wasting is not provided for, and the units have an operating characteristic of the MLSS concentration gradually increasing until a critical value is reached. At this point the clarifier becomes overloaded, and the solids are flushed out of the system. The flocculant culture must then be reestablished. An effluent solids concentration vs time chart generally shows fairly low values up to the time when the clarifier becomes overloaded. A sharp increase in effluent solids concentration occurs for a short period, and then the value drops off and gradually improves as the culture is reestablished. Complete cycle times are often of the order of 1 month.

A major problem with package extended-aeration units is that they are usually used for small subdivisions, motels, or similar establishments. Professional operation is often nonexistent and at best minimal, which increases the natural instability of the process. In addition, the discharge is often into a small stream with little capacity to absorb such solids loadings.

Properly designed and operated extended-aeration units produce excellent effluents. The advantage of low cell yield is obvious, but remembering the stoichiometry of biological processes, we recognize that more oxygen must be supplied; i.e., as yield decreases the fraction of the substrate converted to CO_2 increases. In large plants it is generally more economical to operate with a higher yield and lower oxygen requirement because the cell solids can sometimes be utilized to produce methane in another process (see Chap. 10).

8-7 STEP AERATION

Step aeration was introduced by Gould[35] as an attempt to even out the organic load in a conventional (nominal plug-flow) activated-sludge plant. As is shown in Fig. 8-9, the influent waste stream is split and introduced at a number of points along the length of the tank. In effect, this approach gives an approximation of a CFSTR. In some cases particularly where industrial wastes are involved, an old plant may be modified quite inexpensively by this

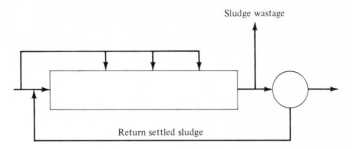

FIGURE 8-9
Schematic diagram of step-aeration activated-sludge plant.

procedure. A similar concept was developed with respect to diffused aeration of long narrow tanks. Because the highest oxygen demand is at the head end of the tank, more diffusers are placed there. The number per foot is gradually decreased along the tank length. This modification, called *tapered aeration*, is now considered good design and is generally used in nominal plug-flow processes.

8-8 HIGH-RATE ACTIVATED SLUDGE

This modification of the basic activated-sludge process was also introduced in New York City by Gould.[35] The fact that it could be introduced reflects the relatively loose effluent quality standards then in existence and, perhaps more importantly, the relatively unresponsive receiving-water system. High-rate processes are systems with a Θ_c value in the range where little flocculation occurs. Growth and organic removal rates are high on a unit microbial mass basis but not necessarily on an overall basis. The systems characteristically have low MLSS concentrations, generally of the order of 600 to 1000 mg/l. Removal of the soluble organics remains high (as predicted from Figs. 8-5 and 8-8), but most of the cell solids produced are in the effluent.

The process was developed by overloading conventionally designed processes. Because the loading rates were higher (of the order of 2 g COD/g MLSS-d or greater) and the effluent met imposed requirements, the modification was quite economically feasible.

8-9 BIOSORPTION OR CONTACT STABILIZATION

Discussion up to this point has centered on the removal of soluble organic material from wastewater. Most wastes contain nonsoluble organic material ranging from colloidal to particles of greater than 1 mm in size. Breakdown of these particles takes place after they have been absorbed onto floc particles,

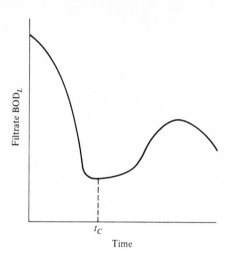

FIGURE 8-10
Batch-contact-stabilization test results.

and this phenomenon is the fundamental basis of the contact-stabilization form of the activated-sludge process.

If a batch experiment is run using a mixed colloidal-soluble wastewater and an activated-sludge culture, a plot of the filtrate organic concentration will usually be similar to Fig. 8-10. During the initial period, soluble organics are converted to cell material and organic solids are adsorbed onto floc surfaces. At some critical time t_c, a minimum value of filtrate to BOD is reached. At this time the solubilization of sorbed solids begins to increase the filtrate organic concentration.

Eventually, these organics are metabolized, and the filtrate concentration decreases. This conceptual picture can be used to describe most soluble-nonsoluble wastewater mixtures, but the importance of the process varies with the waste. For example, if the solids are difficult to hydrolyze, the peak associated with resolubilization might be very low. On the other hand, if the solids were extremely easy to hydrolyze, the peak would be high (assuming high solids concentration), and the time scale would be compressed.

In cases where the concentration profile in the region of t_c is fairly flat, the aeration-tank detention time can be considerably decreased by separating the solids (MLSS + adsorbed organic solids), and using a second tank (the reaeration basin) to carry out the hydrolysis and conversion of the adsorbed solids, i.e., the part of Fig. 8-10 to the right of t_c. A schematic diagram of this process is shown in Fig. 8-11.

Contact stabilization was developed by Ulrich and Smith[36] and Eckenfelder and Grich[37] independently during the 1950s. Applications of the concept have not always been successful because of the need for a low organic concentration associated with t_c and a reasonably long time period in this low-concentration

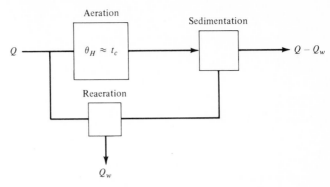

FIGURE 8-11
Schematic diagram of contact-stabilization process.

region. The designer must remember that time in the clarifier is also important. Clearly, this process cannot be taken off the design shelf and slapped in anywhere, but used only where there is experimental evidence that it will work. Perhaps one of the most distressing aspects of the process is knowing what will happen if the designer puts in a safety factor to ensure a satisfactory detention time.

Busch[4] has noted that the best candidates for successful application of the contact-stabilization process are wastewaters containing very low soluble organic concentration. Examples would include certain industrial wastes and municipal wastewaters carried through very long sewers. In the latter case, one might expect that the soluble fraction is removed in the sewer.

8-10 PURE-OXYGEN-ACTIVATED SLUDGE

Use of oxygen rather than air in activated-sludge systems was first investigated in 1948 by Okun.[38] Practical application of the concept was first made in 1969 at Batavia, N.Y.[39] An obvious advantage of using pure oxygen is the increase in oxygen-concentration gradients, both between the gas and liquid phase and within the liquid phase. Oxygen is not ordinarily a rate-limiting material, and therefore modifications to conventional operations procedures must be made to take advantage of the higher potential rates. The most obvious modification is to utilize higher cell concentrations. Higher overall organic removal rates associated with higher cell concentrations make use of the increased oxygen transfer rates. Shorter hydraulic residence times, smaller aeration tanks, and because of the longer cell residence times needed to produce the high cell concentrations, low excess-sludge-production rates are the beneficial results. Because most treatment processes are overdesigned with respect to hydraulic

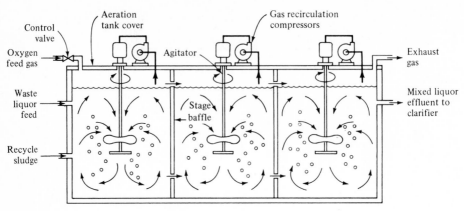

FIGURE 8-12

Schematic diagram of Unox process. (*Union Carbide Corporation.*)

residence time Θ_H, there is no satisfactory information available on the actual potential savings associated with tank construction costs.

The most commonly used pure-oxygen-activated-sludge system is the proprietary Unox process of Union Carbide Corporation (Fig. 8-12). This system consists of an aeration tank divided into three or more sections by baffles, with mechanical mixing in each section. Initial studies of Unox process operation utilized MLSS values between 7000 and 10,000 mg/l, but recent design MLSS values[39] used have varied between 3000 and 5000 mg/l. Sludge-settling rates reported have, in general, been high. If this proves to be unique to pure-oxygen systems, they will have a distinct advantage over conventionally aerated processes. A possible explanation is the difference in cell residence times used, however. Humenick and Ball[40] have suggested that lower shearing rates in pure-oxygen-activated sludge result in larger, better settling flocs.

Rate and yield values reported for pure-oxygen systems are identical with those that use air.[40, 41] Thus the advantages of the use of pure oxygen are the higher potential oxygen transfer rates that allow the use of higher cell concentrations and the possibly better sludge settling. Both advantages can result in decreased tankage volume.

8-11 AERATED LAGOONS

When cell recycle is not used, the overall reaction rates are low because the cell concentration in the aeration basin is limited to the direct yield, $Y_{obs}(C_i - C_b)$. As a result, the aeration basin must be considerably larger than in conventional activated-sludge processes. Process design follows exactly the same steps, i.e., determination of rates and stoichiometry as before, however.

Aerated lagoons were developed as a modification of sewage lagoons (hence, the name) rather than activated sludge. Consequently, little attention has been given to developing either kinetic or hydraulic specifications for their design. In most cases, mixing is not great enough to eliminate the buildup of sludge deposits in basin corners and between aeration units. These deposits are limited in size by fluid turbulence and rarely cause great difficulty. Control or elimination of the deposits can be achieved by adding additional aeration-mixing equipment.

Most aerated lagoons are designed with floating-surface turbine aerators (see Fig. 4-8).

8-12 AEROBIC DIGESTION

Disposal of waste-activated sludge is a significant problem because of the quantities produced and the biological activity of the material. In most cases, some form of "biological digestion" is utilized to decrease the quantity of biodegradable material, and aerobic digestion[42] is one such process. In essence, the process is an unfed activated-sludge system. Because the only source of energy available is endogenous respiration, the culture effectively consumes itself. Effluents from aerobic digesters are highly nitrified, and the solids reduction $(1 - X_{out}/X_{in})$ is of the order of 50 percent. Hydraulic residence time equals the cell residence time because there is no recycle, and values in use range from 15 to 25 d.

Design of aerobic digesters should be on the same principles as activated-sludge processes. Net oxygen requirements must be determined, and these will be related to the sludge concentration and the endogenous respiration rate. Mixing requirements must also be checked to make sure that aeration will also satisfy mixing criteria.

8-13 CENTRIFUGAL SCREENS

Tchobanoglous[43] has suggested a schematic modification of the activated-sludge process that appears very promising. If centrifugal screens (or any other suitable mechanisms) are used to remove solids from the aeration-tank effluent, the solids loading imposed on the secondary clarifiers will be decreased. Centrifugal-screen effluent does not meet most effluent standards, but if the screen effluent undergoes sedimentation, an excellent quality effluent is produced. Screens are relatively inexpensive and offer a simple method of altering systems with overloaded secondary clarifiers (Fig. 8-13).

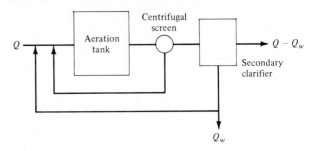

FIGURE 8-13
Placement of centrifugal screen as proposed by Tchobanoglous.[43]

8-14 NITRIFICATION

When nitrogen-containing organic material is oxidized in a biological wastewater treatment process, ammonia-valence nitrogen is released into solution. This nitrogen is available as a nitrogen and energy source for the nitrifying bacteria briefly discussed in Chap. 6 and poses an oxygen-demand threat to receiving waters. Two groups of nitrifying bacteria exist: *Nitrosomonas*, which oxidizes ammonia nitrogen to nitrite, and *Nitrobacter*, which oxidizes nitrite to nitrate. Both groups are chemautotrophs, and the two reactions can be written

$$v_1[NH_3] + v_2[CO_2] + v_3[O_2] \xrightarrow{\textit{Nitrosomonas}} v_4[\text{cells}] + v_5[NO_2^-] \qquad (8\text{-}33)$$

$$v_6[NO_2^-] + v_7[CO_2] + v_8[O_2] \xrightarrow{\textit{Nitrobacter}} v_9[\text{cells}] + v_{10}[NO_3^-] \qquad (8\text{-}34)$$

Data reported in the literature on the stoichiometry and rates of nitrification have been summarized by Lawrence and McCarty[20] and are presented in Table 8-2. Because the nitrifying organisms are usually part of the mixed bacterial population of an activated-sludge process, nitrification can be carried out in the aeration tank provided the cell residence time Θ_C is greater than k_g^{-1} for nitrite oxidation. Alternatively, a separate tank can be provided as shown in Fig. 8-14.

Table 8-2 DATA ON NITRIFICATION[20]

	Y^*, g cells/gN	k_a, d^{-1}	k_n, d^{-1}	K_m, mg/l
Ammonia oxidation	0.05	0.33	6.6	1
Nitrite oxidation	0.02	0.14	7.0	2

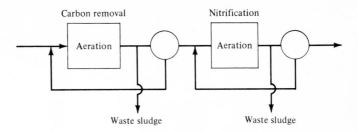

FIGURE 8-14
Two-sludge–activated-sludge nitrification scheme.

EXAMPLE 8-1   Treatment of an industrial wastewater has been studied using bench and pilot scale treatment plants. Data from these studies are summarized as follows:

Waste characteristics

Ultimate BOD	400 mg/l (soluble)
Organic N	60 mg/l
PO_4^{3-}	30 mg/l
pH	7.0
T	20°C

Process kinetics and stoichiometry

k_0	2.0 d^{-1}
K_m	5 mg/l
Y_{obs}	$0.4/(1 + 0.06\Theta_C)$

Settling-rate data

$$v_H = 2.7 - 0.00026X \text{ cm/min}$$

For a constant flow rate of 10^6 l/d (0.264 mgd), determine suitable design values for MLSS, Θ_H, Θ_C, and oxygen transfer rate for both the activated-sludge process and a separate nitrification process. Maximum allowable effluent BOD and NH_3–N concentrations are 10 and 0.5 mg/l, respectively.

Activated-sludge aeration tank   Assuming all the influent N and P are available for growth, carbon will be the limiting nutrient. Making a mass balance on carbon (as ultimate BOD) around the aeration tank for the limiting condition gives

$$R_0 = \frac{Q(10 - 400)}{V} = -\frac{X}{Y_{obs}\,\Theta_C}$$

A mass balance around the aeration tank on cells gives

$$R_g = \frac{Y_{obs} k_0 CX}{K_m + C} = \frac{X}{\Theta_C}$$

Substituting the given coefficient values and $C = 10$ mg/l into the cell-mass-balance expression results in an expression involving the minimum allowable Θ_C value.

$$\Theta_{C, min} = \frac{K_m + C}{Y k_0 C - b(K_m + C)}$$

$$= \frac{5 + 10}{0.4(2)(10) - 0.06(5 + 10)}$$

$$= 2.1 \text{ d}$$

Therefore, any Θ_C value greater than 2.1 d will result in a satisfactory effluent BOD_L. In order to maintain a stable, well-flocculated sludge, the MCRT should be 3 d or greater, and this factor becomes the constraint on Θ_C.

Aeration-tank volume and oxygen uptake rate In practice, solutions for all Θ_C and X combinations can be easily made with the computer. Here four values of Θ_C and five values of X will be used to demonstrate the concepts. Values of aeration-tank volume are calculated directly from the organic material mass balance for chosen Θ_C and X values. Oxygen uptake rates are developed as described in Chap. 7.

$$R_{O_2} = R_0(1 - 1.41 Y_{obs})$$

Secondary-clarifier-thickener design According to the given expression, the settling rate is a function of X. The settling velocity drops to zero at a MLSS concentration of about 10,500 mg/l. This then is the maximum underflow concentration. For convenience in this example the number of underflow flux rates studied will be limited to four (0.2, 0.4, 0.8, 1.6 cm/min). Results of the flux analysis are shown in Fig. 8-15.

Table 8-3 AERATION-TANK VOLUME AND OXYGEN UPTAKE RATE

	$\Theta_C = 3$ d $Y_{obs} = 0.34$ $C = 4.8$ mg/l		$\Theta_C = 6$ d $Y_{obs} = 0.29$ $C = 2.0$ mg/l		$\Theta_C = 10$ d $Y_{obs} = 0.25$ $C = 1.3$ mg/l		$\Theta_C = 15$ d $Y_{obs} = 0.21$ $C = 0.9$ mg/l	
X, mg/l	V_A, m³	R_{O_2}, mg/l-d	V_A, m³	R_{O_2}, mg/l-d	V_A, m³	R_{O_2}, mg/l-d	V_A, m³	R_{O_2}, mg/l-d
2000	202	1018	346	680	498	518	629	447
4000	101	2037	173	1360	249	1037	314	895
6000	67	3071	115	2046	166	1555	210	1337
8000	50	4114	87	2704	125	2065	157	1789
10,000	40	5143	69	4310	100	2582	126	2230

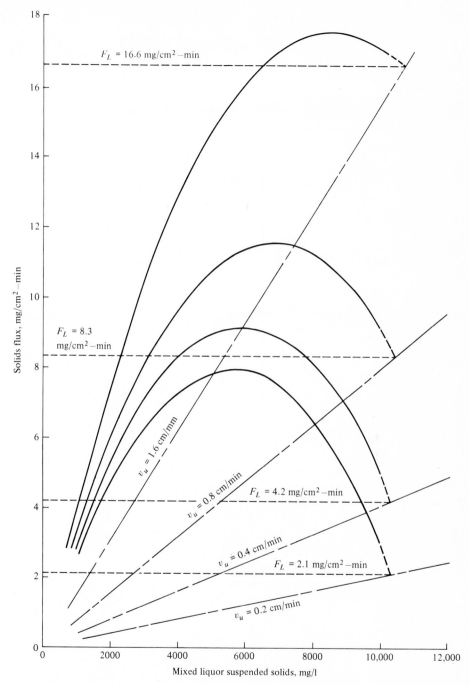

FIGURE 8-15
Solids-flux curves for Example 8-1.

A mass balance on the secondary clarifier gives the following equations:

$$A = \frac{(1 + \alpha)(Q - Q_w)X}{F_s}$$

$$\alpha = \frac{1 - \Theta_H/\Theta_C}{X_u/X - 1}$$

where A is the clarifier area. In most cases $Q_w \ll Q$, and excellent approximations are given by

$$A = \frac{(1 + \alpha)QX}{F_s}$$

$$\alpha = \frac{X}{X_u - X}$$

Table 8-4 can now be developed for the clarifier area and recycle ratio necessary.

Clarifier volumes presented in Table 8-4 are based on an arbitrary average depth of 2.5 m. This value is in the normal range used for clarifiers in the United States. Some flexibility in the choice of depth is possible, although the availability of equipment is a constraint. This constraint is important because of capital cost ramifications and more importantly because of the effect of clarifier detention time Θ_{H_C}. Because of the active nature of the microbial culture separated out in the secondary clarifier, problems occur if the detention time is too long. Maximum detention time should be less than $1\frac{1}{2}$, preferably less than 1 h. Detention times for this problem are shown in Fig. 8-16.

Total system volume data are presented in Fig. 8-17. Only values for $v_u = 1.6$ cm/min are shown because of the extremely limited range of concentrations possible at lower underflow flux values. Total activated-sludge-system volume is minimized for a 3-d MCRT and an MLSS concentration of 6000 mg/l, although the total system volume changes by less than 1 percent if the MLSS concentration used is 4000 mg/l.

Two additional considerations should be made at this point. First, limiting oxygen transfer rates should be compared to those given in Table 8-3. Oxygen-transfer coefficients $(K_L a)$ are usually reported in the range of 0.002 to 0.003 s^{-1}. For a $C_{sat} - C_{O_2}$ value of 6 mg/l, a transfer rate between 1040 and 1550 mg/l-d can be expected. The particular equipment chosen and results of pilot plant studies will provide a more exact estimate in particular cases. Here oxygen uptake rates are plotted against cell concentration for the five Θ_C values used. Lines showing oxygen uptake rate limitations are dashed in Fig. 8-18. Apparently, oxygen transfer rate will be a greater constraint than clarifier detention time in this problem.

The second consideration involves nitrification. If a 10- or 15-d MCRT is used, considerable nitrification can be expected. Thus, the aeration-tank-plus-clarifier volume at these residence times can be compared to the total tankage of "two-sludge" systems using a shorter activated-sludge MCRT.

Table 8-4 SECONDARY-CLARIFIER PARAMETERS

X, mg/l	v_u = 0.2 cm/min			v_u = 0.4 cm/min			v_u = 0.8 cm/min			v_u = 1.6 cm/min		
	A, m²	V_C,† m³	α	A, m²	V_C,† m²	α	A, m²	V_C,† m³	α	A, m²	V_C,† m³	α
10,000	6940	17,350	20	3470	8675	20	1756	4390	20	878	2195	20
8000	1110	2776	3.2	555	1388	3.2	281	702	3.2	140	351	3.2
6000	462	1155	1.33	231	578	1.33	117	292	1.33	60	149	1.33
4000	214	535	0.62	107	268	0.62	54	135	0.62	40	100	0.87
2000	81	203	0.23	41	102	0.23	25.5	94	0.36	33	83	0.74

† Clarifier volumes are based on an arbitrary depth of 2.5 m.

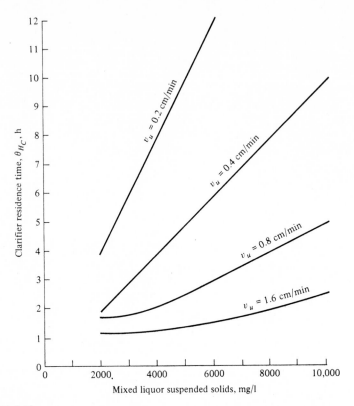

FIGURE 8-16
Secondary-clarifier residence time as a function of underflow rate and MLSS concentration in Example 8-1.

Nitrification The first consideration is to determine the amount of ammonia leaving the activated-sludge process. An assumption has been made that for $\Theta_C = 10$ d nitrification is complete. At the other two MCRTs, effluent nitrogen will be assumed to be in the ammonia form.

$$\text{Entering nitrogen concentration, } C_{Ni} = 60 \text{ mg/l}$$

$$\text{Nitrogen assimilated in activated-sludge process} = \tfrac{14}{113}(Y_{obs})(C_i - C)$$

$$\text{Nitrogen assimilated} = \begin{cases} 16.6 \text{ mg/l} & \text{at } \Theta_C = 3 \text{ d} \\ 14.3 \text{ mg/l} & \text{at } \Theta_C = 6 \text{ d} \end{cases}$$

$$r_{\theta NH3} = \frac{1}{\Theta_{C_{NH3}}} = \frac{k_{\theta NH3} C_{NH3}}{K_{NH3} + C_{NH3}}$$

where C_{NH_3} = ammonia-nitrogen concentration, mg/l N

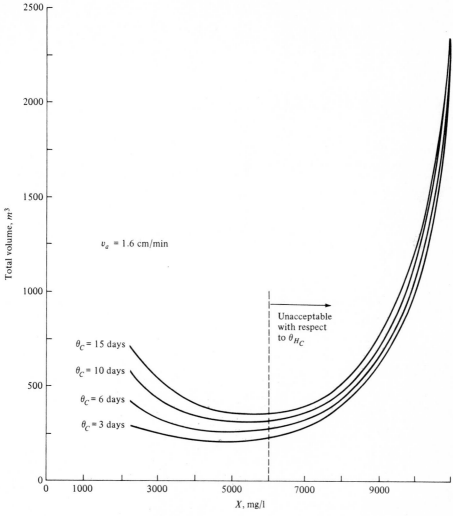

FIGURE 8-17
Total aeration-tank and clarifier volume as a function of X and θ_C.

and the other subscripts are for the ammonia-limiting situation. Using coefficient values from Table 8-2,

$$\frac{1}{\Theta_{C_{NH3}}} = \frac{0.33(0.5)}{1 + 0.5} = 0.11 \text{ d}^{-1}$$

Thus, the assumption that ammonia conversion is essentially complete at $\Theta_C = 10$ d in the aeration tank appears satisfactory. Using the values given for nitrite

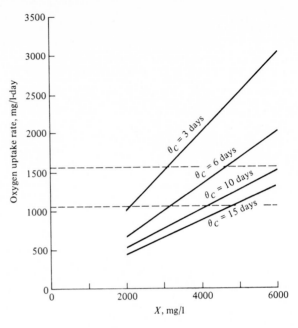

FIGURE 8-18
Oxygen uptake rate as a function of X and θ_C.

in Table 8-2 allows calculation of the nitrite concentration

$$C_{NO_2^--N} = \frac{2}{0.14\Theta_C - 1} = \frac{2}{0.14(9.1) - 1} = 7.3 \text{ mg/l}$$

Thus, at $\Theta_{C_{NH_3}} = 9.1$ d the nitrite-nitrogen concentration is over 7 mg/l and the potential oxygen demand is about 8 mg/l. This value would satisfy the total effluent oxygen-demand requirement for a process operated at 9 d Θ_C or greater. Further reduction in NH_3 and NO_2^- concentrations can be made by extending the MCRT of the activated-sludge process or by adding on a nitrification reactor as shown in Fig. 8-14.

Process calculations are similar to those made for activated sludge except for determining the oxygen uptake rate. An approximate set of stoichiometric expressions are given as follows:

$$NH_3 + \tfrac{3}{2}O_2 \rightarrow NO_2^- + H_2O + H^+$$
$$NO_2^- + \tfrac{1}{2}O_2 \rightarrow NO_3^-$$

Cells are not included in the expressions because of the low cell-yield values. Thus the expressions slightly overestimate the oxygen demand.

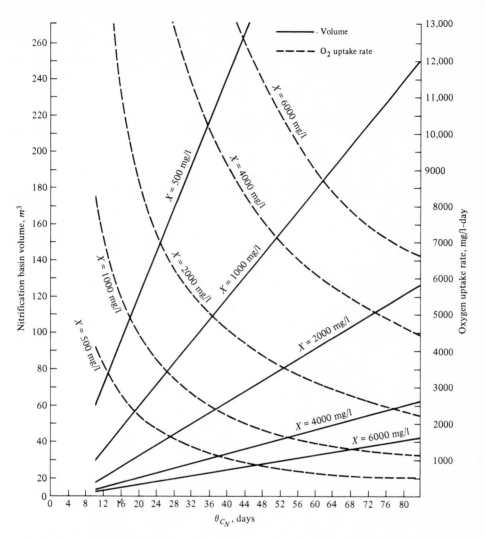

FIGURE 8-19
Nitrification-basin volume and oxygen uptake rate as a function of θ_{C_N} for activated sludge $\theta_C = 3$ d.

For a given Θ_C value, the steady-state ammonia and nitrite-nitrogen concentrations can be calculated, and this allows calculation of the aeration-tank volume, sedimentation-tank volume, recycle ratio, and hydraulic detention times as before. Oxygen uptake rate must be calculated as follows:

$$O_2 \text{ uptake rate} = \frac{32}{14}\frac{Q}{V}\left\{\frac{3}{2}\left(C_{NH_{3i}} - C_{NH_3}\right) + \frac{1}{2}\left[C_{NH_3} - \left(C_{NH_3} + C_{NO_2^-}\right)\right]\right\}$$

In this example problem, the settling rate will be assumed to follow the same relationship as for the activated-sludge mixed liquor. Hence, only MLSS values less than 6000 mg/l and a $v_u = 1.6$ cm/min will be used. In determining rates, the total MLSS concentration must be used because there is no way of differentiating the types of cells.

Nitrification-basin volume and the corresponding oxygen uptake rates are shown in Fig. 8-19 for a range of MLSS concentrations and an activated-sludge process MCRT of 3 d. When longer activated-sludge process MCRTs are used, less nitrogen is assimilated and the nitrification process must do more work. Oxygen transfer rate limitations will constrain MLSS concentrations to approximately 1000 mg/l and force Θ_C values to be maintained above 60 d. The minimum nitrification reactor volume will be 180 m^3, giving a hydraulic residence time Θ_{H_N} of 4.3 h.

The extremely long MCRTs that are necessary to maintain a flocculant sludge in a nitrification process will result in other operating problems. One such problem is a high inert solids fraction with a significant amount of protein and other surface-active material being released into the mixed liquor. Foaming is a common problem under these conditions. One solution to this problem is to choose the single-reactor system. If a 10-d MCRT does not produce a satisfactory effluent with respect to nitrification, a longer MCRT could be used. At $\Theta_C = 10$ d the nitrifier cell concentration predicted by the model would be 109 mg/l at an MLSS concentration of 4000 mg/l. These small additions to the solids concentration would not change the sedimentation-tank design and can be neglected in the overall calculations. Very similar conclusions on the advantages of one-sludge systems have been reached in pilot plant studies at Contra Costa County in California.   ////

8-15 ANAEROBIC BACTERIAL DENITRIFICATION

Many bacteria are able to utilize nitrate as an electron acceptor as well as oxygen. This fact was briefly mentioned in Chap. 6 and will be further developed here. Two types of nitrate reduction occur: assimilatory, in which nitrate is reduced to the ammonia valence and incorporated into organic molecules, and dissimilatory, in which the end product is molecular nitrogen. Because not all the bacterial population present in a denitrifying culture carry out both the reduction of nitrate and the reduction of nitrite, the processes should be considered separately.

$$NO_3^- + organic \rightarrow cells + NO_2^- + CO_2 \qquad (8\text{-}35)$$

$$NO_2^- + organic \rightarrow cells + N_2 + CO_2 \qquad (8\text{-}36(8\text{-}36)$$

In most cases, significant quantities of nitrite are not observed, and the process can be effectively described by an overall expression. The organic material

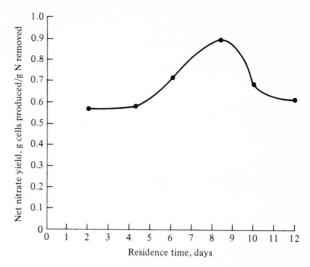

FIGURE 8-20

Nitrate yield as a function of residence time (θ_c) for a system using methanol as a carbon and energy source.[45]

usually must be added to the reaction system because organics are ordinarily removed from wastewaters prior to nitrification.

In most cases, only one form of nitrogen is available, and this has important implications with respect to system stoichiometry. Because energy is expended in the reduction process, cell production from a given quantity of organic material will be less in systems when nitrate is the nitrogen source than in systems where ammonia nitrogen is available. In aerobic systems, a corresponding increase in the oxygen demand occurs, and this relationship has been quantified by Lewis and Busch.[44] Where anaerobic conditions exist (as are necessary for denitrification), the corresponding situation is lower cell yields and a higher fraction of NO_3^- being reduced to N_2.

A major difference between denitrification processes and conventional biological systems is that cell yield does not decrease with Θ_c throughout the operating region. As available nitrogen decreases and the system becomes increasingly nitrogen limited, the bacteria begin to produce considerable quantities of polysaccharides. Because of the polysaccharide, production cell mass yield (dry weight) actually increases with Θ_c up to the point where maintenance-energy requirements force net production to decrease. An example of this phenomenon is shown in Fig. 8-20.

Nearly all the available data on denitrification of wastewater are based on the use of methanol as the energy and carbon source.[25, 26, 45-49] Methanol has been the least expensive organic available and has the further advantage of producing low cell yields. Recent price increases from about $0.085 to $0.85/kg

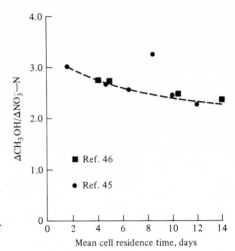

FIGURE 8-21
Methanol consumption as a function of
MCRT in bacterial denitrification.

have made methanol a good deal less desirable. Low-cost methods of providing energy such as use of endogenous respiration and low-nitrogen industrial wastes must be developed. Basic properties of the denitrification process have been worked out, and a number of systems are in operation. Data are not available that would allow development of yield functions, i.e., a general expression related to Fig. 8-20. Thus design must be based on pilot plant studies to a considerable degree. The range of cell yields reported is between 0.4 g cells/g N removed to 0.9 g cells/g N removed, however. Methanol consumption decreases on a unit basis with cell residence time (Fig. 8-21). Values reported in the literature are ranged between 2.2 and 3.2 g CH_3OH/g N removed. These values do not account for methanol necessary to maintain anaerobic conditions. Thus, in systems where oxygen is entering, an additional methanol demand is exerted.

Nitrogen reduction rate has been reported to obey the Monod model in a variety of physical systems.[25, 26, 46, 50] Saturation-coefficient (K_m) values reported are of the order of 0.08 mg/l. Moore and Schroeder[25] and Kaufman[26] reported a maximum removal rate value k_N of 0.31 d^{-1} for a chemostat operating at 5.6 d residence time. In a previous paper,[45] Moore reported removal rate values as high as 0.9 d^{-1}. These higher values were derived from a chemostat at residence times of 2 d. Engberg[46] reported similar values of removal rate as a function of cell residence time in a system with cell recycle, and other workers[51, 52] have reported values in this range also. Maximum removal rate is not ordinarily found to be a function of cell residence time or influent nutrient concentration, although Grady and his coworkers[24] have reported similar results in aerobic systems. One possibility is the depletion of a trace element. Another possible conclusion is that for a given cell residence time the steady-state nitrogen concentration is not a single-valued function. These results illustrate the need

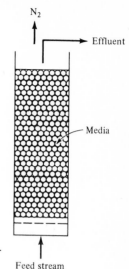

FIGURE 8-22
Schematic diagram of packed-bed denitrification process.

for further research in this area. We should note that working processes have
been and will continue to be designed and constructed on the basis of the
limited knowledge available. A major consideration should be in providing a
flexible process, i.e., one that can be modified with operating experience. In most
cases, the easiest method of providing flexibility is to allow for a variable cell
age. This procedure is normal in activated-sludge process design and will be
quite naturally applied to fluidized-culture denitrification processes.

Cell separation in actively denitrifying cultures is often made difficult by
gas generation in the clarifier. Gas bubbles attach to flocs and bring them to
the surface. In most cases, the introduction of a slight turbulence will correct
the problem by separating gas bubbles from floc particles.

Packed-bed Dentrification

This system falls between suspended-culture and film-flow systems and is discussed
here because an overall view of denitrification is provided. The process was
suggested by McCarty in 1966 (Ref. 48) and has been studied by other
workers.[50, 53, 54] Physically, packed-bed denitrification systems consist of upflow
beds packed with a media of suitable porosity and surface characteristics
(Fig. 8-22). Denitrifying bacteria grow on the media surface and gradually
increase in mass and volume until the pores are nearly filled with cells.
Hydraulic flow rates are low with actual velocities in the range of 0.01 to 0.1
cm/s, which allows cells to accumulate in units until they plug. Thus head loss
becomes a limiting operational variable. Requa[50] reported on kinetic studies
with laboratory-scale packed beds using Raschig rings for a packing medium.

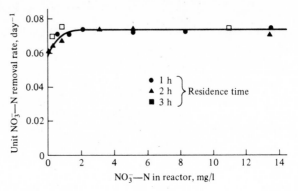

FIGURE 8-23
Unit nitrate removal rate as a function of nitrate-nitrogen concentration in a packed-bed reactor.

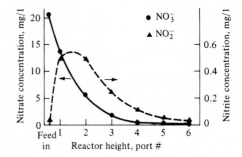

FIGURE 8-24
Nitrate and nitrite concentration in a packed-bed reactor.

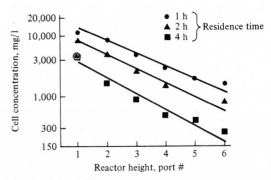

FIGURE 8-25
Cell concentration in a packed-bed reactor.

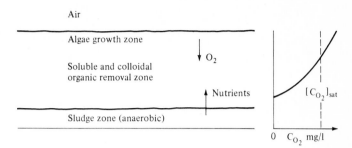

FIGURE 8-26
Schematic diagram of an oxidation pond.

His results are presented in Figs. 8-23 to 8-25. The low maximum removal rate (0.07 d^{-1}) may result from approximations made in his analysis. Removal rate was independent of hydraulic residence time as would be expected. Thus the major influence of applied flow rate is on the time of operation before plugging occurs.

8-16   OXIDATION LAGOONS

Three major biological processes occur simultaneously in oxidation lagoons or ponds. Organic solids accumulating on the bottom are broken down by anaerobic bacterial degradation, soluble and colloidal organics are removed from suspension by aerobic bacterial metabolism, and the nutrients released in these two processes are utilized by algae growing on the surface (Fig. 8-26). The limitation on organic removal rate which can be obtained in such a system is the rate at which the algae can produce oxygen through photosynthesis $(CO_2 + H_2O \xrightarrow{\text{light}} CH_2O + O_2)$. This latter rate is a function of light intensity and temperature. Algae photosynthesis uses light in the visible to near-infrared wavelength range (4–10,000 Å). Light in this range is absorbed by clouds, and therefore photosynthesis is greatest on clear, warm days.

Because process design must consider the limiting conditions, a range of light intensity and organic removal rate combinations must be considered. For example, municipal sewage flows 24 h a day, and therefore oxygen must be available at night when photosynthesis is not occurring. One of two situations will exist: (1) Enough oxygen is stored during the day through supersaturation of the lagoon to supply oxygen through the night, and (2) the lagoon will go anaerobic during the night. Nighttime temperatures are lower, and therefore conversion rates decrease after sundown. Most municipal wastewaters decrease in both flow rate and organic concentration during the night, and this greatly decreases the oxygen demand also. Industrial wastes must be considered on a case-by-case basis. If an industry operates only one shift, there may not be any nighttime flow, for example.

Similar considerations must be made with respect to cold-weather conditions. During the winter, the rate of photosynthesis may decrease to a negligible value due to cold weather. Under those conditions, the organic conversion rate will also become very low and the lagoons will become storage tanks rather than biological reactors. Sudden increases in temperature during the spring may result in organic conversion rates increasing rapidly. Odor problems may occur during these periods just as in any overloaded pond.

When the organic loading rate exceeds the ability of the algal population to provide oxygen, anaerobic conditions develop. Sulfate-reducing organisms are then able to produce sulfides, including H_2S, and odor problems may result. Two groups of photosynthetic bacteria, the green and purple sulfur bacteria, use H_2S as an electron donor in the synthesis of cell material from CO_2, however.

$$CO_2 + H_2S \xrightarrow{\text{infrared light}} CH_2O + H_2O + 2S$$

$$3CO_2 + 2S + 5H_2O \xrightarrow{\text{infrared light}} 3CH_2O + 4H^+ + 2SO_4^{2-}$$

If conditions are such that the photosynthetic sulfur bacteria can thrive, odor problems can be controlled. The process effluent will contain a high BOD, however. Adequate information is not available to enable an engineer to design a stable process using the photosynthetic bacteria. Thus this type of lagoon is not suitable even in cases where there is no effluent.

Process Design

Because of the complexities of the process and the limited operational control possible with lagoons, process design must lean heavily on experience and judgment. For example, the designer should be concerned about the buildup

Table 8-5 LOADING RATES AND BOD REMOVALS OF OXIDATION PUMPS

Type of pond	Loading rate, g/m²-d		Percent BOD$_5$ removal	
	Average	Cold	Average	Cold
Aerobic	9.0–13.5	6.7	80–95	70
Facultative	2.8–5.6	1.7	80–95	70
Facultative-aerated	4.5–11.2	3.4	80–95	70

of sludge deposits and short circuiting. Sludge deposits should be spread over as much of the pond bottom as possible, and this can be accomplished by providing multiple inlets. Short circuiting can be controlled by using baffles or ponds in series. The two problems are coupled to a certain extent because having an inlet near the outlet would certainly increase short circuiting. Thus multiple outlets may also be needed in some cases.

Maximum conversion rates could be predicted using organic concentration profiles developed from tracer studies, but this approach would not provide accurate enough results to predict rates on a 24-h basis. In addition, the variables associated with photosynthesis must be considered, and thus the analytical problem becomes far more complex. A sophisticated mathematical approach to lagoon design seems quite a bit like shooting flies with an elephant gun. Several useful approaches are in the literature, however. Marais[55] and Oswald and his coworkers[56-58] have incorporated factors such as incident solar-radiation cell production and loading rate into the design process. Their work has contributed a great deal to the understanding of these systems.

Otte[59] has summarized typical loading rates and BOD_5 removal percentages reported in the literature. His values are presented in Table 8-5.

The BOD removals reported in Table 8-5 are often filtrate BODs; that is, suspended solids are removed prior to analysis. Because of the amount of algae present in oxidation ponds (often as high as 200 mg/l), considerable amounts of organic material are released from ponds with discharges. Removal of this material is difficult and requires either large amounts of coagulants[60] or the use of a harvester of some type. To date, only one harvester[61] has been developed that seems suitable, but economic limitations may preclude its general application. In summary, there seems to be an excellent case for the use of a standard no-discharge requirement on photosynthetic oxidation ponds. Friedman[62] has found that flocculation is greatly enhanced at pH values above 9.0. He suggests taking advantage of the cyclic pH variation in oxidation ponds by operating on a fill-and-draw basis. During the late afternoon when pH values are between 9 and 11 in most ponds, $Mg(OH)_2$ can be added. The pH of the pond (or a section of the pond) must be raised to 11 in order for a $Mg(OH)_2$ precipitate to form. After sedimentation the product can be drawn off. In most cases, the quantity of $Mg(OH)_2$ necessary will be less than 10 mg/l.

PROBLEMS

8-1 A parameter known as the sludge volume index (SVI) has been used for many years as a measure of sludge settleability. As a rule of thumb, a sludge is considered in good condition if the SVI is below 100. The index is defined as the volume (ml) occupied by 1 g of sludge after settling for 30 min. Determine SVI for the sludges in the following table, and draw a conclusion about how the index should be applied.

Initial volume, ml	Initial concentration, mg/l	Settled volume, ml
1000	100	10
1000	1000	100
1000	10,000	1000
1000	100,000	1000

8-2 A CFSTR activated-sludge process is to be designed using the pilot plant data in the following table. Determine the aeration-tank size necessary, the required surface area of the secondary clarifier, air requirement, and sludge wastage rate to meet an effluent ultimate BOD requirement of 20 mg/l. Wastewater flow rate is expected to be 9.5×10^6 l/d. The pilot plant had an integral clarifier, and wasting was from the mixed liquor.

PILOT PLANT DATA

Aeration-tank volume: 500 l
Flow rate: 100 l/h

Phase 1: Waste rate = 150 l/d

Date	COD_i, mg/l	COD_o, mg/l	MLSS, mg/l	X_e, mg/l	pH	DO, mg/l
3/4	1071	19	5560	5	6.9	5.2
3/6	1122	15	5870	9	6.9	5.4
3/8	1151	16	6050	10	6.9	5.4
3/10	1059	18	5620	4	6.9	5.3
3/12	1007	19	5300	5	6.9	5.1
3/14	1123	21	5882	12	6.9	5.3
3/16	1153	16	6064	10	7.5	5.3
3/28	1061	14	5584	3	6.9	5.1
3/30	1203	10	6342	5	6.9	4.7
3/22	1147	14	6040	8	6.6	5.3
3/24	1128	13	5941	6	6.8	5.6
3/26	1015	11	5351	4	6.8	5.5
3/28	1031	15	5411	5	7.0	5.1
3/30	1028	13	5400	4	7.1	5.3
4/1	1019	16	5348	7	6.9	5.6
4/3	1136	17	5967	9	6.9	5.1

Maximum settling-velocity data, 4/3

MLSS, mg/l	1000	1500	3000	5967	7500	9000	12,000
V_H, cm/min	0.60	0.75	1.25	1.27	0.85	0.35	0.22

Phase 2: Waste rate = 100 l/d

Date	COD_i, mg/l	COD_o, mg/l	MLSS, mg/l	X_e, mg/l	pH	DO, mg/l
4/18	1090	16	7731	7	7.0	4.7
4/20	1075	14	7641	8	7.1	4.8
4/22	1062	17	7530	9	7.0	4.8
4/24	1051	12	7490	6	7.0	4.7
4/26	1033	10	7366	5	7.1	4.6
4/28	1103	9	7871	5	7.2	4.7
4/30	1006	9	7185	4	7.0	4.9
5/2	1121	8	8021	5	7.0	4.7
5/4	1105	7	7911	3	7.2	4.8
5/6	1119	8	8003	4	7.4	4.7
5/8	1096	6	7840	4	7.1	4.6
5/10	1090	9	7765	5	7.0	4.5
5/12	1052	8	7511	5	7.0	4.3
5/14	1071	12	7625	7	7.1	4.5
5/16	1067	11	7601	6	6.9	4.7
5/18	1082	9	7714	5	7.0	4.7
5/20	1009	8	7218	4	7.0	4.6

Maximum settling-velocity data, 5/20

MLSS, mg/l	1200	2400	4800	7218	10,800	14,400	17,900
v_H, cm/min	0.38	0.68	1.10	1.35	1.20	0.46	0.40

Phase 3: Waste rate = 200 l/d

Date	COD_i, mg/l	COD_o, mg/l	MLSS, mg/l	X_e, mg/l	pH	DO, mg/l
6/4	1062	17	4317	9	6.8	5.5
6/8	1047	18	4261	12	6.4	5.6
6/10	1028	21	4148	14	6.6	5.9
6/12	1064	16	4340	10	6.6	5.7
6/14	1028	19	4154	13	6.7	5.8
6/16	1018	17	4127	11	6.6	5.6
6/18	1014	23	4100	16	6.8	5.4
6/20	1009	14	4095	18	6.5	5.3
6/22	1063	18	4316	9	6.6	5.2
6/24	928	19	3758	12	6.3	5.1
6/26	950	18	3850	11	6.3	5.6
6/28	979	22	3981	9	6.7	6.0
6/30	980	20	3975	6	6.5	5.4
7/2	921	17	3756	10	6.5	5.3
7/4	954	21	3889	8	6.6	5.5
7/6	996	18	4027	11	6.5	5.6
7/8	987	16	4009	8	6.6	5.5
7/10	928	20	3757	7	6.6	5.9
7/12	954	19	3861	11	6.5	5.5

Maximum settling-velocity data, 7/12

MLSS, mg/l	1000	1700	2570	3861	5800	7720
V_H, cm/min	0.55	0.73	0.95	1.05	0.86	0.40

8-3 A wastewater has the following characteristics:

$$BOD_L = 550 \text{ mg/l} \qquad NO_3{}^- \text{-}N = 0 \text{ mg/l}$$

$$N_{total} = 29 \text{ mg/l} \qquad P_{total} = 7 \text{ mg/l}$$

$$N_{organic} = 12 \text{ mg/l} \qquad Fe = 0.4 \text{ mg/l}$$

$$NH_3\text{-}N = 16 \text{ mg/l} \qquad SO_4{}^{2-} = 0.5 \text{ mg/l}$$

$$NO_2{}^- \text{-}N = 1 \text{ mg/l} \qquad k_0 = 2 \text{ day}^{-1}$$

$$K_m = 20 \text{ mg/l}$$

A typical bacterial composition is given as follows:

	% dry weight
Carbon	50
N	12–15
P	3
S	1
Fe	0.4

Using this data and the yield expression

$$Y = \frac{0.4}{1 + 0.06\Theta_c} \quad \frac{g \text{ cells}}{g \text{ BOD}_L \text{ removed}}$$

determine for a CFSTR

(a) The Θ_C value corresponding to stoichiometric excess of all materials other than BOD_L

(b) The nutrient balance necessary to satisfy effluent requirements of $BOD_L = 20$ mg/l, $N_{total} = 2.0$ mg/l, $P_{total} = 0.5$ mg/l at $\Theta_C = 3$ d.

8-4 An industrial wastewater has the following characteristics:

Design Q	10⁸ l/d
COD	1420 mg/l
Plateau BOD	720 mg/l
Ultimate BOD	1200 mg/l
Total N	60 mg/l
Total P	8 mg/l
pH	7.2
T	20°C

A continuous-flow stirred-tank aerobic biological process without recycle has been used to determine the necessary kinetic coefficients. Data from these studies are as follows:

Volume of reactor = 10 l

Q, l/d	Ultimate BOD, mg/l	Cell concentration, mg/l
3.33	4.0	420
1.67	1.8	370
1.11	1.2	336
0.56	0.6	252

The available aeration device has a reported $K_L a$ value of 0.12 min^{-1}.

(a) Assuming that the design of the secondary clarifier will not be a problem, determine the acceptable volume and MLSS concentration region as a function of Θ_C and aeration capacity. Assume $C_{sat} - C = 6$ mg/l.

(b) What additional information would you need to determine if a contact stabilization process should be used to treat this wastewater? Explain how the information is used.

(c) As the design Θ_C value becomes greater, several process variables change. Clearly there are cost implications in these variations. List the variables that change with Θ_C in two columns—those that result in increasing cost and those that result in decreasing cost. Justify your choices.

8-5 A recent (1975) proposal for an integrated biooxidation nitrogen-removal process is shown in Fig. 8-27. For a wastewater contining organic N: (1) Anaerobic

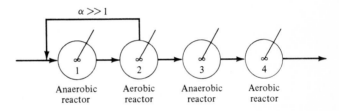

FIGURE 8-27
Nitrogen-removal scheme proposed by Barnard.[63]

reactor 1 partially oxidizes organic material and reduces NO_3 recycled from reactor 2; (2) aerobic reactor 2 is operated at high Θ_C (extended aeration) to ensure nitrification; (3) reactor 3 is anaerobic and NO_3 in effluent is denitrified by endogenous respiration; (4) aerobic reactor 4 provides final removal of residual organics and provides aerobic effluent.

(a) Will the process work in practice? Explain!

(b) Where should settling tanks be inserted into the schematic?

(c) If your answer to (a) is yes, estimate qualitatively the economic and operating advantages and problems that will result.

8-6 Three schematics for phosphorus removal which have been proposed in the literature are shown in Fig. 8-28. Explain the advantages of each and recommend a best choice. (Note phosphorus is removed by adding lime and precipitating as calcium phosphate or by adding alum and precipitating as aluminum phosphate.)

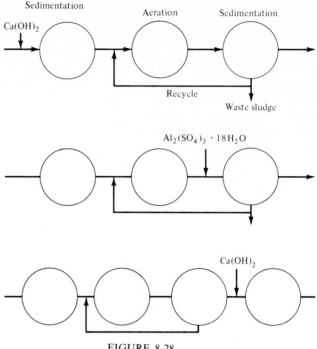

FIGURE 8-28
Proposed schematics for phosphorus removal.

REFERENCES

1 MURPHY, K. L., and P. L. TIMPANY: Design and Analysis for an Aeration Tank, *J. Sanit. Eng. Div., ASCE*, vol. 93, SA5, p. 1, 1967.

2 BUSCH, A. W.: Theory and Design of Bench Scale Units for Bio-Oxidation Studies, *Water Sewage Works*, vol. 106, p. 254, June 1959.

3 MCKINNEY, R. E., et al.: Design and Operation of a Complete Mixing Activated Sludge System, *J. Sanit. Eng. Div., ASCE*, vol. 30, SA3, p. 287, 1958.

4 BUSCH, A. W.: "Aerobic Biological Treatment of Wastewater," Olygodynamics Press, Houston, Tex., 1968.

5 MCKINNEY, R. E.: Closure—Mathematics of Complete Mixing Activated Sludge, *J. Sanit. Eng. Div., ASCE*, vol. 89, SA3, p. 71, 1963.

6 TAPALSHAY, J. A.: Total Oxidation Treatment of Organic Wastes, *Sewage Ind. Wastes*, vol. 30, p. 652, 1958.

7 BUSCH, A. W., and H. N. MYRICK: Food-Population Equilibria in Bench Scale Bio-Oxidation Units, *Water Pollut. Control Fed.*, vol. 32, p. 949, 1960.

8 MCKINNEY, R. E.: Mathematics of Complete Mixing Activated Sludge, *J. Sanit. Eng. Div., ASCE*, vol. 88, SA3, pp. 87, 81, 1962.

9 JENKINS, D., and W. E. GARRISON: Control of Activated Sludge by Mean Cell Residence Time, *Water Pollut. Control Fed.*, vol. 40, 1968.

10 METCALF, L., and H. P. EDDY: "American Sewerage Practice," vol. III, 3d ed., McGraw-Hill Book Company, New York, 1935.

11 PHELPS, E. B.: "Stream Sanitation," John Wiley & Sons, Inc., New York, 1944.

12 MONOD, J.: "Recherches sur la Croissance des Cultures Bacteriennes," Hermann & Cie, Paris, 1942.

13 SELNA, M. W.: Response of Activated Sludge Process to Quantitative Transient Loadings, M.S. thesis, University of California, Davis, 1973.

14 ATKINSON, B.: "Biochemical Reactors," Pion Press, London, 1974.

15 WILKE, C. R., and P. CHANG: *AIChE J.* vol. I, p. 264, 1955.

16 SHERWOOD, T. K., and R. C. RIED: "The Properties of Gases and Liquids," McGraw-Hill Book Company, New York, 1958.

17 PIRT, S. J.: The Maintenance Energy of Bacteria in Growing Cultures, *Proc. R. Soc. London, Ser. B*, vol. 163, p. 224, 1965.

18 VAN UDEN, N.: Yield and Maintenance Analysis in the Chemostat, *Arch. Mikrobiol.*, vol. 62, p. 34, 1968.

19 AGARDY, F. J., R. D. COLE, and E. A. PEARSON: Kinetic and Activity Parameters of Anaerobic Fermentation Systems, SERL Rep. 63-2, Sanit. Eng. Res. Lab., University of California, Berkeley, 1963.

20 LAWRENCE, A. W., and P. L. MCCARTY: A Unified Basis for Biological Treatment Design and Operation, *J. Sanit. Eng. Div., ASCE*, vol. 96, SA3, 1970.

21 GUJER, W., and D. JENKINS: The Contact Stabilization Activated Sludge Process— Oxygen Utilization, Sludge Production and Efficiency, *Water Res.*, vol. 9, p. 553, 1975.

22 GRADY, C. P. L., and R. E. ROPER: A model for the Bio-Oxidation Process Which Incorporates the Viability Concept, *Water Res.*, vol. 8, p. 471, 1974.

23 LEE, S., A. P. JACKMAN, and E. D. SCHROEDER: A Two State Microbial Growth Kinetics Model, *Water Res.*, vol. 9, 1975.

24 GRADY, C. P. L., JR., and D. R. WILLIAMS: Effects of Substrate Concentration on the Kinetics of Natural Microbial Populations in Continuous Culture, *Water Res.*, vol. 9, p. 171, 1975.

25 MOORE, S. F., and E. D. SCHROEDER: The Effect of Nitrate Feed Rate on Denitrification, *Water Res.*, vol. 5, p. 445, 1971.

26 KAUFMAN, G.: Kinetics of Methanol Limited Denitrification, M.S. thesis, University of California, Davis, 1974.

27 WEDDLE, C. L., and D. J. JENKINS: The Viability and Activity of Activated Sludge, *Water Resour.*, vol. 5, p. 621, 1971.

28 MCKINNEY, R. E.: Discussion of Kinetics of Aerobic Removal of Organic Matter, *J. Sanit. Eng. Div., ASCE*, vol. 90, SA5, p. 59, 1964.

29 BISOGNI, J. J., and A. W. LAWRENCE: Relationships between Biological Solids Retention Time and Settling Characteristics of Activated Sludge, *Water Res.*, vol. 5, p. 753, 1971.

30 METCALF AND EDDY, INC.: "Wastewater Engineering," McGraw-Hill Book Company, New York, 1972.

31 MILBURY, W. F., W. O. PIPES, and R. B. GRIEVES: Closure to Compartmentalization of Aeration Tanks, *J. Sanit. Eng. Div., ASCE*, vol. 92, SA5, p. 103, 1966.

32 KOUNTZ, R. R., and C. FORNEY: Metabolic Energy Balances in a Total Oxidation Activated Sludge System, *Sewage Ind. Wastes*, vol. 31, p. 819, 1959.

33 GAUDY, A. F.: Innovations in Secondary Treatment, *Ind. Water Eng.*, vol. 9, pp. 6, 24, 1972.

34 OBAYASHI, A. W., and A. F. GAUDY: Aerobic Digestion of Extracellular Microbial Poly-saccharides, *Water Pollut. Control Fed.*, vol. 45, p. 1584, 1973.

35 GOULD, R. H.: Economical Practices in the Activated Sludge and Sludge Digestion Processes, *Sewage Ind. Wastes*, vol. 31, p. 815, 1959.

36 ULRICH, A. A., and M. W. SMITH: The Biosorption Process of Sewage and Waste Treatment, *Sewage Ind. Wastes*, vol. 23, p. 1248, 1951.

37 ECKENFELDER, W. W., and E. R. GRICH: High Rate Activated Sludge Treatment of Cannery Wastes, *Proc. 10th Ind. Waste Conf.*, p. 549, 1955.

38 OKUN, D. A.: System of Bio-Precipitation of Organic Matter from Sewage, *Sewage Works J.*, vol. 21, p. 763, 1949.

39 FEDERAL WATER QUALITY ADMINISTRATION: Investigation of the Use of High Purity Oxygen Aeration in the Conventional Activated Sludge Process, *Rept.* 17050 DNW 05/70, Washington, D.C., 1970.

40 HUMENICK, M. J., and J. E. BALL: Kinetics of Activated Sludge Oxygenation, *Water Pollut. Control Fed.*, vol. 46, p. 735, 1974.

41 SHERRARD, J. H., and E. D. SCHROEDER: Cell Yield and Growth Rate in the Activated Sludge Process, *Water Pollut. Control Fed.*, vol. 45, p. 1889, 1973.

42 LAWTON, G. W., AND J. D. NORMAN: Sludge Digestion Studies, *Water Pollut. Control Fed.*, vol. 36, p. 495, 1964.

43 TCHOBANOGLOUS, G.: Personal communication, Davis, Calif., March 1974.

44 LEWIS, J. W., and A. W. BUSCH: BOD Progression in Soluble Substrates—The Quantitative Error due to Nitrogen as a Nitrogen Source, *Proc. 19th Ind. Waste Conf.*, 1964.

45 MOORE, S. F., and E. D. SCHROEDER: An Investigation of the Effect of Residence Time on Denitrification, *Water Res.*, vol. 4, p. 685, 1970.

46 ENGBERG, D. J.: Kinetics and Stoichiometry of Bacterial Denitrification as a Function of Cell Residence Time, *Water Res.*, vol. 9, 1975.

47 MULBARGER, M. C.: Nitrification and Denitrification in Activated Sludge Systems, *Water Pollut. Control Fed.*, vol. 43, p. 2059, 1971.

48 MCCARTY, D. L.: Feasibility of the Denitrification Process for Removal of Nitrate Nitrogen from Agricultural Drainage Waters, Report to the Department of Water Resources, State of California, Fresno, Calif., June 30, 1966.

49 TAMBLYN, T. A., et al.: Bacterial Denitrification of Agricultural Tile Drainage, presented at American Geophysical Union Meeting, San Francisco, Calif., Dec. 16, 1969.

50 REQUA, D. A., and E. D. SCHROEDER: Kinetics of Packed Bed Denitrification, *Water Pollut. Control Fed.*, vol. 45, p. 1696, 1973.

51 STANSELL, H. D., R. K. LOEHR, and A. W. LAWRENCE: Biological Kinetics of Suspended Growth Denitrification, *Water Pollut. Control Fed.*, vol. 45, p. 249, 1973.

52 JOHNSON, W. K., and G. J. SCHROEPFER: Nitrogen Removal by Nitrification and Denitrification, *Water Pollut. Control Fed.*, vol. 36, p. 1016, 1964.

53 ENGLISH, J. N., et al.: Denitrification in Granular Carbon and Sand Columns, *Water Pollut. Control Fed.*, vol. 46, p. 28, 1974.

54 SUTTON, P. M., K. L. MURPHY, and R. N. DAWSON: Low-temperature Biological Denitrification of Wastewater, *Water Pollut. Control. Fed.*, vol. 47, p. 122, 1975.

55 MARAIS, G. R.: Dynamic Behavior of Oxidation Ponds, *2d Int. Symp. Waste Treatment Lagoons*, R. E. McKinney (ed.), University of Kansas, Lawrence, 1970.

56 OSWALD, W. J.: Fundamental Factors in Stabilization Pond Design, in W. W. Eckenfelder and B. J. McCabe (eds.), "Advances in Biological Waste Treatment," Pergamon Press, New York, 1963.

57 OSWALD, W. J., and H. B. GOTAAS: Photosynthesis in Waste Treatment, *Trans. Am. Soc. Civ. Eng.*, Paper No. 2849, vol. 122, 1957.

58 OSWALD, W. J., et al.: Designing Waste Ponds to Meet Water Quality Criteria, *2d Int. Symp. Waste Treatment Lagoons*, R. E. McKinney (ed.), University of Kansas, Lawrence, 1970.

59 OTTE, G. B.: Design of Waste Stabilization Ponds, M.S. thesis, University of California, Davis, 1974

60 CALIFORNIA DEPARTMENT OF WATER RESOURCES: Removal of Nitrate by an Algal System, Rep. No. 1303ELY, 4/71-7, April 1971.

61 DODD, J. C.: Harvesting Algae with a Paper Precoated Belt-type Filter with Integral Dewatering and Drying, Ph.D. dissertation, University of California, Davis, 1972.

62 FRIEDMAN, A. A.: Seminar on Algae Removal from Oxidation Ponds, University of California, Davis, June 28, 1975.

63 BARNARD, J. L.: Biological Nutrient Removal without Addition of Chemicals, *Water Res.*, no. 9, p. 485, 1975.

9

BIOLOGICAL FILM-FLOW PROCESSES

Trickling filters were briefly described in Chap. 1, and their oxygen-transfer characteristics were discussed in Chap. 4. These discussions will be extended in this chapter to a more complete physical description, theoretical model development, and design criteria presentation. Rotating-biological-disk (RBC) processes, and relatively new film-flow systems will also be discussed.

9-1 TRICKLING FILTERS

A number of physical configurations are used in trickling filters, some of which would be expected to significantly affect operational characteristics. Standard-rate filters are subjected to hydraulic loading rates of 900 to 9000 l/m²-d (1 to 10 mgad) and organic loading rates of 110 to 370 g BOD_5/m³-d (300 to 1000 lb BOD_5/acre-ft-d).[1] In addition, standard-rate filters are usually loaded intermittently and normally do not incorporate effluent recycle. High-rate filters employ recycle in all cases and have hydraulic loading rates in the range of 9000 to 27,000 l/m²-d (10 to 30 mgad), including recycle for rock-media filters, and up to 40,000 l/m²-d (45 mgad) for plastic-media filters. High-rate trickling-

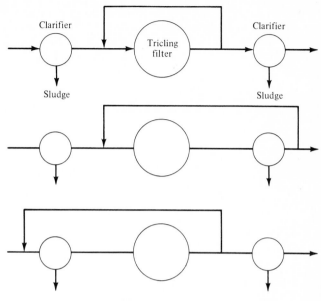

FIGURE 9-1
Common trickling-filter-recycle configurations.

filter organic loadings are of the order of 370 to 1850 g BOD_5/m^3-d (1000 to 5000 lb BOD_5/acre-ft-d). The cited benefits of recycle are flow equalization, improved distribution of flow over the media, improved filter maintenance, prevention of the development of pychoda flies, and improved organic re-movals.[2,3] Recycle ratios used are generally in the range of 1 to 5, and several common configurations are shown in Fig. 9-1. It should be noted that two of the configurations shown in Fig. 9-1 affect the sedimentation tanks.

A Microscopic View of the Trickling Filter

The trickling filter is an inherently nonsteady-state system. Cell growth occurs on the solid media on a somewhat continuous basis, and therefore, the depth and possibly the configuration of the biological slime is constantly changing. As slime depth increases, the rate of organic uptake would be expected to increase also, at least up to some limiting depth. Atkinson[4] and Kornegay and Andrews[5] have demonstrated that this expectation is real and that maximum removal rates occur in the slime-thickness range of 70 to 100 μm. If the slime close to the media is subjected to anaerobic conditions, gas (H_2S and possibly CH_4) and organic acids would be expected to be produced and could cause pieces of slime to slough off by pushing the slime away from the media or by decreasing the local pH and killing the lower layer of cells. A more probable situation is starvation of the lower levels of cells as the slime becomes thicker, which probably results in sloughing as hydraulic shear on the surface increases. The effect of shear on the slime becomes greater as thickness increases, and thus a maximum filter thickness will be associated with a given flow rate. These sloughing mechanisms would be expected to be coupled together to some degree in actual systems and result in a "patchy" type of microbial surface and continuous sloughing of cells in high-rate filters.

If we consider an idealized section of filter (Fig. 9-2) and assume that over a reasonable period of time approximate steady-state conditions exist, a useful if not predictive model can be developed. A mass balance on a soluble component on a cross section of the liquid film at a point z gives

$$N_y \, dA - (dV)R_l = N_{y+dy} \, dA \qquad (9\text{-}1)$$

where $N_y = y$ direction mass flux
   $R_l = $ liquid-phase reaction rate

The liquid-phase reaction rate would be nonzero in most cases due to the presence of sloughed and recycled cells, but the relative importance of the term has not been established.[6] Because reactions also occur throughout the slime layer, a general assumption has been made that liquid-phase transport is not limiting, however.

Reactant material reaching the liquid-slime interface diffuses into the slime, and the result is a coupled diffusion-reaction problem similar to the one for flocs presented in Chap. 8. Atkinson and his coworkers[4,7-9] have developed an effectiveness-factor model similar to one presented earlier:

$$N_y \, dA - (dV)R_s = N_{y+dy} \, dA$$
$$\mathscr{D}_s \frac{d^2C}{dy^2} - R_s = 0 \qquad (9\text{-}2)$$

where R_s is the reaction rate in the slime layers and would be expected to be of the same form as other biological rate expressions.

$$R_s = -\frac{k_0 \, CX}{K_m + C}$$

In the slime, the cell concentration should be constant, allowing the rate to be expressed as a function of substrate concentration alone.

$$R_s = -\frac{k_0^* \, C}{K_m + C} \qquad (9\text{-}3)$$

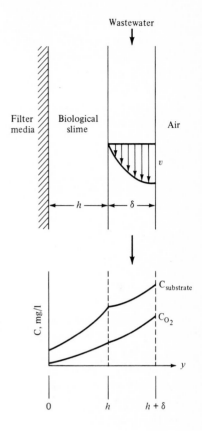

FIGURE 9-2
Idealized microscopic view of a trickling
filter.

where k_0^* has units of mg/s-cm^3. Substituting (9-3) into Eq. (9-2) and rewriting
in terms of the dimensionless terms,

$$C' = \frac{C}{C_b} \qquad Y = \frac{y}{h} \qquad B = h\left(\frac{k_0^*}{\mathscr{D}_s K_m}\right)^{1/2} \qquad \text{and} \qquad \beta = \frac{C_b}{K_m}$$

$$\frac{d^2 C'}{dY^2} = \frac{B^2 C'}{1 + \beta C'} \tag{9-4}$$

Appropriate boundary conditions are

$$Y = \begin{cases} 1 & C' = 1 \\ 0 & \dfrac{dC'}{dY'} = 0 \end{cases}$$

Organic substrate-concentration profiles are not of particular interest, although oxygen-concentration profiles would be. Profiles are of considerable difficulty to develop, and an effectiveness-factor approach is both simpler and more flexible in application. Thus we can write

$$N_y = -\mathscr{D}_s \left.\frac{dC_b}{dy}\right|_{y=h} \tag{9-5}$$

$$N_y = E \frac{hk_0^* C_b}{K_m + C_b} \tag{9-6}$$

Atkinson, Davies, and How[9] have developed expressions for the effectiveness factor.

$$E = \begin{cases} 1 - \dfrac{\tanh B}{B}\left(\dfrac{\phi}{\tanh \phi} - 1\right) & \text{for } \phi \leq 1 \tag{9-7}\\[2em] \dfrac{1}{\phi} - \dfrac{\tanh B}{B}\left(\dfrac{1}{\tanh \phi} - 1\right) & \text{for } \phi \geq 1 \tag{9-8} \end{cases}$$

$$\phi = \frac{B(C_b/K_m)}{\sqrt{2}(1 + C_b/K_m)}\left[\frac{C_b}{K_m} - \ln\left(1 + \frac{C_b}{K_m}\right)\right]^{-1/2} \tag{9-9}$$

We should note that the term B is nearly identical to the term $k_2 V_p/A_p$ in Eqs. (8-19) to (8-21), and that in both cases, there is virtually no diffusion rate limitation for values of the parameter B, or $k_2 V_p/A_p$, below 0.5. We could develop a set of curves similar to Fig. 8-6 for film-flow reactors, but because the results would be qualitatively identical, there is no need to do so. Instead, Eqs. (9-7) to (9-9) should be used to develop a conceptual view of the trickling-filter reaction system.

In systems treating wastewaters having a high fraction of soluble organic materials, we would expect that the C_b/K_m ratio was initially high (i.e., at the point of application). In most cases, we can extend this reasoning to eliminate consideration of diffusion limitations in the upper areas of trickling filters. As the wastewater flows downward, organic concentrations are reduced, and the probability of diffusion rate becoming the controlling variable increases significantly. At slime depths of 0.07 cm, which Kornegay and Andrews[5] suggested were near optimal, and K_m values between 0.01 and 10 mg/l, we find that B varies between 1 and 12. Organic materials in wastewaters are far more likely to have K_m values near 10 than 0.01, and thus we conclude that as concentrations are reduced (i.e., as flow downward in a trickling filter progresses) diffusion control within the slime layer increases.

There is no way in which the complete Atkinson model can be directly applied to trickling-filter design because flow patterns and filter thicknesses change both with time and position. Certain conclusions can be made from their analysis, however. Because liquid-phase organic concentrations decrease to the point where diffusion control of removal rates is quite probable, organic removals in trickling filters may be limited in practical terms. This may be why trickling filters are seemingly insensitive to moderate changes in design (e.g., depth) or operation (e.g., hydraulic loading rate).

A second conclusion can be derived from consideration of the effect of recycle on the system. If the liquid-phase reaction (i.e., the reaction due to suspended cells) is not significant, recycle will decrease the slime-liquid interface organic concentration C_b and thus decrease the transport rate into the reaction region. Removal rates should decrease with increasing values of the recycle ratio under these circumstances. Reports in the literature vary considerably with respect to the effect of recycle,[2,3,10] probably due to other effects. Effluent recycle improves distribution of the wastewater over the filter surface, and this should improve performance up to some critical value. Additionally, taking the recycle stream off ahead of the secondary clarifier, as is often the practice, tends to recycle cells, and this would increase the liquid-phase reaction rate and possibly negate dilution effects.

Diffusion control may be a partial explanation of the decreasing rates encountered as removal of organics progresses. Eckenfelder and O'Conner[11] noted this phenomenon and assumed that the decreasing rates were due to changing substrate characteristics. Differentiation between the two mechanisms would not be possible from ordinary input-output data.

9-2 DEVELOPMENT OF A DESIGN EQUATION

Because of the complexity of Eqs. (9-6) to (9-9), there is little chance of using them directly for design purposes. Coefficient evaluation is, alone, an insurmountable problem. These equations can be used to predict behavior of the trickling filter, as was discussed above, or to develop a useful (i.e., simpler) design model, an approach suggested by Atkinson.[4]

High-rate trickling filters, particularly where municipal sewage is the wastewater being treated, operate in a region where the effectiveness factor E is approximately proportional to the interface concentration C_b. If we assume that the slime "phase" diffusion controls and there is no significant concentration gradient in the liquid film, Eq. (9-6) can be rewritten

$$N_y = \frac{fhk_0^* C_b^2}{K_m + C_b} \qquad (9\text{-}10)$$

where f is a proportionality factor with units of $m^{-1}l^{-3}$. For ranges of C_b normally encountered, Eq. (9-10) is approximately linear and has a slope of nearly fhk_0^*.

Making a mass balance on a small section of the liquid film, as shown in

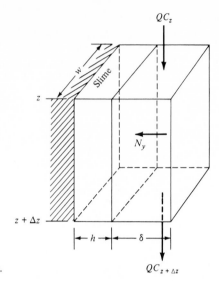

FIGURE 9-3
Idealized section of trickling-filter film.

Fig. 9-3, and assuming steady state exists, results in Eq. (9-11):

$$QC_z - N_y w \, \Delta z = QC_{z+\Delta z}$$

$$-Q \frac{dC}{dz} = N_y w = \frac{fhwk_0^* C_b^2}{K_m + C_b} \qquad (9\text{-}11)$$

where w is the wetted perimeter. Using the linearized form of Eq. (9-10),

$$\frac{dC}{dz} = -\frac{fhwk_0^* C_b}{Q} \qquad (9\text{-}12)$$

or

$$\frac{C_0}{C^*} = \exp\left(-\frac{fhwk_0^* z}{Q}\right) \qquad (9\text{-}13)$$

where C^* is the entering concentration (after the addition of recycle), Q includes recycle, and C_0 is the filter-effluent concentration.

Expressions similar to Eq. (9-13) have commonly been used in trickling-filter design.[12-14] Eckenfelder[14] has suggested using Eq. (9-14):

$$\frac{C_0}{C^*} = \exp\left[-KZa_v^{1+m}\left(\frac{A}{Q}\right)^n\right] \qquad (9\text{-}14)$$

where A is the cross-sectional area of the trickling filter (i.e., Q/A is the hydraulic loading rate) and a_v is the media specific surface area, l^{-1}. The term Q/hw in Eq. (9-13) can be interpreted as a loading rate, and arguments for a direct relationship between Q/hw and Q/A can be made. The coefficient K in Eq. (9-14) is a pseudorate coefficient. In the model presented, the coefficient is a rate coe-

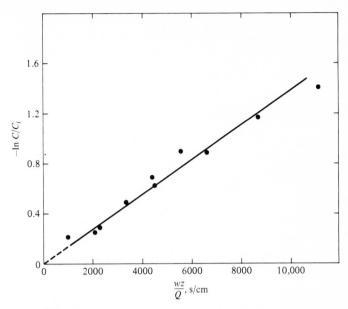

FIGURE 9-4
Organic removal in a high-rate trickling filter with plastic media.

fficient, but in practice, it is a curve-fitting coefficient. In essence, this is also the fate of the term f or possibly fhk_0^* in Eq. (9-13). Using reasonable values of the coefficients as was done in Chap. 8, we can estimate that f should be in the range 0.006 to 0.01 l/mg.

The maximum rate of removal k_0^* can be estimated from the bacterial density of the slime and the maximum removal rate of organics by bacterial suspensions. A reasonable range of values for k_0^* is 0.2 to 0.5 mg/l-s. Values of the parameters h and w are difficult to estimate. For simple geometries, wetted perimeter can be calculated as was done in Chap. 4, but for rock filters, the number becomes an average value. Slime-layer thickness varies considerably in any trickling filter, particularly with depth. Equation (9-13) was developed using a constant (i.e., average) value of slime-layer depth, and is, therefore, somewhat suspect in this regard. Average values should give reasonable results in most cases, however. Values should be of the order of 0.1 to 0.5 cm.

Data presented by Eckenfelder[15] for plastic-media trickling filters are presented in Fig. 9-4. The original data was in terms of loading rate (that is, Q/A), but for the plastic media, the wetted perimeter per unit area is approximately 1 cm/cm^2, and the data can be converted to the form presented here. A value of the function $fhk_0^* = 1.4 \times 10^{-4}$ cm/s can be calculated from the slope of the curve. This value is in the range predicted above.

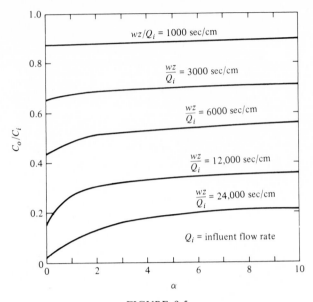

FIGURE 9-5
The effect of recycle on effluent quality.

In practice, the exponential coefficients would be expected to vary with the wastewater characteristics and the pattern of flow over the slime. This is the primary reason why so many correlations have been presented in the literature for trickling-filter performance as a function of various parameters[12-14,16,17] and also the reason for the exponents n and m in Eq. (9-14). There is nothing wrong with using these empirical correlations where they have been justified, but this means that justification must be made for each usage. In addition, extension of the functions beyond the experimental region is extremely dangerous. Finally, we should remember that when double exponential functions are used (for example, $\ln x = y^{-n}$), data scatter is also reduced, often to the point that false functional relationships appear acceptable. In developing design criteria from pilot plant data, this fact presents difficult problems because rarely is there sufficient data for statistical testing.

The effect of recycle on effluent quality should be noted. Assuming that Eq. (9-13) holds over a range of recycle ratios between 1 and 10, and using the value of fhk_0^* from Fig. 9-4, a family of curves of C_o/C_i (not C_o/C^*) for selected values of wz/Q can be developed (Fig. 9-5). In discussing the effects of recycle, the point must be made that both the fraction of the entering organic concentration removed, $1 - C_o/C^*$, (Fig. 9-6) and the overall fraction removed, $1 - C_o/C_i$, decrease with increasing recycle. This is not true with all currently used models (see Prob. 9-2).

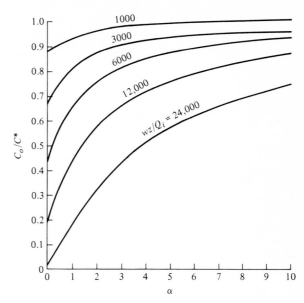

FIGURE 9-6
The effect of recycle ratio α on organic removal in a trickling filter.

An indication of the range of acceptable organic loadings cannot be obtained from Eq. (9-13) or similar models. Two major limitations must be considered. If the incoming concentration is extremely high, the oxygen-transfer capacity of the film may be exceeded, and a portion of the trickling filter may go anaerobic. If this condition persists, sulfide reduction and production of organic acids will result, and decreased treatment and odors will be observed. The second limitation is associated with the development of an extremely thick slime layer. Anaerobic conditions will develop in this case also, particularly at the slime-support interface, but plugging of portions of the bed may result due to sloughing. Eckenfelder[15] noted that the organic load on the trickling filter of Fig. 9-4 was 9.3 BOD_5 g/m^3-d at 25°C. Increasing the loading rate to approximately 30 g BOD_5/m^3-d resulted in greatly increased slime production (1.6 to 8.5 g/m^3-d), effluent-quality deterioration, and anaerobic conditions in the system.

9-3 MAXIMUM INFLUENT CONCENTRATION

An approximation of the maximum allowable organic concentration C^*_{max} and the maximum organic loading rate can be obtained by considering the maximum oxygen flux rate. In Chap. 4, the probable oxygen flux to a liquid film in a trickling filter was estimated to be approximately 3.3×10^{-5} mg/cm^2-s. Using the same coefficient values as before and equating flux to N_y in Eq. (9-10), the maximum BOD_L concentration in the wastewater, C^*_{max}, entering the filter should be of the order of 410 mg/l. This is well above the value which occurs in high-rate filters treating municipal wastes. The value is conservative because flow is not continuous, and oxygen can pass directly into the slime layer once the pulse of flow has passed. Thus, the actual maximum concentration is probably much higher. Using 410 mg/l as the entering organic concentration C^* and calculating organic loading rates in terms of mass/volume-time at conventional loading rates results in an approximate loading rate value of 4.1 kg/m³-d (11,100 lb BOD_5/acre-ft-d). This value is higher than can normally be used because slime production rates become limiting. Thus, in most cases, oxygen transfer is not the limiting variable as long as airflow rates through the filter bed are satisfactory even where incoming organic concentrations are extremely high, as in the case of cannery wastes.

9-4 AIRFLOW RATE

Temperature differential between the ambient air and the air within a trickling filter is the primary driving force for airflow. If the ambient air is hotter than that in the trickling filter, the flow will be downward, and if the ambient air is cooler, airflow will be upward. The natural tendency is for air to be cooled by evaporation in the trickling filter, but because of fluctuations in the ambient air and wastewater temperatures, airflow in both directions can be expected.

Clearly, if air flows in both directions, stagnation periods must also occur. If these are prolonged, serious problems can result. Another consideration must be resistance to airflow and airflow distribution. Conventional trickling-filter design procedures provide for adequate airflow and distribution in most cases. Application of conventional criteria to unconventional systems has led to problems in some cases, however.[18]

The driving force resulting from temperature differences can best be described as a pressure difference between two air columns with different temperatures. This pressure difference is termed *draft* and is normally given in units of length (e.g., centimeters).

$$H = \frac{1}{\rho_w} \int_0^Z d(\rho_{ac} - \rho_{ah})z \qquad (9\text{-}15)$$

where H = draft

Z = filter depth

ρ_{ac}, ρ_{ah} = cold and hot air densities, respectively.

Ambient air density will be constant with depth, but the air density within the filter will vary with position. Remembering that as environmental conditions change, temperature and density profiles also change from the steady-state values determined here, and we can develop a steady-state model using average densities.

Air can be approximated by the perfect gas laws; thus we can write

$$pV = nRT$$

$$\rho_a = \frac{n}{V} (29) = \frac{29p}{RT}$$

where p = air pressure
  V = volume
  n = number of moles
  R = gas constant
  T = absolute temperature

Substituting this latter expression into Eq. (9-15) gives

$$H = \frac{29}{\rho_w R} \int_0^z d\left(\frac{p}{T_c} - \frac{p}{T_h}\right) z \qquad (9\text{-}16)$$

where T_c and T_h are the cold and hot temperatures, K. Using average values for p/T_c, Eq. (9-16) becomes

$$H = \frac{29pz}{\rho_w R} \left(\frac{1}{T_c} - \frac{1}{T_h}\right) \qquad (9\text{-}17a)$$

$$H = 0.353\left(\frac{1}{T_c} - \frac{1}{T_h}\right) z \qquad (9\text{-}17b)$$

Where H is in centimeters of water, z is in centimeters, and temperature is in kelvins. Estimation of the average air temperature in the filter is difficult, even assuming steady-state conditions. Two simple approximations can be made: (1) assuming air and water temperatures are the same and (2) using the log-mean temperature

$$\frac{T_a - T_w}{\ln (T_a/T_w)} = T_m \qquad (9\text{-}18)$$

where T_m = log-mean temperature
  T_a, T_w = air and water temperatures, respectively

The latter method is both more realistic and more conservative and thus should be generally used.

Head Losses in Trickling-Filter Airflow

Friction, form, and expansion-contraction losses are the primary sources of airflow head loss in trickling filters. A rough schematic, such as Fig. 9-7, can be used to set up the problem. Assuming that flow is upward, the head losses are

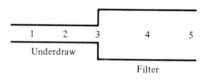

FIGURE 9-7
Airflow schematic diagram for a trickling
filter.

(1) entrance loss, (2) friction in underdrain, (3) expansion loss (contraction when flow is downward), (4) friction loss in filter, and (5) exit loss. Airflow will generally be laminar in a trickling filter, and we can use a classical expression for friction loss:

$$h_L = f' \frac{L}{d_e} \frac{v^2}{2} \qquad (9\text{-}19)$$

where L = length
$\quad d_e$ = equivalent diameter
$\quad v$ = velocity, cm/s

h_L is in centimeters of water, and the friction factor f' is defined by Eq. (9-20):

$$f' = \frac{0.0255}{N_{Re}} \qquad (9\text{-}20)$$

Form losses, such as contractions and expansions, are functions of velocity and the area ratios. For a sudden expansion, the head loss is given by Eq. (9-21):

$$h_L = K_1 \frac{v^2}{2g} \frac{\rho_A}{\rho_w} \qquad (9\text{-}21)$$

where h_L is in centimeters of water, and K is a function of the upstream-to-downstream area ratio A_1/A_2.

Table 9-1 FORM-LOSS COEFFICIENTS†

A_1/A_2	K_1	K_2	K_1	K_2	K_3
0.1	4.8		5.2		
0.2	3.2	6.2	4.1	2.1	12.0
0.3	1.9		3.2		
0.4	0.6	4.8	2.3	1.6	8.0
0.5			1.6		
0.6		3.1	1.0	1.0	4.1
0.7			0.6		
0.8		1.2		0.4	1.3

† h_L in cm water, v in sm/s, ρ in g/cm³.

Contraction losses can be described by expressions similar to (9-21).

$$h_L = \begin{cases} K_2 \dfrac{v_1^2}{2g} \dfrac{\rho_a}{\rho_w} & \text{sudden contractions} \\[3mm] K_3 \dfrac{v_1^2}{2g} \dfrac{\rho_a}{\rho_w} & \text{sharp-edged orifices} \end{cases} \qquad (9\text{-}22)$$

Values of the loss coefficients are given in Table 9-1.
Exit losses are given by the change in velocity head.

$$h_{ex} = \frac{v_1^2}{2g} \frac{\rho_a}{\rho_w} \qquad (9\text{-}23)$$

Entrance losses are a special case of the sudden contraction. Values of the coefficient K_{en} are between 2.2 and 3.2.

Because the airflow velocity is unknown, incorporation of draft and airflow head loss into the design procedure is not straightforward. One method is to obtain data on wastewater and ambient air temperature as a function of time. The worst conditions are chosen for calculation of available draft. This value would be an average over perhaps the 2-h period with the smallest value of $T_a - T_i$ (i.e., the minimum draft). A curve of head loss vs airflow rate is then constructed. The curve is entered on the head-loss coordinate at a value equal to the draft, and the airflow rate is read off the flow coordinate. Airflow rates can be used to estimate oxygen-transfer capacities as shown in the following example.

EXAMPLE 9-1 An industrial wastewater with an ultimate BOD concentration after primary settling of 700 mg/l and a flow rate Q_i of 100 l/s (2.3 mgd) is to be treated in a plastic-media trickling filter. The maximum effluent ultimate BOD concentration is 30 mg/l. Air- and wastewater-temperature differences have been studied, and the worst situation occurs during the later summer period when

Table 9-2 **MODEL TRICKLING-FILTER OPERATING DATA**

$C_i = 300$ mg/l, $A = 21$ cm^2, $w = 20$ cm

z, cm	C_z, mg/l	
	$Q = 0.1$ cm^3/s	$Q = 0.18$ cm^3/s
25	140.0	190.0
50	64.6	125.0
75	33.1	86.7
100	17.3	44.6

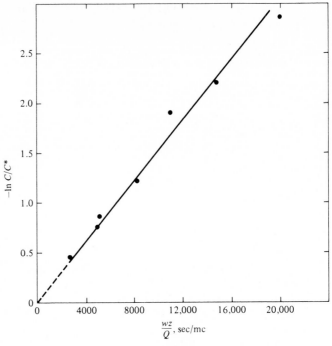

FIGURE 9-8
Organic removal data for Example 9-1.

wastewater temperature is 30°C and the ambient air temperature is 28°C for several hours each day. Wastewater organic removal kinetics have been studied on a 1-m deep, cylindrical laboratory bench-scale basis, and the data are given in Table 9-2.

Process sizing   Calculate and plot (Fig. 9-8) $\ln (C/C^*)$ versus wz/Q.

$-\ln (C/C^*)$	wz/Q, s/cm
0.46	2,800
0.76	5,000
0.88	5,600
1.24	8,300
1.54	10,000
1.91	11,100
1.20	15,000
2.85	20,000

From the slope of the curve,

$$f h k_0^* = 1.5 \times 10^{-4} \text{ cm/s}$$

Using 300 mg/l as a conservative maximum C^* value, the minimum recycle ratio can be calculated:

$$700Q_i = 300(1 + \alpha)Q_i$$

$$\alpha_{min} = 1.33$$

Use

$$\alpha = 2$$

Design

$$C^* = \frac{700}{1 + \alpha} = 233 \text{ mg/l}$$

$$\frac{wz}{Q} = \frac{1}{1.5 \times 10^{-4}} \ln \frac{30}{233} = 13{,}700 \text{ s/cm}$$

$$Q = (1 + \alpha)Q_i = 300 \text{ l/s} = 3 \times 10^5 \text{ cm}^3/\text{s}$$

$$wz = 4.1 \times 10^9 \text{ cm}^2$$

$$= 4.1 \times 10^5 \text{ m}^2$$

A choice must now be made between depth z and area. Because experience has shown that hydraulic loading rates for plastic-media trickling filters can be of the order of 40,000 l/m²-d, an area (gross cross section) of 650 m² will be used. This corresponds to a diameter of 28.8 m (94 ft). For the media used in the experiments, and presumed used in the actual process, the wetted-perimeter-to-area ration is $\frac{20}{21}$ cm/cm². Thus we can calculate w.

$$w = \tfrac{20}{21}(650)(100 \text{ cm/m})^2 = 6.19 \times 10^6 \text{ cm}$$

$$z = \frac{4.10 \times 10^9}{6.19 \times 10^6} = 662 \text{ cm} = 6.6 \text{ m}$$

This is an extremely deep trickling filter, even for one using plastic media. Pumping costs would be high because the flow is relatively large. A head-loss specification seems appropriate (it may not always be so, however), and we will choose 1 m. The tradeoff between area and depth also allows better airflow. Therefore

$$w = 4.1 \times 10^7 \text{ cm}$$

$$A = \frac{21}{20} \frac{4.1 \times 10^7}{10^4} = 4305 \text{ m}^2$$

For one filter the diameter would be 74.0 m (242 ft), which is quite high. Two filters should be specified, each having an area of 2200 m² and a diameter of 53 m. Total area is then 4400 m², and the design hydraulic loading rate is 5890 l/m²-d (6.3 mgad). This value is below the conventional high-rate trickling-filter range of 9000 to 40,000 l/m²-d. The reason is the high influent ultimate BOD (700 mg/l) and the low acceptable head loss.

Determine minimum airflow rate For the purposes of this example, the trickling filter will be assumed to be essentially open, i.e., having no airflow resistance in the underdrain. Thus the entire resistance will be in the pores of the trickling filter.

$$T_m = \frac{30 - 28}{\ln \frac{30}{28}} = 29°C = 302 \text{ K}$$

$$\text{Draft} = 0.353(\tfrac{1}{302} - \tfrac{1}{303})(100)$$

$$= 3.86 \times 10^{-4} \text{ cm water}$$

$$h_L = \text{entrance loss} + \text{friction loss} + \text{exit loss}$$

$$= 3.2 \frac{v^2}{2g} \rho_A + \frac{0.0255}{N_{Re}} \frac{L}{d_e} \frac{v^2}{2} + \frac{v^2}{2g} \rho_A$$

$$\mu_{29°C} = 1.85 \times 10^{-4} \text{ g/cm-s}$$

$$\rho_{A, 29°C} = 0.00117 \text{ g/cm}^3$$

$$h_L = 3.2 \frac{v^2}{2g} 0.00117 + 0.0255 \frac{(1.85 \times 10^{-4})Lv^2}{2d_e^2(0.00117)} + \frac{v^2}{2g} 0.00117$$

$$d_e = 4R_H = 4(\tfrac{21}{20}) = 4.2 \text{ cm}$$

$$L = 100 \text{ cm} \qquad \text{(approximate mixing length)}$$

$$h_L = (2.78 \times 10^{-6})v^2 + 0.0114v$$

$$= \text{draft}$$

$$v^2 + 4111v - 139 = 0$$

$$v = 0.034 \text{ cm/s}$$

$$\text{Airflow rate} = (v)(A)(\text{porosity})$$

$$= (0.034)(4.752 \times 10^7 \text{ cm}^2)(0.94)$$

$$= 1.52 \times 10^6 \text{ cm}^3/\text{s}$$

$$= 1.52 \text{ m}^3/\text{s}$$

Air is 23 percent oxygen by weight. Thus

$$\text{Mass O}_2 \text{ flow rate} = (1.52 \text{ m}^3/\text{s})(10^6 \text{ cm}^3/\text{m}^3)(0.00117)(0.23)$$

$$= 409 \text{ g/s}$$

$$\text{O}_2 \text{ demand} = (0.7 \text{ g/l})(100 \text{ l/s})(1 - 1.42Y)$$

Y is not known and cannot be calculated from the data. For trickling filters, a value of 0.28 or less is reasonable. Thus

$$\text{O}_2 \text{ demand} = 0.7(100)(0.6) = 42 \text{ g/s}$$

Oxygen-transfer efficiency must be greater than 10.3 percent. This is a high efficiency. In effect it means that 10.3 percent of the entering oxygen will be used. The net effect will be that the flux of oxygen into the filter will be decreased because near the end of the filter the O_2 concentration in the air will be reduced by 10.3 percent. Because the wastewater is heating the air, the airflow will be countercurrent (i.e., upward), and the low concentration will be at the entrance. An estimate of the upper-layer O_2 flux and flux needed can be made using the O_2 depletion and the value for maximum O_2 transfer rate determined earlier:

$$N_{O_2,\,max} \approx (0.89)(3.3 \times 10^{-5}\ \text{mg/cm}^2\text{-s}) = 2.9 \times 10^{-5}\ \text{mg/cm}^2\text{-s}$$

And using Eq. (9-10),

$$N_{O_2,\,necess.} = 0.6\ \frac{1.5 \times 10^{-4}(0.233)^2}{0.1 + 0.233}$$

$$= 1.47 \times 10^{-5}\ \text{mg/cm}^2\text{-s}$$

Thus the flux in the upper layer is satisfactory even under the worst airflow rate conditions. If the recycle ratio was below a value of 0.65, the oxygen flux would not satisfy the predicted maximum uptake rate.   ////

9-5 PHYSICAL FACTORS IN TRICKLING-FILTER DESIGN

This discussion of trickling-filter principles has consistently assumed that wastewater and airflow distribution are uniform. Some care must be taken to ensure that this is actually the case. Wastewater distribution can be adequately handled by geometrically spacing the nozzles in a rotary distributor or making sure that head losses to all nozzles are the same where stationary or fixed sprays are used. Controlling airflow distribution presents a problem in underdrain design. Either head loss in the underdrains (the open portion) must be negligible (i.e., for the airflow) or a network of channels must be provided for concentric sections of the filter. This is ordinarily not a significant problem, but in the deep, large-diameter trickling filters now being used for municipal waste treatment, problems can occur.[18] Because of the light weight of the plastic media used in these trickling filters, underdrains with large air spaces above the flow line are easily constructed without structural problems.

9-6 ROTATING BIOLOGICAL CONTACTORS

Rotating biological contactors (RBCs) are continuous-flow units composed of a cylindrical bottomed tank and a number of large lightweight disks rotating through the liquid. A typical unit is shown in Fig. 9-9. Materials used for the disks include expanded polystyrene and corrugated polyethylene. Disk diameters range from 2.5 to 3.5 m (9 to 12 ft), and rotational speeds used are 1 to 2 r/min.

FIGURE 9-9   Styrofoam disks used in RBC units. (Biosystems Division, Autotrol Corporation.)

The disks are submerged to a depth just below the drive shaft. Disk thickness is 1 to 2 cm, and center-to-center spacing is 30 to 40 cm.

Conceptually, RBC units are similar to trickling filters. A microbial slime develops on the disk surface. Organics diffuse into or are absorbed by the slime during the submerged portion of the cycle, and biooxidation reactions are carried out throughout the cycle. The major portion of the oxygen transfer occurs during the period when the slime is in direct contact with air. This may result in higher oxygen transfer rate coefficients than can be obtained in trickling filters. Presently, oxygen transfer rate information for RBC processes is not available, greatly limiting the ability of the designer.

Because of structural limitations and drive control, RBC units are most easily constructed as modular units. In most cases, two to six modules are placed in series, depending on the strength of the wastewater and the degree of treatment required. Sloughing of slime from the disks is a continuous process, and secondary clarification is necessary. In some cases, intermediate clarifiers are placed between modules.

Application of RBC units to industrial wastewaters has been extensive. A wide variety of food-processing wastes including meat, vegetables, and fish are currently being treated by this method. Textile, chemical manufacturing, and pulp and paper wastes with BOD_5 values up to 13,000 mg/l have also been treated with RBC units.[20]

Capital costs of RBC systems reported in the literature[20, 21] have been about 20 percent higher than those for activated-sludge processes. Annual cost differences are considerably less because of the lower operating and maintenance costs of the RBC system relative to activated sludge.

Conclusions on the advantages and disadvantages of the process have varied. Antoine and Hynek[20] concluded that RBC processes were stable, versatile, and competitive with activated sludge. Their studies included a wide variety of industrial and municipal wastewaters. Thomas and Koehrsen,[21] working with distillery wastewaters, concluded that the activated-sludge process was more stable when subjected to shock loads, provided better overall removals, and was less expensive on both a capital outlay and annual cost basis.

Reaction Model for the RBC Process

Removal rate expressions for RBC processes should be similar to those developed for trickling filters. Rotational speed may have a minor influence on the rates because agitation in the tank results in increased oxygen transfer but otherwise will not be a rate-controlling variable. If organic concentrations are assumed to be homogeneous in the cross section of the tank (Fig. 9-10), an expression can be written for the movement organics into the slime that is virtually identical to Eq. (9-10).

$$N_z = \frac{fhk_0^* C_b^{\,2}}{K_m + C_b} \qquad (9.24)$$

where the nomenclature is the same as for the trickling-filter model. The mass rate of removal per disk face is then

$$M_z = \frac{fhk_0^* C_b^2 A_s}{K_m + C_b} \qquad (9-25)$$

where M = mass of BOD_L removed per unit time
A_s = submerged area of one side of disk

Because the disks are closely spaced, an approximation of a continuous or homogeneous removal rate function can be made which is analogous to a plug-flow reactor.

$$V_L \frac{dC_b}{d\Theta_H} = -M_z \qquad (9-26)$$

where V_L is the actual liquid volume (tank volume − disk volume) per disk.

Integrating Eq. (9-26) gives an expression for effluent concentration from an RBC unit.

$$K_m\left(\frac{1}{C_b} - \frac{1}{C_{bi}}\right) + \ln \frac{C_{bi}}{C_b} = \frac{fhk_0^* A_s \Theta_H}{V} \qquad (9-27)$$

where C_{bi} is the influent BOD_L concentration.

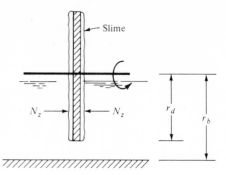

FIGURE 9-10
Flux of organics into slime on RBC disk.

In most situations the effluent BOD concentration will be specified and design variable will be Θ_H. Because RBC units have standard dimensions, the number of disks is directly proportional to total liquid volume and

$$n_d = \frac{Q\Theta_H}{2V_L} \qquad (9\text{-}28)$$

Equation (9-28) provides a theoretical number of disks and is based on the assumption that the liquid-phase reaction is homogeneous. The actual design number will be somewhat greater because disk surfaces occur at discrete intervals. As data are accumulated on RBC performance, an efficiency factor e can be developed that accounts for the difference between the model and the actual RBC system.

$$n_d = e\,\frac{Q\Theta_H}{2V_L} \qquad (9\text{-}29)$$

Oxygen Transfer Rate Limitations

Oxygen transfer to the biological slime will limit the removal rate for RBC systems. A satisfactory approximation is to consider the reaction rate to be constant over the entire surface of each disk face. The oxygen concentration at the disk-slime interface will be 0 for the limiting case. An oxygen mass transfer rate can be defined for this situation.

$$M_{O_2,\,z} = K_L\,a C_{sat}(A - A_s) \qquad (9\text{-}30)$$

where A is the total disk area.

The limiting condition is defined as a situation where $M_{O_2,\,z} = M_z$. Serious operational problems may not occur unless this situation exists for a number of disks.

Design Factors for RBC Systems

Process design of RBC systems is usually based on hydraulic and organic loading rates per unit area. Typical hydraulic loading rates are in the 40 to 60 l/m²-d (1.0 to 1.5 gal/ft²-d) range. Maximum organic loading rates in use are of the order of 210 g BOD_L/m²-d. This value would be used in the first stage of a series of RBC units. Average values over a series of stages are of the order of 55 g BOD_L/m²-d.

Wilkey and Friedman[22] reported removal rates as high as 97 g COD/m²-d in studies of effects of transient loading on RBC units. They also found that effluent quality decreased sharply with increasing magnitude of the transient organic loading. Response of RBC units to transients can be characterized by sharp increases in effluent organic and suspended-solids concentrations. Because the culture is fixed to the disk surfaces, recovery from shock or transient loadings is fast, i.e., of the same order of time as the transient.

PROBLEMS

9-1 The idealized expression for the contact time of fluid running over an inclined plate (of angle Θ with the horizontal) is

$$t_c = D\left(\frac{3\mu}{\rho g \sin \Theta}\right)^{1/3}\left(\frac{b}{Q}\right)^{2/3}$$

where D = length of plate
$\quad b$ = width of flow surface
$\quad Q$ = volumetric flow rate

(a) Derive the above expression.
(b) Justify use of Eckenfelder's[11] expression

$$t_c = K\frac{Da_v^m}{Q^n}$$

where a_v = specific surface, l^{-1}.
(c) Eckenfelder[11] gives values of m and n in the area of 0.7. Does this seem reasonable? Explain!

9-2 Prediction of effluent quality is an important part of design. Currently, several formulas are in use. Among the most commonly used are those listed below.

NRC formula[16]

$$\frac{C_i - C_o}{C_i} = \frac{1}{1 + 0.12[QC_i/(V\varepsilon)]^{1/2}}$$

where C_i = influent BOD_5 [assume 0.67 (ultimate BOD), mg/l]
    C_o = effluent BOD_5, mg/l
    Q = volumetric flow rate, l/s
    V = filter media volume, m^3
    ε = recirculation factor

$$= \frac{1 + \alpha}{(1 + 0.1\alpha)^2}$$

Velz formula[1]

$$\frac{C_o}{C_i^*} = e^{-KD}$$

where D = filter depth, m
    K = coefficient = 1.13 m^{-1}

and C_i^* is the applied concentration of BOD_5, that is,

$$C^* = \frac{C_i + \alpha C_o}{1 + \alpha}$$

Galler and Gotaas[19]

$$C_o = \frac{2.58 C^{*1.19}(1 + \alpha)^{0.28}(Q/A)^{0.13}}{(1 + 3.3D)^{0.67}T^{0.15}}$$

where C_o = effluent BOD_5, mg/l
    C^* = applied BOD_5, mg/l
        = $C_i(1 + \alpha)$
    α = recycle ratio
    Q = influent flow rate, l/s
    D = depth, m
    A = cross-sectional area, m^2
    T = °C

Eckenfelder[15]

$$\frac{C_o}{C^*} = e^{-K\left[\frac{A}{(1 + \alpha)Q}\right]^n Da_v^{1+m}}$$

where C_o = effluent BOD_5, mg/l
    C^* = applied BOD_5, mg/l = $(C_i + \alpha C_o)/(1 + \alpha)$
    A = cross-sectional area, m^2
    a_v = media specific surface area, m^{-1} (~ 1 for polygrid)
    D = depth, m
    Q = flow rate, l/s
    K = pseudo rate coefficient, ≈ 0.24
    m, n = 0.75 for polygrid plastic media

For a wastewater with a flow of 50 l/s, a temperature of 20°C, and an influent ultimate BOD concentration of 450 mg/l, plot the BOD fraction remaining predicted by the four models above and by Eq. (9-13), the Atkinson model:

(a) As a function of recycle ratio, using $A = 220$ m^2 and $D = 2$ m

(b) As a function $(1 + \alpha)Q/A$, using $\alpha = 2$ and $D = 2$ m

(c) As a function of depth, using $\alpha = 2$ and $A = 220$ m^2

9-3 The formulas presented in Prob. 9-2 all contain recycle ratio as an implicit or explicit variable. Determine any predicted optimal values and discuss their significance.

9-4 The coefficients associated with the formulations in Prob. 9-2 were developed for specific wastewaters. Search the literature and determine the range of these values that might be encountered.

9-5 Often removal predictions are based on *removable BOD*.[1] Discuss this term in a conceptual fashion. What does it mean?

9-6 The NRC[16] and Galler and Gotaas formulas[1] are based on curve fits. Quite often formulas based upon regression analysis or other curve- fitting techniques are valid in only limited regions. Determine if these two formulations:

(a) Are continuous, positive, and real over the possible range of the parameter values

(b) Are limited to particular types of wastewaters

(c) Can be supported by data reported in the recent literature

(d) Have reasonable standard deviation values for the original data. Why would this be important even if the formulations have high correlation coefficients?

9-7 For the wastewater in Prob. 9-2, estimate the maximum oxygen transfer rate that will be required and the ambient temperatures which will provide satisfactory air-flows. Assume the filter is plastic and is to be 2 m deep, 220 m^2 in cross-sectional area, and that recycle rates between 1 and 4 are under consideration. Use the Atkinson model, assume $K_m = 20$ mg/l and $Y = 0.3$. Oxygen transfer efficiency expected is 8 percent.

REFERENCES

1 METCALF AND EDDY, INC.: "Wastewater Engineering," Mc-Graw-Hill Book Company, New York, 1972.

2 FAIR, G. M., J. C. GEYER, AND D. A. OKUN: "Water and Wastewater Engineering," vol. 2, John Wiley & Sons, Inc., New York, 1968.

3 BABBIT, HAROLD, AND E. R. BAUMAN: "Sewerage and Sewage Treatment," John Wiley & Sons, Inc., New York, 1958.

4 ATKINSON, B.: "Biochemical Reactor Engineering," Pion Press, London, 1974.

5 KORNEGAY, B. H.: Characteristics and Kinetics of Fixed Film Biological Reactors, Ph.D. thesis, Clemson University, Clemson, S.C., 1969.

6 KEHRBERGER, G. J., AND A. W. BUSCH: Mass Transfer Effects in Maintaining Aerobic Conditions in Film Flow Reactors, *Water Pollut. Control Fed.*, vol. 43, p. 1514, 1971.

7 SWILLEY, E. L., AND B. ATKINSON: A Mathematical Model for the Trickling Filter, *Proc. 18th Ind. Waste Conf.*, 1963.

8 ATKINSON, B., AND S. Y. HOW: The Overall Rate of Substrate Uptake by Microbial Films, Part I, *Trans. Inst. Chem. Eng.*, 1974 (in press).

9 ATKINSON, B., I. J. DAVIES, AND S. Y. HOW: The Overall Rate of Substrate Uptake by Microbial Films, Part II, *Trans. Inst. Chem. Eng.*, 1974 (in press).

10 QUIRK, T. P.: Scale Up and Process Design Technique for Fixed Film Biological Reactors, *Water Res.*, vol. 6, p. 1333, 1972.

11 ECKENFELDER, W. W., AND D. C. O'CONNOR: "Biological Waste Treatment," Pergamon Press, New York, 1961.

12 HOWLAND, W. G.: Flow over Porous Media as in a Trickling Filter, *Proc. 12th Ind. Waste Conf.*, p. 435, 1957.

13 SCHULTZE, K. L.: Load and Efficiency of Trickling Filters, *Water Pollut. Control Fed.*, vol. 32, p. 245, 1960.

14 ECKENFELDER, W. W.: Trickling Filter Design and Performance, *J. Sanit. Eng. Div.*, *ASCE*, vol. 87, 1961.

15 ECKENFELDER, W. W.: Closure of Trickling Filter Design and Performance, *J. Sanit. Eng. Div.*, *ASCE*, vol. 89, SA3, p. 65, 1963.

16 SUBCOMMITTEE ON SEWAGE TREATMENT, COMMITTEE ON SANITRARY ENGINEERING: Sewage Treatment at Military Installations, *Sewage Works J.*, vol. 18, 1946.

17 GREAT LAKES–UPPER MISSISSIPPI RIVER BOARD OF STATE SANITARY ENGINEERS: Recommended Standards for Sewage Works, Health Education Service, Albany, N.Y., 1968.

18 TCHOBANOGLOUS, GEORGE: Sacramento, California, Main Treatment Plant, report to the Attorney General, State of California, Sacramento, Calif., July 1973.

19 GALLER, W. S., AND H. B. GOTAAS: Analysis of Biological Filter Variables, *J. Sanit. Eng. Div.*, *ASCE*, vol. 90, SA6, p. 59, 1964.

20 ANTOINE, R. L., AND R. J. HYNEK: Operating Experience with Bio Surf Process Treatment of Food Processing Wastes, *Proc. 28th Ind. Waste Conf.*, 1973.

21 THOMAS, J. L., AND L. G. KOEHRSEN: Activated Sludge–Bio-Disc Treatment of Distillery Wastewater, EPA-6601 2-74-014, Superintendent of Documents, Washington, D.C., 1974.

22 WILKEY, R. C., AND A. A. FRIEDMANN: Response of Rotating Biological Contractors to Shock Loadings, *Proc. 5th Environ. Eng. Sci. Conf.*, University of Louisville, Louisville, K., 1975.

10

ANAEROBIC PROCESSES

Anaerobic processes were introduced in Chaps. 6 and 8. The discussion in Chap. 6 concerned anaerobic metabolism, while that in Chap. 8 involved anaerobic bacterial denitrification. This latter process is a special case because an exogenous electron acceptor, NO_3, is used, and thus there is a direct analogy with aerobic processes. In this chapter we will consider processes that utilize endogenous (internal) electron acceptors. Of specific interest is the fact that methane (CH_4) is a major end product of the fermentation.

10-1 BIOLOGICAL PROCESS CHARACTERISTICS

As in Chap. 6, when organic materials are anaerobically metabolized by bacteria, the final steps involve reoxidation of the reduced nucleotides $NADH^+$ and $NADPH^+$. Because exogenous electron acceptors are not available, ones such as pyruvic acid must be used. The end products of these oxidation-reduction reactions are the oxidized nucleotides and low-molecular-weight alcohols and organic acids (e.g., acetic, butyric, propionic, formic). Facultative bacteria, those that can function under either aerobic or anaerobic conditions, are the predominant species in both aerobic and anaerobic wastewater treatment processes, and their principal end products at neutral pH values and under anaerobic conditions are the volatile (low-molecular-weight) acids. An example of the distribution of end products is given in Table 10-1.

A group of obligate anaerobic bacteria, the methane bacteria, is able to use the volatile acids and simple alcohols as either energy or carbon sources and to produce methane as an end product. Although the group is widespread in nature, there is little known about the metabolism of the methane bacteria. This lack of knowledge is partly due to the difficulty of culturing the organisms and partly due to their diversity in characteristics. Four genera of methane bacteria have been identified, *Methanobacterium* (nonspore-forming rods), *Methanobacillus* (spore-forming rods), *Methanococcus* (micrococci), and *Methanosarcina* (sarcinae, a cocci group that divides at $90°$ angles and grows in cubes of eight cells each).[2] All the methane bacteria can oxidize molecular hydrogen by using carbon dioxide as an electron acceptor.

$$4H_2 + CO_2 \rightarrow CH_4 + 2H_2O \qquad (10\text{-}1)$$

This reaction provides energy but not carbon for growth. Carbon and carbon dioxide [for Eq. (10-1)] must be obtained from the medium or, in the case of CO_2, be produced from the oxidation of organic materials. Each species seems to be remarkably substrate specific, however. *Methanobacterium formicicum* can utilize only H_2, CO_2, and formic acid (CHOOH), for example.[3,4] Volatile acids that have been shown to be substrates for methane bacteria are formic (COOH), acetic (CH_3COOH), propionic (CH_3CH_2COOH), butyric

Table 10-1   ANAEROBIC METABOLISM END PRODUCTS[1]

Product	m mol/$100m$ mol glucose fermented by *E. coli*	
	pH 6.2	pH 7.8
2,3-Butanediol	0.3	0.26
Acetoin	0.06	0.19
Glycerol	1.42	0.32
Ethanol	49.8	50.5
Formic	2.43	86.0
Acetic	36.5	38.7
Lactic	79.5	70.0
Succinic	10.7	14.8
Carbon dioxide	88.0	1.75
Hydrogen	75.0	0.26
Carbon recovered	91.2%	94.7%

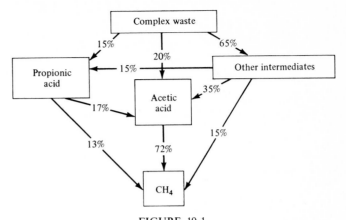

FIGURE 10-1
Sources of methane in anaerobic digestion.[5]

$(CH_3CH_2CH_2COOH)$, valeric $(CH_3CH_2CH_2CH_2COOH)$, isovaleric $[(CH_3)_2$-$CH_2CH_2COOH]$, and caproic $(CH_3CH_2CH_2CH_2CH_2COOH)$. As in the case of other anaerobic bacteria, oxidation of the above compounds is often incomplete. For example, when valeric acid is oxidized by *Methanobacterium suboxydans*, acetic acid and propionic acid are end products. Clearly, the energy production is small, resulting from the oxidation of only one methyl group to an acid, and not surprisingly growth is slow.

Most wastewaters are complex and provide a wide range of organic materials for a bacterial population to oxidize. In the case of methane fermentation, this means that the various species interact. Thus larger volatile acids such as caproic and valeric are split into smaller acids that act as methane precursors. One of the most important reactions in this chain is the fermentation of acetic acid, a reaction which seems to be fairly common among the methane bacteria.[2]

$$
\begin{array}{l}
CH_3 \rightarrow CH_4 \\
| \\
COOH \rightarrow CO_2
\end{array}
\qquad (10\text{-}2)
$$

Jeris and McCarty[5] used radioactive tracer studies to shed some light on the sources of methane in anaerobic sludge digestion. Their conclusions are illustrated in Fig. 10-1. The "other intermediates" box would include valeric acid, which is a source of propionic as well as acetic acid, and formic acid, which is directly involved with CO_2 in the production of methane.

We have not discussed cell production, but clearly, cells are produced. Thus Eqs. (10-1) and (10-2) and Fig. 10-1 are not complete because some of the

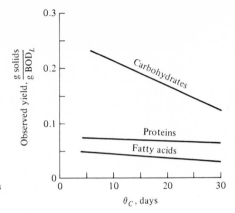

FIGURE 10-2
Yield from methane fermentation as a function of θ_c.

carbon entering the system must end up as cell material. A more correct state-ment of the overall reactions is given by Eqs. (10-3) and (10-4).

$$\text{Organics + nutrients} \xrightarrow{\text{bacteria}} \text{cells + volatile acids + alcohols} + H_2 + CO_2 \tag{10-3}$$

$$\text{Volatile acids + alcohols} + H_2 + CO_2 + \text{nutrients} \xrightarrow[\text{bacteria}]{\text{methane}} \text{cells} + CH_4 + CO_2 \tag{10-4}$$

10-2 CELL PRODUCTION

Because both the acid-forming systems and the methane-forming systems act together in most cases, there is no way in which the cells produced can be separated. Thus, cell yields reported are for both reaction systems combined. Most anaerobic processes used in waste treatment are sludge digesters—processes for breaking down organic solids. Cell residence times are long in these systems, and consequently, yields for anaerobic systems with methane fermentation reported in the literature are very low. A generally reported value is 0.05 g cells/g COD introduced into a digester. Speece and McCarty[6] considered cell yields for various substrates over a range of cell residence times, and their work is summarized in Fig. 10-2.

The high yield values reported by Speece and McCarty[6] for carbohydrates raised considerable controversy at the time but simply reflect normal carbohy-drate metabolism in bacteria. Nearly one-half the energy obtained in the glycolysis–Krebs cycle–electron transport system is from substrate-level phos-phorylation and therefore is available to the bacteria under anaerobic conditions. This situation does not exist in higher organisms where the electron transport system is up to six times more efficient.[7]

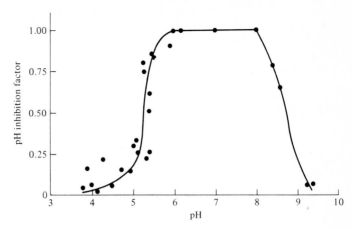

FIGURE 10-3
Inhibition as factors observed at different pH values.[8]

10-3 pH EFFECTS

Because of the organic acid production in the first stage of anaerobic treatment, pH control is a significant problem. If acids are not oxidized as fast as they are produced, their concentration will increase. Eventually the buffering capacity of the wastewater will be exceeded, and the pH of the system will drop. Methane fermentation rates do not vary greatly between pH values of 6.0 to 8.5 but decrease very rapidly outside this range.†

In municipal treatment-plant sludge digesters, the volatile acid concentrations are normally in the range of 200 to 400 mg/l, but if the methane fermentation rate decreases or if for some reason the acid production rate increases sharply, the volatile acid concentration may reach 4000 to 10,000 mg/l. As the concentration increases, the pH begins to drop, usually leveling off in the neighborhood of pH 3. This latter value is controlled by the pH values of volatile acids.

Causes of the acid production–methane fermentation imbalance are varied and not completely understood. Sudden changes in inlet organic concentration, decreases in temperature, or the introduction of toxic materials are commonly reported factors. If the organic concentration suddenly increases, the acid formers are able to respond much faster than the methane formers because of their higher growth rates and the greater amounts of energy derived from the acid-production reactions. High acid concentrations, lower pH values, and decreased

---

† Inhibition of methane fermentation at low and high pH values is shown in Fig. 10-3. The pH inhibition factor is the ratio of the fermentation rate at a given pH to the maximum value for the system.[8]

methane fermentation rates are the result of this situation. When the methane fermentation rate decreases, the relative imbalance increases, and the situation becomes worse. Methane fermentation systems using complex organic media might well be described as metastable systems.

10-4 TEMPERATURE

Decreases in temperature generally affect methane fermentation rates to a greater extent than acid fermentation rates. Thus decreases in temperature can result in instability and eventually process failure. The relationship of temperature and methane fermentation rate is not precisely predictable because available data do not provide enough information. An optimal temperature in the range of 37 to 40°C (98 to 106°F) has been reported by a number of workers,[9-12] and a second optimum with somewhat higher removal rates has been reported in the thermophilic range (55°C). A sharp decrease in methane production at 42°C has been noted by some workers.[9,12] This would not be particularly surprising considering the limited number of bacterial species that produce methane and their known characteristics. Hills'[12] studies are an example of temperature-rate data available, and his results are summarized in Fig. 10-4.

When toxic materials are introduced into an anaerobic treatment process, the effect may be much more noticeable on the methane fermentation step. The result is instability as in the previous two cases. Metal ions, for example, $Na^+, K^+, Ca^{2+}, Mg^{2+}$, are known to be toxic at certain concentrations. Normally, these toxic concentrations are of the order of 1000 mg/l or greater. Values reported by Speece and McCarty[6] are given in Table 10-2.

When a treatment process has become unstable or "gone sour," problems other than the cessation of treatment develop. Hydrogen sulfide (H_2S), organic acids, and other odorous compounds are produced in sufficient quantities to

Table 10-2 EFFECT OF COMMON IONS ON METHANE FERMENTATION[6]

	Concentration, mg/l		
Ion	Stimulatory	Moderately inhibitory	Strongly inhibitory
Sodium	100–200	3500–5500	8,000
Potassium	200–400	2500–4500	12,000
Calcium	100–200	2500–4500	8,000
Magnesium	75–150	1000–1500	3,000
Ammonia (as N)	50–200	1500–3000	3,000

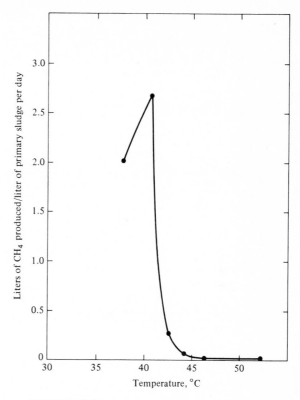

FIGURE 10-4
Methane production rate as a function of temperature.[12]

cause considerable discomfort to people nearby. A sour digester can often be detected by homeowners several miles from the source.

Restoring upset anaerobic treatment processes to stable operation is usually accomplished by stopping the feed and raising the pH. If feeding continues, reducing the volatile acid concentration is considerably more difficult even if the pH is raised to the 6.6 to 7.6 range by chemical addition. Adjustment of the pH is accomplished by adding a base such as lime or ammonia, with lime being the most often used.[12] The calcium ion in lime may cause problems if large quantities are added (see Table 10-2). Ammonia may also be toxic in concentrations of the order of 2 g/l or more, although reliable information on the threshold values is not available. Thus, pH correction may result in problems if considerable care is not taken.

According to McCarty,[13] most heavy metals will precipitate as sulfides prior to toxicity threshold level.

10-5 PROCESS OPERATING PARAMETERS

A number of parameters are normally monitored in anaerobic fermentation processes. Among the most important are alkalinity, methane production, and the CO_2/CH_4 ratio in the gas, in addition to the pH volatile acids concentration and temperature. Alkalinity is measured as the amount of acid needed to lower the system pH to the methyl orange end point (about pH 4) and, for convenience, is reported in units of calcium carbonate (MW = 100). The higher the alkalinity, the greater the buffering capacity of the system. Therefore, alkalinity is in general a more sensitive parameter than pH. A decrease in system alkalinity may signal an unbalance long before the pH begins to decrease.

Carbonate-bicarbonate equilibrium is an important factor in the buffering system at near-neutral pH values.

$$[H^+] = K_1 \frac{[H_2CO_3]}{[HCO_3^-]} \qquad K_1 = 4.41 \times 10^{-7} \qquad \text{at } 20°C \qquad (10\text{-}5)$$

Adding carbon dioxide to the system, as occurs in bacterial metabolism, increases the carbonic acid (H_2CO_3) concentration and the hydrogen-ion concentration. In aerated systems excess CO_2 is purged out of the liquid through the introduction of "clean" gas, but in anaerobic fermentations, CO_2 will accumulate if methane fermentation rates are low.

Neutralized volatile acids act as a buffer, i.e., also add to the alkalinity. Thus the total alkalinity can be estimated from Eq. (10-6) (Ref. 13):

Total alkalinity (mg/l $CaCO_3$) = bicarb alkalinity

$$+ 0.85(0.833)(\text{total volatile acids concentration}) \qquad (10\text{-}6)$$

where 0.833 is a unit conversion factor and 0.85 accounts for the fact that only 85 percent of the total volatile acids are accounted for by titrating to pH 4.5.

Methane production, like alkalinity, is a sensitive parameter. Normally, methane is reported as cubic meters per hour or cubic feet per hour, and some interpretation and judgment must be used in deciding what the reported values mean. If the feed has decreased sharply, a decrease in production rate would be expected, for example. If a methane production rate decrease is accompanied by a decrease in alkalinity and an increase in the volatile acids concentration, a significant problem has developed. Quite often an increase in the feed rate of organics will result in a sharp increase in volatile acid fermentation. If there is enough alkalinity, there will not be a significant decrease in pH, and methane fermentation will stay constant and then slowly rise to meet the new situation. Thus, a change in one variable does not necessarily mean the system is upset, and consideration of a number of parameters is important.

Carbon dioxide–methane ratio is similar to methane production rate in meaning. In most processes, the ratio should be about 0.7 on a volumetric basis (40 percent CO_2, 60 percent methane), but any given waste will have a particular associated value. A sudden decrease in the ratio indicates a decrease in acid fermentation (where most of the CO_2 is generated) and probably can be traced to a decrease in the organic feed rate. An increase in the ratio normally results from an increase in acid fermentation rate relative to the methane fermentation rate. An increase in organic feed rate would cause this, and if the alkalinity is adequate, no difficulty will result. If the increase in the ratio is due to a decrease in methane fermentation rate, problems have clearly developed.

Other parameters are of importance in particular types of anaerobic treatment processes. For example, in digesters where solids destruction is the goal, the condition of the effluent solids is important. Low volatile solids content, good drainability, and stability are the criteria normally used. Because anaerobic solids digestion was until very recently the only anaerobic process used in waste treatment, much of the available data are in terms of these criteria. For example, early work on temperature effects was reported as temperature vs time necessary for complete digestion.[9]

10-6   FERMENTATION RATES

Anaerobic fermentation rates are considerably lower than those for aerobic systems. Because of the necessity of maintaining a balanced reaction system, one in which overall methane and acid fermentation rates are equal, researchers have nearly always studied the overall system rather than the individual steps. Methane fermentation is assumed to be rate controlling because the methane bacteria grow so much slower than the acid formers. Generation times of methane bacteria are greater than 12 h under optimal conditions,[15] while facultative bacteria growing under optimal conditions have generation times as low as 0.3 h.[2,16] Under anaerobic conditions, facultative bacteria would still be expected to grow much faster than the methane bacteria.

Reported values of k_0, the maximum organic removal rate, are usually of the order of 0.2 g COD/g cells-d where soluble substrates are used. McCarty

Table 10-3   METHANE FERMENTATION RATE DATA

Substrate	Maximum rate, kg/d	Temperature, °C	Ref.
Methanol	0.50	35	6
Formate	0.33	35	6
Acetate	0.20	35	6
Acetate	0.24–0.50	35	17
Acetate	0.24	25	17
Propionate	0.13	35	6
Propionate	0.36	25	17
Butyrate	0.37	35	17
Primary and activated sludge	0.31	37	18

and his coworkers have studied a number of substrates, mostly volatile acids, and their data are summarized in Table 10-3 with that of other workers.

The low-rate coefficient values are a major reason for the paucity of data and also can be used in determining applicability of the process. Because the rate coefficients are so low, obtaining steady-state values takes a great deal of time. When a researcher faces the problem of low rates and an unstable system, he usually decides to work on something else.

10-7 APPLICATION OF ANAEROBIC PROCESSES

In determining whether or not an anaerobic treatment process should be used in treating a particular waste, several factors related to conversion rates and stoichiometry must be considered. As has been noted, cases where large changes in the feed characteristics occur over short time intervals present problems for anaerobic treatment processes. If flow or organic mass rate equalization is feasible, these changes may not be important, but if flow rates are high, this is often not possible. Seasonal wastes such as those from canneries are generally not suitable for anaerobic treatment because start-up times are long. Where a food-processing industry operates throughout the year, anaerobic processes may be well suited, however.

Wastewaters containing large quantities of soluble fats are difficult to treat by anaerobic processes. Fat breakdown by β oxidation (Fig. 6-12) produces large amounts of reduced pyrimidine nucleotides ($NADH^+$). Because of the difficulty in reoxidizing the nucleotides, fat breakdown is relatively slow under anaerobic conditions as evidenced by layers forming in unmixed digesters or secondary separation tanks.

Because of the temperature sensitivity of anaerobic processes using methane fermentation, temperature control is necessary. Most processes are maintained at approximately 35°C (95°F), slightly below the optimal temperature, and must be heated. Methane produced can be used to heat the system if the organic concentration of the wastewater is sufficiently high. Heat value of methane is approximately 9420 cal/l at 20°C (1060 Btu/ft^3). For the gas mixture, this corresponds to approximately 6500 cal/l (740 Btu/ft^3). Buswell and Mueller[19] reported that the relationship below could be used to predict methane production.

$$C_nH_aO_b + \left(n - \frac{a}{4} - \frac{b}{2}\right)H_2O \rightarrow \left(\frac{n}{2} - \frac{a}{8} + \frac{b}{4}\right)CO_2 + \left(\frac{n}{2} + \frac{a}{8} - \frac{b}{4}\right)CH_4 \quad (10\text{-}7)$$

For carbohydrates, Eq. (10-7) predicts 0.35 l CH_4 produced per gram of COD or ultimate BOD removed (5.62 ft^3/lb). We should note that cell production is not included in the expression, and therefore, the actual value must be lower than 0.35 l/g. Metcalf and Eddy[14] suggests the use of an expression derived from Eq. (10-7).

$$M_{CH_4} = 0.35(\eta QC_i - 1.42R_g V) \quad (10\text{-}8)$$

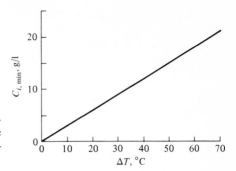

FIGURE 10-5
Minimum BOD$_L$ concentration to sustain operating temperature in anaerobic processes (100 percent heat-transfer efficiency assumed).

where M_{CH_4} = methane production rate, l/t,
η = conversion
1.42 = conversion factor for cells to BOD$_L$

The growth rate is quite small, and in most cases, Eqs. (10-8) and (10-7) will normally give similar values.

If we assume the growth rate term in Eq. (10-8) is negligible, the minimum waste BOD$_L$ concentration based on sustaining a stated operating temperature can be calculated. A curve of $C_{i,\,min}$ versus ΔT, assuming 100 percent efficiency, is presented in Fig. 10-5. From this curve we can conclude that most sludges encountered in wastewater treatment processes are quite good candidates for anaerobic treatment. Wastewaters must be quite high in organic content before anaerobic treatment can be self-sustaining, however. Some industrial wastewaters are both high in organic concentration and temperature. Among this group are a number of food-processing wastes.

10-8   LOADING RATES

Traditionally, loading rate (mass of COD or solids/d-unit volume) has been a design parameter for anaerobic processes.[20,21] There is not a direct relationship between volumetric loading and process operation because current models for biological processes are all single valued. Thus a process should be able to accommodate any volumetric loading as long as the residence time is satisfactory. As was noted in Chap. 8, the assumption of a single-valued response may be in error, and at this time, reliance on field experience is recommended. A possible second reason for loading rate limitations is related to mixing. If mixing were inadequate, actual residence times would be below predicted values in parts of the tank. This would result in a local drop in pH, a situation that would probably cause the entire system to become unstable. Mixing problems would be most pronounced in solids digestion.

Zablatsky and Baer[22] reported that a loading rate of 8000 g undigested solids/m^3-d resulted in stable operation at a 10-d residence time. The solids in their study were a mixture of primary and activated sludge and had a volatile-solids fraction of approximately 0.7. Conventional loading rates for high-rate digesters are of the order of 1600 to 6400 g/m^3-d of volatile solids with residence times between 10 and 20 d.

EXAMPLE 10-1 An anaerobic process is being considered for treating a food-processing waste with the following characteristics:

COD, mg/l	9500
Ultimate BOD, mg/l	9200
Nitrogen, mg/l:	
Organic	160
Ammonia	20
pH	4.5
Acidity, mg/l $CaCO_3$	490
Temperature	26°C
Flow rate, l/s	5.6

The source of the low pH and acidity is known to be organic acids in the C_3 to C_5 size range.

Make a first estimate of the feasibility of anaerobic treatment, approximate tank size, and net methane production rate.

Consideration of pH correction If the low pH and the acidity are due to fermentable organic acids, pH control should not be a problem. Providing a system in which the organic acids are converted as fast as they enter will result in a "balanced system." The only problem will be during the start-up period. Pilot plant data will be necessary to predict process performance.

Approximate tank size The type of waste (low pH, some acidity) may result in an anaerobic treatment process that is somewhat temperamental. Thus a conservative residence time value should be considered, at least as a first estimate. A 15-d residence time in a completely mixed system would probably be appropriate. A second tank for cell separation will also be necessary and will also have a 15-d residence time.

Approximate tank volume is

$$V_1 = V_2 = 5.6(8.65 \times 10^4 \text{ s/d})(15 \text{ d})$$
$$= 7.26 \times 10^6 \text{ l}$$
$$= 7.26 \times 10^3 \text{ m}^3 \ (2.6 \times 10^5 \text{ ft}^3)$$

Four tanks, each 8 m deep and 24 m in diameter, will be satisfactory. Tank sizes chosen are based on standard practice, availability of materials, and the flexibility provided by four units.

Pilot plant studies should be run at three residence times, 10, 20, and 30 d, to establish whether or not 15 d is really suitable. Ideally, kinetic data would be developed for design and operation of the system. This is usually a difficult job.

Approximate net methane production Using Eq. (10-8) and an approximate yield of 0.06 g cells/g BOD allows determination of an estimate of methane production rate:

$$Rg = kgX = -YQ\,\Delta C$$
$$= -Y\eta QC$$

Metcalf and Eddy[14] suggests using $\eta = 0.8$.

$$M_{CH_4} = 0.35[0.8(5.6)(9.2)][1 - 1.42(0.06)]$$
$$= 13.200 \text{ l/s}$$

The sludge will have to be heated from $2.6°C$ to operating temperature, approximately $37°C$.

$$\text{Energy required} = (5.6 \text{ l/s})(37 - 25)(1000 \text{ cal/l-°C})$$
$$= 67,200 \text{ cal/s}$$
$$= 7.2 \text{ l CH}_4/\text{s}$$
$$\text{Net methane production} = 6 \text{ l/s} \qquad ////$$

10-9 ANAEROBIC DIGESTION

Organic solids from either primary sedimentation or from biological treatment processes are both highly putrescible and difficult to treat aerobically. Anaerobic biodegradation of these solids has been used since the early days of wastewater treatments. The earliest anaerobic system for solids stabilization was the septic tank. In the early part of the century, the Imhoff tank was developed.[21] Gas generation had been observed to be associated with anaerobic degradation and collection, and use of the gas was begun before 1930 (Ref. 23). As the overall sophistication of wastewater treatment plants increased, sludge stabilization or digestion was set aside as a separate process. Initially, digesters were simply closed tanks into which sludge was pumped and stored for periods of 30 to 120 d. Fair and Moore,[9] Morgan,[24] and Torpey[25] reported that digestion rate could be greatly increased by maintaining the digester contents at higher temperatures and by mixing. These two innovations were the basic ingredients of the modern high-rate digester.

In digestion there are three products of interest: the gas, the supernatant liquor, and the digested solids. Supernatant liquor is returned to the biological treatment process, and therefore, can be "biologically unstable." Digested solids must be suitable for final disposal, however. In most cases, digestion criteria include sludge drainability and stability. If drying is slow or if odors associated with further biodegradation are produced, digestion is not complete. As has been noted, most of the data published on digestion prior to 1955 used completeness of digestion as a control parameter.

In small communities, final disposal of digested sludge can often be accomplished by draining and drying on sand beds. The dried sludge is useful as humus or can be disposed of in a sanitary landfill. Larger communities often have space limitations that make drying economically unfeasible. Vacuum filtration, centrifugation, and incineration are often used. Digested municipal sludge has a heat value of approximately 6200 cal/g dry volatile solids. Theoretically, incineration can be self-sustaining if volatile-solids concentrations of the order of 10 percent or greater are provided. Economical operation requires solids concentrations of the order of 30 percent, however.

Most modern sludge-digestion systems make use of both high-rate (mixed and heated) digesters and heated unmixed processes (Fig. 10-6). The latter systems serve to separate the digested solids from the supernatant liquor as well as provide additional time to degrade volatile solids.

Mixing

Mixing can be accomplished either by recycling a portion of the gas or by mechanical stirring. In either case the process is difficult because of the viscosity of the material. If mechanical mixing is used, the best method is to pump from the center and distribute the solids to the outer edge. Gas mixing should be through several sprayers distributed through the tank. Heating provides some mixing also but not enough to thoroughly mix the contents of a digester.

A major problem associated with digester heating is caking of sludge on the heat-exchanger lines. This both reduces the heat-transfer coefficient and the flow cross section. Most heat exchangers use hot water, and if the water is kept below 57°C (138°F) caking is not a problem.

Torpey's studies[18] indicated that the maximum rate of digestion, i.e., the washout rate, was 0.31 d^{-1}. This would correspond to a chemostat (which is what a high-rate digester is) residence time of 3.2 d. In practice, full-scale digesters rarely function well at residence times less than 12 d. This may be due to mixing problems and resultant short circuiting, or due to the fact that as maximum rates are approached, system stability decreases. Standard-rate digesters are normally designed using residence times of 20 d or greater.

Primary digester

Secondary digester

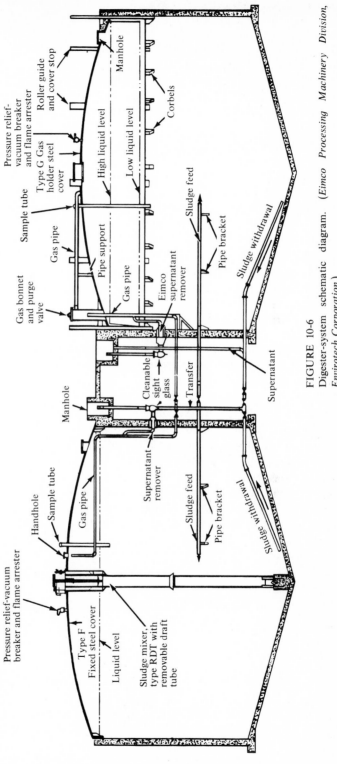

Pressure relief-vacuum
breaker and flame arrester

Handhole

Sample tube

Gas pipe

Type F
Fixed steel cover

Liquid level

Sludge mixer,
type RDT with
removable draft
tube

Manhole

Supernatant
remover

Cleanable
sight glass

Transfer

Sludge feed

Pipe bracket

Sludge withdrawal

Supernatant

Pressure relief-
vacuum breaker
and flame arrester

Roller guide
and cover stop

Manhole

Sample tube

Type G Gas
holder steel
cover

High liquid level

Low liquid level

Corbels

Gas pipe

Pipe support

Gas bonnet
and purge
valve

Gas pipe

Eimco
supernatant
remover

Sludge feed

Pipe bracket

Sludge withdrawal

FIGURE 10-6
Digester-system schematic diagram. (*Eimco Processing Machinery Division, Envirotech Corporation.*)

10-10 ANAEROBIC CONTACT PROCESS

Schroepfer et al.[26] and Steffens[27] developed a modification of the conventional digestion process for use with meat packing or other high soluble–low solids wastewaters. They proposed placing a solids separation tank on the process effluent line and recycling the solids. This results in a flow sheet similar to activated sludge. In order to provide a satisfactory conversion rate in such a system, the solids residence time must be considerably longer than the hydraulic residence time. An ordinary sedimentation tank cannot be used to separate the cells from the liquid because gas generation continues in the separator and lifts cells to the surface. A degasification system must be included. Gentle stirring can be used to separate bubbles from floc particles. The process is difficult to control and has been used in only a limited number of cases.

10-11 ANAEROBIC PACKED BEDS

The problem of cell separation in the anaerobic contact process led to the development of the anaerobic packed bed by McCarty and his coworkers,[28,29] and the process is now being used for some food-processing wastes.[30] The physical system is the same as that described for packed-bed denitrification processes in Chap. 8. Bacteria grow on the media surfaces and in the pores. Gas bubbles generated by flocs may cause the particles to rise for a short distance, but separation usually occurs on collision with the packing. Thus most of the cells are in the lower portions of the packed bed. This region has correspondingly higher reaction rates and accounts for most of the organic conversion.

Kinetic data are not available at the present time that will satisfactorily allow prediction of organic and cell concentrations in anaerobic packed beds. Young and McCarty[29] worked with 15.2-cm-diameter, 183-cm-high columns packed with 2.5 to 4 cm nominal-size rock media. They used loading rates as high as 2.4 mg COD/l-min. Their results are summarized in Table 10-4.

Table 10-4 ANAEROBIC PACKED-BED DATA OF YOUNG AND McCARTY[29]

	Volatile acids	Protein carbohydrate
Loading, mg COD/l-min	0.59	0.59
Waste COD, mg/l	3000	3000
Effluent suspended solids, mg/l	7	90
Θ_C, d	665	84
Θ_H, d	1.5	1.5
Effluent COD, mg/l	470	130
Effluent BOD, mg/l	470	280
Yield	0.02	0.16

The major advantage of using packed beds is the separation of cell and hydraulic residence times, and the major disadvantages are the problems associated with flow distribution, heating, and the fact that plugging will eventually occur. Hydraulic flow rate is limited by the scour velocity, but this is not a very severe limitation. Wastewater suspended-solids concentration is also a limiting parameter. The plugging rate increases with increasing suspended-solids concentration. For this reason, anaerobic packed beds appear to be best suited to treating wastewaters with high soluble organic contents.

Increasing the cell age, i.e., retaining the cells longer than the liquid, usually increases process stability, and the anaerobic packed bed appears to follow this general rule. Young and McCarty's[29] experience with laboratory-scale systems led to the conclusion that shut-down–start-up situations will be far less difficult with packed beds than with either digesters or anaerobic contact units.

Flow distribution is a problem in any packed-bed reactor. Anaerobic packed beds will be much larger than packed beds used in the chemical industry, and as a consequence, flow-distribution problems will be intensified. As cells accumulate in pores, increased channelization will develop, and these problems may severely limit the time period between cleaning.

Process-cleaning techniques have yet to be worked out. Backwashing of a unit the size of a conventional trickling filter appears out of the question. There is the additional problem of maintaining satisfactory reaction rates in the start-up period following cleaning. Possibly a portion of the cells removed will need to be reintroduced into the system.

Flow distribution and cleaning problems can be minimized by using modular units similar in concept to sand-filtration systems in potable-water treatment. This type of design will allow much greater operational flexibility and control, also.

10-12 ANAEROBIC PONDS

Two types of anaerobic ponds or lagoons are in use, those for sludge and those for soluble wastes. Sludge lagoons have been used for over 70 years in municipal waste treatment systems and recently have been introduced as a manure-disposal process for dairies, feedlots, and chicken ranches.[31–33] Most sludge lagoons are shallow ditches or ponds that are intermittently filled with sewage sludge or manure. No temperature control is provided, and the first stage of digestion can be expected to predominate. Nuisance odors are a common problem, and location of sludge lagoons can be expected to be well known to anyone living nearby. Cattle manure is more stable than swine or chicken manure or municipal sewage sludge and, consequently, causes fewer problems. The reason for this difference would appear to be that cattle manure is relatively stable due to the digestion process of ruminants.

Deep anaerobic lagoons have been used primarily for treatment of meat-packing wastes, although examples exist of their successful application for other wastes. Lagoon depths used vary from 2.8 to 6 m and residence times from 7 to 80 d. Reasonably deep ponds with a minimum feasible area-to-volume ratio are necessary to conserve heat. Organic loading rates of 0.16 to 0.32 g COD/l-d have been used with meat packing (0.01 to 0.02 lb/ft^3-d). Organic removals reported are usually less than 80 percent, although values as high as 90 percent have been observed.[31] In most cases, a secondary pond must be used also. Suspended solids are efficiently removed in the anaerobic pond, and the secondary pond can be a shallow aerobic rather than a facultative system.

Anaerobic ponds are particularly well suited to treating meat-packing wastewaters. These wastes usually contain a considerable amount of grease that, along with paunch manure, forms a fairly tight scum layer on the pond surface. The scum layer helps control odor problems that would be expected from anaerobic degradation of high protein wastes. The sulfur bacteria, those that use H_2S as an energy source, also play an important role in odor control in anaerobic ponds.

EXAMPLE 10-2 A soluble industrial wastewater has the following characteristics:

$$BOD_L = 6700 \text{ mg/l}$$

$$T = 25°C$$

$$\text{Alkalinity} = 850 \text{ mg/l as } CaCO_3$$

$$pH = 7.8$$

$$Q = 3.5 \text{ l/s}$$

Laboratory experiments have been performed with an anaerobic packed bed having a porosity of 0.9, a diameter of 30 cm, and a hydraulic residence time of 24 h. The experiments were run over a 120-d period during which the effluent characteristics consistently improved. Change in effluent quality per day became very slow after 100 d, and the data given for BOD_L concentration were averaged over the last 10 d. At the end of the experiment, the suspended and attached solids were measured at each sampling port, and a pseudo-homogeneous MLSS value $\bar{X}$ was determined.

Port no.	Z, cm	BOD_L, mg/l	$\bar{X}$, mg/l	pH	Gas, l/h	CH_4, l/h	T, °C
1	10	5960	4550	7.4			37
2	40	4150	2870	7.3			37
3	70	3050	1849	7.4			37
4	100	2275	1256	7.5			37
5	130	1780	803	7.6			37
6	160	1510	469	7.6			37
					12.23	7.95	

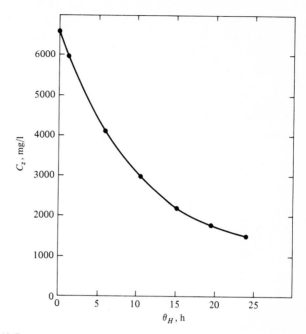

FIGURE 10-7
Ultimate BOD as a function of position in a packed-bed anaerobic reactor.

Develop design criteria for the waste treatment plant with a soluble BOD_L effluent of 50 mg/l.

Estimate the kinetic coefficients   For a soluble waste we can expect K_m to be of the order of 1.0 to 10 mg/l, and therefore the system should be operating near maximum rate throughout the length of the unit. Although a tracer curve has not been furnished, we will assume plug-flow conditions approximately exist.

$$\frac{dC}{d\Theta} = \frac{-k_0\,CX}{K_m + C} \approx -k_0\,\bar{X}$$

Plot C versus Θ and calculate $1/X$ $(\Delta C/\Delta\Theta)$ for 30-cm increments using tangents to the curve (Fig. 10-7).

Z, cm	$-\dfrac{1}{X}\dfrac{\Delta C}{\Delta\Theta}$, d^{-1}
10	2.6
40	2.2
70	2.4
100	2.6
130	2.3
160	2.6
$\sum = 14.7$	$\bar{k}_0 = 2.45$

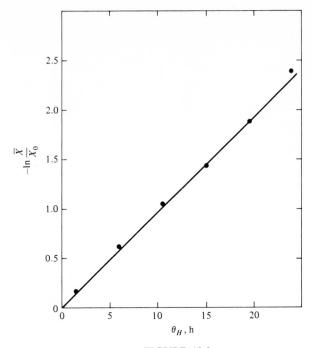

FIGURE 10-8
Cell mass as a function of position.

There is no direct method of predicting the solids concentration because the value at any point keeps increasing with time and because the gradient is the opposite from normal, i.e., decreases with increasing Θ.

We will assume that at 120 d the experimental units were close to a situation in which the cells lost in the effluent were very close to the amount produced per unit time. This would mean that the concentration profiles in the experimental system would be the same as in the prototype. Plot $\ln(X_z/X_0)$ versus Θ after obtaining $X_o = 5300$ mg/l from an arithmetic plot.

From this curve (Fig. 10-8), we can derive the empirical expression

$$\bar{X}_z = \bar{X}_o\, e^{-0.0833\Theta}$$

where Θ represents a time of travel in the reactor.

$$\frac{dC}{d\Theta} = -k_0\, X_o\, e^{-0.2\Theta}$$

$$C - C_o = \frac{k_0\, X_o}{2}(e^{-2\Theta} - 1)$$

This expression fits the data very well but should not be extended to other systems. Thus use of this expression as a design equation is dependent upon being able to reproduce the experimental system on a larger scale. The design hydraulic residence time can now be calculated.

$$\Theta = \frac{-1}{2} \ln\left[\frac{2}{k_0 X_0}(C - C_o) + 1\right]$$

$$= \frac{-1}{2} \ln\left[\frac{-2(6650)}{2.45(5300)} + 1\right] = 1.9 \text{ d}$$

Gas production From the experimental data we can estimate the gas production per liter of wastewater:

$$Q_{exp} = \frac{V_r}{\Theta} = \frac{113}{24} = 4.71 \text{ l/h}$$

Methane production rate $= 7.95 \text{ l/h } (Q_{exp} \Delta C)^{-1}$

$$= \frac{7.95}{4.71(5.19)} = 0.328 \text{ l/g-BOD}_L$$

ΔC prototype $= 6.65$ g/l
Methane rate ≈ 2.2 l/l of wastewater
  Heat value $\approx 20,724$ cal/l of wastewater
 Heat needed $= 1000 \ \Delta T = 12,000$ cal/l

Therefore, 8000 cal/l or 28,000 cal/s will be available for other uses in the treatment plant. ////

PROBLEMS

10-1 An industrial wastewater from an organic chemical plant is to be treated by high-rate anaerobic digesters followed by no-discharge ponds. Bench-scale studies have been run using 5-l reactors and residence times of 4, 8, 10, 16, 20, and 30 d, and the data are summarized below. The wastewater heat capacity is 1.05 cal/g-°C, and the design flow rate is 3 l/s (68,500 gpd). Determine the total tankage volume that should be used and the excess energy available in kilocalories per day. Assume heat losses from the digesters are 10 percent and that the wastewater temperature is 27°C.

	Reactors					
	1	2	3	4	5	6
$\Theta_H = \Theta_C$, d	4	8	10	16	20	30
C_i, mg BOD/l	9700	9700	9700	9700	9700	9700
C, mg BOD/l	9200	315	220	109	82	51
CH_4 production, ml/d		1860	1520	960	780	532
Gas production, ml/d	200	3005	2190	1340	1083	718
pH	3.9	6.9	7.2	7.5	7.6	7.6
Alkalinity		376	625	729	732	760
Temperature, °C	37	37	37	37	37	37
X, mg/l	62	763	713	605	541	395

10-2 Review the literature on anaerobic treatment from 1955 to date. Determine the extent of knowledge on effluent quality from anaerobic processes and what quality might be expected from a modern, well-operated process.

10-3 Early studies on the anaerobic packed-bed process resulted in the conclusion that the systems are more stable than conventional high-rate systems. Would you expect this to be true under conditions of highly variable influent flow rate or concentration? Explain and justify your answer.

10-4 Experiments on anaerobic treatment of a soluble organic industrial wastewater have resulted in the following data:

$$r_0 = \frac{-0.86C}{670 + C} \quad \text{at } 37°C$$

$$Y = 0.08e^{-0.017\theta_c}$$

where C is the ultimate BOD. Similar experiments on aerobic treatment of the same wastewater have resulted in the following data:

$$r_0 = \frac{-1.4C}{17 + C} \quad \text{at } 20°C$$

$$Y = 0.4e^{-0.05\theta_c}$$

If the wastewater BOD_L concentration, flow rate, and temperature are 6300 mg/l, 520 m³/d and 20°C, respectively, determine which of the configurations below is most suitable. If necessary consider an anaerobic treatment process followed by an activated sludge process. Assume that digester mixing consumes 10 percent of the methane produced and that secondary clarifier areas are equal. If possible,

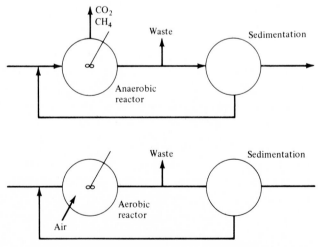

FIGURE 10-9
Alternative process configurations for Prob. 10-4.

include current power and construction costs in your calculations. Final effluent BOD concentration is to be 20 mg/l. Waste solids can be thickened to 1.2 percent by weight.

10-5 In recent years methane fermentation has been suggested as a commercial or home energy source. Use the data presented in the following table to determine the number of animals necessary to provide fuel for a commuter to travel 65 km (40 m)/d in an automobile that consumes 1.25 kc/km (15 mi/gal gasoline) and the area necessary to keep the animals.

	Animal†		
	Chicken	Swine	Beef cattle
Waste volume, l/animal-d	0.19	7.5	34
Waste, BOD_L g/l	150	60	21
Average mass/animal, kg	1.5	90	410
Pen area/animal, m^2	0.2	1.4	4.7

† Average figures. Particular values depend on age, feed, and activity of animals.

REFERENCES

1 WOOD, W. A.: Fermentation of Carbohydrates, in I. C. Gunsalas and R. Y. Stanier (eds.), "The Bacteria," vol. II, Academic Press, Inc., New York, 1961.

2 STANIER, R. Y., J. L. INGRAHAM, AND E. A. ADELBERG: "The Microbial World," 4th ed., Prentice-Hall, Inc., Englewood Cliffs, N.J., 1976.

3 LEMANNA, C., AND M. F. MALLETTE: "Basic Bacteriology," 3d ed., The Williams & Wilkins Company, Baltimore, 1965.

4 THIEMAN, K. W.: "The Life of Bacteria," 2d ed., The Macmillan Company, New York, 1963.

5 JERIS, J. S., AND P. L. MCCARTY: Biochemistry of Methane Fermentation, *Proc. 17th Ind. Waste Conf.*, May 1963.

6 SPEECE, R. E., AND P. L. MCCARTY: Nutrient Requirements and Biological Solids Accumulation in Anaerobic Digestion, *Proc. 1st Int. Conf. Water Pollut. Res.*, 1962.

7 SMITH, LUCIELLE: Cytochrome Systems in Aerobic Electron Transport, in I. C. Gunsalas and R. Y. Stanier (eds.), "The Bacteria," vol. II, Academic Press, Inc., New York, 1961.

8 CLARK, R. H., AND R. E. SPEECE: The pH Tolerance of Anaerobic Digestion, *Proc. 5th Int. Conf. Water Pollut. Res.* 1971.

9 FAIR, G. M., AND E. W. MOORE: Time and Rate of Sludge Digestion and Their Variation with Temperature, *Sewage Works J.*, vol. 6, pp. 3–13, 1934.

10 MALINA, J. F.: The Effect of Temperature on High Rate Digestion of Activated Sludge, *Proc. 16th Ind. Waste Conf.,* May 1961.

11 GOULECKE, C.: Temperature Effects on Anaerobic Digestion of Raw Sewage Sludge, *Sewage Ind. Wastes,* vol. 30, p. 1225, 1958.

12 HILLS, D. J., AND E. D. SCHROEDER: Temperature Effects on the Rate of Methane Fermentation, *Water Sewage Works,* vol. 6, p. 46, July 1969.

13 MCCARTY, P. L.: Anaerobic Waste Treatment Fundamentals, part two, *Public Works,* p. 123, October 1964.

14 METCALF AND EDDY, INC.: "Wastewater Engineering," McGraw-Hill Book Company, New York, 1972.

15 ANDREWS, J. F., R. D. COLE, AND E. A. PEARSON: Kinetics and Characteristics of Multi-Stage Methane Fermentation, SERL Rept. No. 64-11, Sanit. Eng. Res. Lab., University of California, Berkeley, 1964.

16 INGRAHAM, J. L.: Temperature Relationships, in I. C. Gunsalas and R. Y. Stanier (eds.), "The Bacteria," vol. IV, Academic Press, Inc., New York, 1961.

17 LAWRENCE, A. W., AND P. L. MCCARTY: Kinetics of Methane Fermentation, Tech. Rept. 75, Department of Civil Engineering, Stanford University, Stanford, Calif., 1967.

18 TORPEY, W. N.: Loading Failure of a High Rate Digester, *Sewage Ind. Wastes,* vol. 27, p. 121, 1955.

19 BUSWELL, A. M., AND H. F. MUELLER: Mechanics of Methane Fermentation, *J. Ind. Eng. Chem.,* vol. 44, p. 550, 1952.

20 GREAT LAKES–UPPER MISSISSIPPI RIVER BOARD OF STATE SANITARY ENGINEERS: Recommended Standards for Sewage Works, Health Education Service Albany, N.Y., 1968.

21 CLARK, J. W., W. VIESSMANN, AND M. J. HAMMER: "Water Supply and Pollution Control," 2d ed., International Textbook Company, Scranton, Pa., 1971.

22 ZABLATSKY, H. R., and G. T. BAER: High Rate Digester Loadings, *Water Pollut. Control Fed.,* vol. 43, p. 268, 1971.

23 METCALF, L., and H. P. EDDY: "Sewerage and Sewage Disposal," McGraw-Hill, New York, 1930.

24 MORGAN, P. F.: Studies of Accelerated Digestion of Sewage Sludge, *Sewage Ind. Wastes,* vol. 26, p. 462, 1954.

25 TORPEY, W. H.: High Rate Digestion of Concentrated Primary and Activated Sludge, *Sewage Ind. Wastes,* vol. 26, p. 479, 1954.

26 SCHROEPFER, G. J., W. J. FULLER, A. S. JOHNSON, N. R. ZIEMKE, and J. J. ANDERSON: The Anaerobic Contact Process as Applied to Packinghouse Waste, *Sewage Ind. Wastes,* vol. 27, p. 460, 1955.

27 STEFFENS, A. J.: The Treatment of Packinghouse Wastes by Anaerobic Digestion, in B. J. McCabe and W. W. Eckenfelder (eds.), "Biological Treatment of Sewage and Industrial Wastes," vol. II, Reinhold Publishing Corporation, New York, 1958.

28 MCCARTY, P. L.: Anaerobic Treatment of Soluble Wastes, in E. F. Gloyna and W. W. Eckenfelder, (eds.), "Advances in Water Quality Improvement," University of Texas Press, Austin, 1966.

29 YOUNG, J. C., and P. L. MCCARTY: The Anaerobic Filter for Waste Treatment, *Proc. 22d Ind. Waste Conf.,* May 1967.

30 CARLSON, D. A.: Recent Developments in Anaerobic Waste Treatment, *Potato Waste Treatment Symp.,* Pacific Northwest Water Laboratory, Corvallis, Ore., 1968.

31 DORNBUSH, J. N.: State of the Art—Anaerobic Lagoons, *Proc. Int. Symp. Waste Treatment Lagoons,* R. E. McKinney (ed.), University of Kansas, Lawrence, 1970.

32 LOEHR, R. C.: "Agricultural Waste Management," Academic Press, Inc., New York, 1974.

33 HART, S. A., and M. E. TURNER: Lagoons for Livestock Manure, *Water Pollut. Control Fed.,* vol. 37, p. 1578, 1968.

11

PROCESS SELECTION AND SYSTEM SYNTHESIS

Selection of processes for treatment of a particular water or wastewater is based upon product-water requirements, influent-water characteristics, and, where alternatives are available, cost. In development of a treatment system consisting of a train of individual processes, the interaction of steps must be considered, and this is often a distinct economic factor.

Product-water requirements are, in most cases, related to use. Domestic water supplies should meet the U.S. Public Health Service Drinking Water Standards.[1] Boiler-water supplies must be low in minerals, particularly those that cause scale. Cooling waters must not cause scaling, and there must also be control of the growth of microorganisms in the pipes and cooling tower. Requirements placed on wastewater treatment processes are usually based on receiving-water quality. In selecting and designing either water or wastewater processes and systems, the engineer has the responsibility to ensure that requirements can be met. In cases where requirements are inappropriate, the engineer has the responsibility to propose improvements or to design a system that meets his personal standards. Most wastewater discharges are into public waters, either surface or ground, and because of this fact the engineer must always be concerned with the public's interest as well as a client's.

11-1 WASTEWATERS

Wastewater effluent requirements in the United States are set by federal and state agencies (usually in cooperation). Natural or background conditions in the receiving water are used to determine selected parameter values for a given stream or lake, and these in turn are used to set effluent quality criteria. In some cases wastewater effluent requirements can affect the choice of water supplies. For example, discharges into the Sacramento–San Joaquin Delta must have TDS concentrations less than 500 mg/l. Groundwaters in this area are often in excess of this value and therefore cannot be used for domestic supplies. A typical set of effluent requirements used for municipal discharges in the Central Valley of California is presented in Table 11-1.

The use of natural or background conditions to set criteria seems both wise and fair. If there is a major weakness, it is that the parameters used to develop criteria are tuned to predict acute problems and that the criteria do not provide satisfactory protection from long-term or chronic problems. As methods of determining these situations become available, controls can be added, however.

The impact of a set of criteria on process selection can be illustrated from Table 11-1. Consistent production of an effluent with a BOD less than 20 mg/l or a COD less than 60 mg/l is difficult. Trickling-filter effluents rarely meet this criterion, and many activated-sludge processes meet it only intermittently. In

Table 11-1 SELECTED DISCHARGE REQUIREMENTS FOR SACRAMENTO, CALIFORNIA

Constituent	30-d average	7-d average	Daily maximum
BOD$_5$:			
mg/l	30	45	90
lb/d	17,000	25,000	50,000
Total suspended matter:			
mg/l	30	45	90
lb/d	17,000	25,000	50,000
Settleable matter, ml/l	0.1		0.2
Total coliform organisms, MPN/100 ml	23		500
Total fixed nitrogen:			
mg/l			15
lb/d			8500

The discharge shall not cause the dissolved-oxygen concentration in the Sacramento River to fall below 5.0 mg/l.

The discharge shall not increase turbidity of the receiving waters by more than 10 percent over background levels.

The discharge shall not cause surface-water temperatures to rise more than 4°F above the natural receiving waters at any time.

essence, effluent oxygen-demand criteria similar to those in Table 11-1 force the designer to choose activated sludge (if he is confident of being able to design a process that has few operational problems) or go to a trickling filter with some form of further treatment. A significant fraction of trickling-filter effluent BOD is due to suspended solids, and either coagulation or sand filtration will normally provide a product that meets the criteria of Table 11-1. Either of the tertiary treatment processes will add significantly to both capital and operating costs, however. An alternate approach would be to use coagulation and adsorption rather than biological treatment. Process stability would be much improved, relative to activated sludge, and the effluent criteria could be met. Economic feasibility would rule out this choice in most cases because of the cost of sorbent. Oxidation ponds could also be considered if enough land were available to allow operation without a surface discharge.

Dissolved-solid limitations on effluents may restrict the use of chemical addition in wastewater treatment processes. Where groundwaters are used, TDS concentrations are often high before use. Municipal use increases the TDS concentration by 200 mg/l or more, and addition of coagulants in treatment may well cause the effluent to exceed effluent requirements.

Inclusion of nutrient limitations in effluent requirements is becoming more common. Usually these criteria are written in terms of total nitrogen or total phosphate. Here it is extremely important that the designer understand the stoichiometry of biological processes. Nutrient uptake by the cells increases as the cell age (θ_C) decreases.[2] Thus, if the nutrient concentrations are of the right magnitude, an activated-sludge process can often be operated in such a manner that additional treatment is unnecessary. For example, if the BOD_L/N ratio in a wastewater is greater than 16, the nitrogen can be assimilated by the cell growth using a θ_C of 4 d. Higher BOD_L/N ratios will require longer residence times (lower yields) for good stoichiometric operation. Wastes with lower BOD_L/N ratios will need additional treatment (e.g., denitrification). Domestic sewage has BOD_L/N ratios in the range of 6.5 to 12.1 in most cases, and therefore, conventional wastewater treatment will not remove all the nitrogen.

Other characteristics of wastewaters affect process selection also. Many biodegradable organics are toxic at high concentrations. Plug-flow processes will consistently be upset when these materials are introduced into the treatment system. A CFSTR activated-sludge process (or a trickling filter with a high recycle rate) will often be unaffected by either continuous influx or slugs of toxic organics because of the dilution factor provided. When the toxic material is conservative, i.e., does not react, a different situation exists. A slug of the material may have satisfactory dilution, but a continuous influx will cause an accumulation in the reactor until the reactor concentration is equal to the influent concentration. If the influent concentration is above the toxicity threshold, a CFSTR activated-sludge process will have operational problems.

11-2 INDUSTRIAL WASTEWATERS

Industrial wastewater often is the combined product of a number of different processes within a manufacturing operation. For example, a metal-plating plant may produce materials coated with copper, nickel, chrome, or zinc. Each plating process involves a number of steps involving cleaning, rinsing, plating, and rinsing again, and each step produces a wastewater. Copper and zinc plating usually makes use of the respective cyanide salts, and the waste from the plating and rinsing steps contains both toxic heavy metals (Cu^{2+} and Zn^{2+}) and the very toxic cyanide ion (CN^-). Treatment of these wastes involves a number of steps, and mixing of the plating wastes with the cleaning wastes would make the problem more difficult.

Cyanide can be oxidized by chlorine under high-pH conditions. The cyanide treatment process then involves raising the pH to approximately 11 by addition of lime or sodium hydroxide and oxidation with excess chlorine and strong agitation (to prevent precipitation of calcium or sodium cyanide salts). End products of the oxidation include molecular nitrogen and carbonate ion. After the cyanide has been removed, the wastewater can be combined with metal-cleaning solutions containing heavy metals and grease. The mixture is usually quite acidic, and the pH must be raised to a neutral or slightly alkaline value before the metals can be precipitated.

Chromium-plating wastes present a similar specialized problem. Precipitation is simplest if the chromium is in the trivalent form. Unfortunately, the plating operation uses hexavalent chromium. Reduction to the trivalent state by $FeSO_4$, $NaHSO_3$, or SO_2 at $pH = 3$ is followed by neutralization and precipitation.[2]

Clearly, if all the wastes from a metal-plating operation are mixed, wastewater treatment will be far more difficult. Cyanide and chromium concentrations will be lower, resulting in slower reaction rates, and the entire waste volume would have to be raised and lowered in pH for the cyanide and chromium reactions. Similar situations exist in many industries. Quite often the wastewater streams that present the greatest problems are relatively small in size. If these streams are segregated from overall flow, methods of treatment can be considered that would be economically unfeasible otherwise.

11-3 INTERACTION OF SYSTEM COMPONENTS

Components of any treatment system are interrelated. We have already pointed out that the manner of operation of an activated-sludge aeration basin affects the flocculation of the sludge. A second example is the relationship between sludge digestion and other forms of treatment. If precipitation is used to improve removal of organic solids in the primary or secondary clarifiers, cation toxicity may occur in the digester. In situations where chemical addition is considered, effects on digestion should be estimated and taken into consideration. A similar situation exists with respect to relatively nontoxic inorganic solids such as grit, soil, and clay. These materials are normally removed in a special sedimentation unit, the grit chamber. If a grit chamber is not provided and significant amounts of these materials are in the wastewater, accumulation in the digester may occur. Fermentation may not be directly hindered, but mixing and pumping problems may occur as well as a significant change in effective volume.

Tertiary treatment by filtration is strongly affected by the properties of the floc overflowing the clarifier. Floc size and strength are a function of the type of treatment process and the manner in which it is operated, and therefore, filtrability of an effluent is related to previous treatment. An extreme example is the case of an activated-sludge process that has been "taken over" by filamentous organisms. Decreased sedimentation rates would be expected, and, consequently, greater quantities of solids would be washed out of the secondary clarifier. Rapid blinding of the filter would be the result. Similar problems exist in filtering algae. Filter cleaning is often much more difficult under these conditions also.

Examples of the interrelationship between processes with organic waste-waters have been furnished in discussing nutrient removal and metal-plating wastewater treatment. In the case of nutrient removal, phosphate can be precipitated by the addition of lime. The lime provides calcium for the precipitate and also raises the pH to some extent, although perhaps excess lime must be added before ammonia stripping becomes possible. If denitrification is to be used, the order of processes must be reversed. Phosphate is necessary for the growth of both the nitrifying and denitrifying bacteria.

In the case of the metal-plating wastes, the order of the treatment steps is important from both an efficiency and cost point of view. A further note may be added because pH adjustment of the wastewater would normally be necessary before discharge, and partial treatment (precipitation of metal ions) could be achieved by simple neutralization. Removal of cyanide and chromium would be far more difficult unless they were segregated and treated individually, however.

11-4 MIXING WASTEWATERS AND REGIONAL PLANTS

Care must be taken in mixing industrial wastewaters with either other industrial wastes or municipal sewage. In addition to toxicity, the designer must consider the different reaction rates involved, variation in flow rates, and the possibility of diluting a wastewater constituent to an undetectable or untreatable concentration.

Many biodegradable industrial wastewaters have very low reaction rates. Mixing these wastewaters with easily degraded effluents forces the entire volume to be treated at the lower rate, resulting in larger tank volumes and higher capital costs.

Variation in flow rates is particularly important where acclimation of the biological treatment process culture to an industrial wastewater is necessary. Many industries shut down on weekends or seasonally. The change in wastewater constituents may result in regular process upset.

Mixing a low-volume industrial wastewater containing a toxic material with a high-volume flow may reduce the concentration to virtually undetectable values. The mass output rate is unchanged, of course. When the material is one that is concentrated in the food chain, a dangerous condition results.

Regionalization of treatment plants has been encouraged by both federal and state regulatory agencies in the United States. The primary reasons for this policy have been expected economics of scale and improved process operation. A number of significant dangers exist with regionalization. A change is made from a quasi-distributed discharge to a point discharge. Thus the mass input of organics occurs at a single point, and the receiving stream must respond to a single large input. The problem is most apparent under conditions of plant upset, but effects of nutrient and toxicant discharge will be important also. Wastewater characteristics change in long sewers. Under anaerobic conditions, trace elements such as iron may be precipitated out as sulfides, making treatment difficult. Attitudes toward wastewater treatment improve when the treatment plant is close by. In the case of industrial effluents, the best place for the treatment process is directly next to the production unit. Finally, there is considerable doubt about the relationship between cost (scale economics) and quality of operation and plant size. General observations lead to the conclusion that small plants often function as well as large ones.

11-5   SYSTEM ECONOMICS

Economic feasibility has been discussed in a qualitative sense at several points in the previous chapters. Two types of costs can be considered: the cost to attain a given degree of treatment and the relative costs of two or more processes that do essentially the same job. Overall treatment cost to attain a stated degree of treatment is always a significant problem. This is not because wastewater treatment is expensive in relation to other services or industrial operations but because the public or the company receive no direct benefits for the expenditure. Only the recent bloom of environmentalism has brought sewage treatment into polite conversation, and very few people can connect the taxes they pay for wastewater management with benefits from an improved environment. For this reason, most of the responsibility for setting effluent criteria and, therefore, determining the overall cost of treatment must fall on government. Economic feasibility must include the overall benefit to society, and it is government's responsibility to determine and articulate these objectives.

Economic feasibility of a particular process relative to alternative candidates is dependent upon capital, operating, and maintenance costs but must include consideration of process stability. If a receiving water is particularly sensitive to effluent characteristics, less-stable processes such as activated sludge may be un-economic even if they are considerably cheaper than alternatives such as physical-chemical treatment. Determination of stability is the responsibility of the design engineer. In most cases pilot studies are desirable. These studies must be run with the determination of stability as an objective.

EXAMPLE 11-1   Food-processing wastewater has been characterized and the state Environmental Protection Agency has placed discharge requirements on the wastewater. Both sets of values are presented in the following table:

Parameter	Waste	Required 30-d average	Required 30-d maximum
Ultimate BOD, mg/l	4000		
BOD_5, mg/l		20	40
COD, mg/l	4100	50	100
Total N, mg/l	120	2	3
Total P, mg/l	5		
pH	3.2	6.5–80	
Acidity, mg/l as $CaCO_3$	1200		
Suspended solids (including settleables), mg/l	300	15	30
Grit, mg/l			
Settleable solids, mg/l	100/220†	10	20
Flow rate average, l/s	9.0		
Peak, l/s	10.5		
24-h minimum, l/s	8.3		
Total dissolved solids, mg/l	2400	500	
Volatile dissolved solids, mg/l	2275		

† Obtained by difference between settled and unsettled samples.

The source of the BOD and the acidity is primarily organic acids.

Assuming that oxidation ponds without discharge are not possible, what type of treatment process should be used?

The high organic concentration and flow rate (9 l/s = 205,000 gpd) effectively limit the use of adsorption to polishing a previously treated wastewater. Virtually all the COD can be accounted for in terms of degradable material (BOD), and thus biological treatment would seem appropriate. A residual COD of 100 mg/l could be expected from the data. Quite likely this would be less than 50 mg/l if the settleable solids were removed, as they would be in a treatment plant. Further testing will determine whether this conclusion is valid. If a residual COD greater than 50 mg/l does exist, carbon adsorption of a portion of the flow or filtration to remove fine solids may be necessary.

Acidity and pH of the waste restrict the choice of biological wastewater treatment processes to either a CFSTR activated-sludge or high-rate trickling filter with a suitable recycle rate. The high BOD concentration would present operating problems for a trickling filter, and thus a CFSTR activated-sludge system is the best choice. A quick estimate of the oxygen transfer rate feasible vs the potential BOD_L conversion rate leads to the conclusion that the hydraulic residence time will be controlled by the capacity to transfer oxygen. Residence times of the order of 40 h or more will be necessary. One option would be the use of aerated lagoons, i.e., a chemostat CFSTR system.

The nitrogen restriction should not be a problem. Assuming a yield of 0.3 g cells/g BOD, approximately 150 mg/l N is stoichiometrically necessary. Thus less N is available than necessary, and nitrogen may have to be added to maintain a suitable culture. At $\theta_H = \Theta_C$ 40 h, the yield will probably be closer to 0.5 than 0.3, and this would mean that 250 mg/l of nitrogen would be needed. The actual amount necessary would be less quite probably and should be determined through laboratory and pilot testing. Although phosphorus requirements have not been placed on the effluent, the quantity available seems adequate for biological treatment but should not necessitate special removal procedures.

Virtually all the dissolved solids can be accounted for as BOD. Thus specialized dissolved-solids removal is unnecessary.

A preliminary recommendation can now be made. A waste treatment system consisting of an aerated lagoon followed by a sedimentation tank is the most suitable basic system. Solids settling to the bottom of the sedimentation tank must be removed and separately treated. A simple sedimentation pond without solids removal would result in release of nitrogen as the sludge decomposed. Algae blooms would probably occur in a settling pond, and thus the effluent would be high in BOD, suspended solids, and total N.

Disposal of the solids may present some problems. Anaerobic digestion with partial feedback of the supernatant liquor to the aerated lagoon would be one approach. This would make nitrogen addition unnecessary but would not solve the nitrogen disposal problem. Tertiary treatment of the excess supernatant may be suitable because the volumetric rate would be low.

If tertiary treatment for COD removal *is* necessary, a portion of the settled effluent could be passed through adsorption columns. Another approach would be to use a slow sand filter for all or part of the flow. The latter solution may be necessary because the sedimentation process may not be efficient enough with the low solids loading imposed by an aerated lagoon.

A schematic of the proposed system is given in Fig. 11-1.

Considerable laboratory and pilot testing will be necessary to determine which of the alternatives is most suitable and what the process parameter values should be. These should begin with studies of the rates and stoichiometry of the proposed aerated-lagoon system. Included in these studies will be experiments with sedimentation. Because efficiency of sedimentation is critical, pilot testing

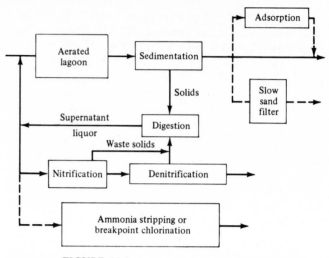

FIGURE 11-1
Alternatives for treatment of wastewater of Example 11-1.

rather than laboratory analysis would be recommended. Coagulation of the suspended solids may be useful if coagulant concentrations used will not cause problems with the digestion process or with final disposal of the sludge.   ////

11-6   WATER TREATMENT SYSTEMS

Water treatment systems have the objective of producing a water suitable for a particular use. Potable-water supplies are the most common examples, but industrial systems often must satisfy more stringent requirements. Bacterial count, presence of particular ions such as Ca^{2+}, Mg^{2+}, $Mg^{2+}Fe^{2+}$, total salt concentration, turbidity, taste, odor, and color provide the classic list of problems which must be dealt with.

In recent years the presence in treated water of carcinogens such as chloroform and certain pesticides has presented a new type of problem in which conventional treatment procedures may result in a decrease in water quality. In this case, chlorination of water containing low concentrations of organics results in the production of chlorinated hydrocarbons. Removal of the organics by carbon adsorption can result in increased bacterial counts by providing regions of high nutrient concentration on the carbon surface.

Potable-water Production Systems

Production of safe drinking water is a problem that has, to a major extent, been satisfactorily dealt with. Recent modifications of the problem, such as improved knowledge about carcinogens in drinking-water supplies, are not technically difficult. The most significant problems in potable-water production involve aesthetic characteristics of the product and the development of less expensive methods of water quality modification. This latter area is becoming increasingly important because of the need to remove specific ions and dissolved solids from water.

11-7 EXPERIMENTAL STUDIES

A brief discussion of the role of experimental studies in water and wastewater treatment seems an appropriate way of completing this chapter and the text. What follows is a statement of the author's beliefs and will, of course, be found incomplete by some and totally irrelevant by others. There are several important distinctions between types of experimental studies that are often ignored. Additionally, the type of information that can be obtained from the various possible approaches is usually rather specific. As a result, much research has been done that does not satisfy the job requirements.

Quite often research is divided into basic and applied. These two definitions seem inappropriate to wastewater treatment, and in fact virtually all the related research should be classified as applied (although to paraphrase Orwell[3]: some is more applied than other). Thus even work using pure organic substrates or pure cultures can be tied to waste treatment problems. Although design numbers being derived from such research is not likely, a better understanding of the mechanisms and operation of waste treatment process is provided (assuming the research is well done).

Here research will be divided by scale: laboratory, pilot, and prototype. Each scale has definite advantages and limitations. For example, flexibility of experimental studies decreases as the scale becomes larger, and usefulness of the variable values in design increases with scale. If we are concerned only with "treatability" of a wastewater, laboratory studies will be adequate. Again, if several alternative treatment procedures are being considered, they can be screened at the laboratory-scale level. The rate and stoichiometric values developed on this scale are usually quite close to those experienced on the prototype scale. Therefore, costs of additives, oxygen requirements, cell yields, and other similar variables can be predicted from laboratory studies. Many treatment problems or necessary conditions can be determined in the same manner. For example, in complex wastes the optimal pH for precipitation or coagulation is rarely the same as for pure solutions. Another example is the determination of a wastewater's toxicity to microorganisms. Clearly, we could go on for a considerable length of time citing examples where laboratory-scale research has a direct use in process design.

Connection of laboratory research on mechanisms of water and wastewater treatment with real-world problems also exists but often presents problems of interpretation. Most of this research is conducted in university laboratories where the participants are not in close contact with designers, regulatory agency staff, or treatment plant operators. Limited contact through conference presentations and journal articles does little to improve the situation because practicing engineers are usually looking for design information, and this is simply not obtainable from most of the work using idealized cultures, substrates, or environmental conditions. What is obtainable is an improved understanding of the mechanisms by which treatment occurs. A general understanding of the response of operating processes to alterations in the feed or environment can be developed. Factors such as which ion or compound is responsible for toxicity or inhibition can be determined and the information applied to real systems. Finally, theories of process design and operation can be tested under carefully controlled conditions.

A significant problem in wastewater treatment is deciding which process modification or new panacea should be used for a particular problem. The proliferation of "new" processes in recent years has left the designers with a wide number of choices and decisions but with little substantial information on which to base a decision. Few presentations of new processes include their constraints and limitations, and quite often these are found by chance at the prototype scale. Laboratory-scale research can eliminate a good deal of this difficulty, often using very simple systems. For example, the introduction of pure-oxygen-activated-sludge systems increased the oxygen transfer rates available to designers. In order to take advantage of this advancement, oxygen utilization rates must be increased, and the only way to do this is to increase cell concentrations. Advantages result because cell yield decreases as the cell residence time is increased, but limitations on the ability of the system to handle solids places an upper limit on the MLSS concentration and hence on the rate. An additional question should be raised as to whether or not the unit or specific removal rate is linear with the MLSS concentration or even the MLVSS concentration.

Theoretical studies offer many advantages and disadvantages also. The most important and difficult feature of this type of research is the need to maintain a relationship with the real world. Theoretical studies on process control of treatment plants have little use unless the systems considered are controllable and the models used represent what is found in the field. As an example, developing a process control model for a chemostat system seems of little real value because "real" processes that do not have feedback are forced to utilize enormous hydraulic residence times, and few control mechanisms other than varying this residence time are possible. Most of the models developed for such processes use short residence times, often orders of magnitude less than necessary, and hence the theoretical system is much more responsive than the real one.

Many of the processes used in wastewater treatment are not mathematically describable enough for application of optimal control modeling. Little is known, at least in a quantitative sense, about the short-term response of activated-sludge processes, and to propose to use a control model at this time does not seem reasonable. Reversing the process can be quite useful, however. Taking the information on transient response that is available, developing a rational model, and subjecting the model to various experiments can eliminate a great deal of difficult laboratory work. As various predictions of the model are checked in the laboratory, the model can be modified and improved. Eventually, a useful model for pilot or prototype systems may be developed, or mechanisms of process upset may be identified in this manner.

Pilot plant research has two purposes: (1) elimination of as many variables connected with the small size of laboratory system as possible and (2) development of design data. Where the second case is the objective, the first is clearly included, but the reverse is not always true. Laboratory systems usually have problems with such items as consistently maintained flow rates and effects of surface-to-volume ratios. Many of these problems can be decreased or eliminated at the pilot scale. Flexibility in the available range of variables and parameters studied is decreased as system size increases, however. For example, temperature control is quite simple in a 10-l tank but considerably more difficult at 1000 l, and while flow rate may be reproducible only to 10 percent at a few milliliters per hour, the range of values feasible in the milliliters per second range may be small. Use of parallel systems, or even controls, becomes difficult as unit size increases also.

Obtaining design data from pilot plant studies must be done with care. Consider the relationship between the final clarifier of an activated-sludge tank and characteristics of the culture. As Busch[4] has noted, cultures operated under different settling conditions quite possibly behave differently. Thus the design settling rate should be taken from the pilot plant studies. More often the pilot plant studies are used to assure treatability, obtain stoichiometric information, and determine effluent quality, and standard overflow rates are used in the prototype. This could be the reason process stability is a problem in many systems.

Rarely is process stability taken into account in pilot plant studies. Often effluent suspended solids, and in some cases filtrate BOD concentrations, fluctuate wildly during pilot plant studies, but when treatability, i.e., biodegradability, is established, the pilot studies are considered complete. If bulking or a filamentous culture develops in the pilot plant, the same conditions will probably occur in the prototype. One purpose of pilot plant studies is to identify problems and their solution prior to construction. On this basis both dischargers and regulatory agencies would be better off if pilot plants were used to demonstrate a system's ability to consistently meet all discharge requirements.

Prototype-scale research often falls into the category of careful observation. Because few modifications in process operation can be made, in most circumstances we are faced with trying to connect perturbation mechanism and response in very complex systems. Where large amounts of data are available, this can be done, of course, and much significant research work has resulted from this approach. In some cases, plant-scale research has involved a significant plant modification, as, for example, the Krauss[5] and contact stabilization processes.[6] As effluent criteria have become more restrictive, the ability of the engineer or plant operator to perform this type of research is decreased, however.

Comparison of the data from a large number of treatment plants can be quite useful also. Ordinarily data must be plotted on log-log paper to show any semblance of correlation, however. While this approach has proved very useful and a number of design equations have resulted from it, the scatter of values normally obtained makes heavy reliance dangerous.

PROBLEMS

11-1   The data shown in Fig. 11-2 were used to develop the NRC formula for trickling-filter design (see Chap. 8).

(a) How does a formulation such as the NRC formula fit with modern discharge requirements?

(b) Determine the standard deviation of the data and discuss its significance.

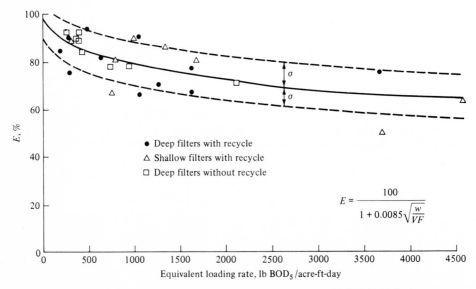

FIGURE 11-2
Original NRC formula data.[7]

11-2   The data presented in the following table were developed from extensive trickling-filter pilot studies with a municipal wastewater. Calculate the hydraulic loading rate including recycled effluent; the fraction removed is based on the actual applied BOD concentration C^*. Plot the data as Cu/C^* vs. $(1 + \alpha)Q/A$ on (*a*) arithmetic scale, (*b*) semilog scale, (*c*) log-log scale, and (*d*) log-log scale using $(Q/A)^{0.7}$. What effect does the scale have on "presentability" of the data? What loading rate would you use to design a process for an effluent requirement of 25 mg/l BOD? What recycle rate?

TRICKLING-FILTER DATA

Depth = 4 m
Diameter = 2 m

Q, l/d	α	C_i, mg/l	C_o, mg/l
3140	0	512	32
	1	507	16
	2	518	19
	4	532	18
	8	501	26
6280	0	517	33
	1	526	39
	2	519	14
	4	541	23
	8	553	24
9420	0	516	38
	1	514	29
	2	507	23
	4	530	16
	8	526	27
12,560	0	541	59
	1	528	28
	2	536	30
	4	514	22
	8	528	24

11-3   List the physical and economic constraints for the principal types of treatment processes.

11-4   A food-processing wastewater has a BOD value ranging from 2100 to 3700 mg/l and a flow rate of 21 l/s for 16 h/d and 150 d/year. Nitrogen content is normally about 130 mg/l as N. Suspended-solids concentrations vary with the crop being processed but may be as high as 1500 mg/l and consist of vegetable pulp. The state regulatory agency has set discharge requirements of 20 mg/l BOD_5 and 20 mg/l suspended solids on the plant. Devise a treatment system for the processor and defend your choice.

11-5   Many semirural mountain areas (e.g., communities of 500–3000) are in fragile environments where wastewater discharges must be of high quality. Discuss the options available to these communities and the financial significance of the choices.

11-6 What noncost constraints are involved in the choice between ammonia stripping and biological nitrification-denitrification?

11-7 A community has contracted to treat the wastewater from three local industries. Characteristics of wastes are given in the following table. Develop a treatment system for the community. Do not size the units, but choose necessary processes.

Characteristic	Domestic	A	B	C
BOD_5, mg/l	270		690	2700
COD, mg/l	450		950	4100
Total solids, mg/l	750	2250	1650	5650
Suspended	125	325	987	1700
Dissolved	625	1925	663	3950
pH	7.8	2.1	6.9	6.3
T, °C	18–24	18–21	28	17
Alkalinity, mg/l	350		250	120
Acidity, mg/l		1750	38	100
Q, l/s	97	0.8	15	38
Cr^{6+}, mg/l		150		
Cu		25		
Grease, mg/l	2	1.5	60	2

REFERENCES

1 SHERRARD, J. H., and E. D. SCHROEDER: Cell Yield and Growth Rate in Activated Sludge, *Water Pollut. Control Fed.*, vol. 45, p. 1889, 1973.
2 NEMEROW, N.: "Liquid Waste of Industry; Theories, Practice and Treatment," Addison-Wesley Publishing Company, Inc., Reading, Mass., 1971.
3 ORWELL, GEORGE: "Animal Farm," Harcourt, Brace and Company, Inc., New York, 1946.
4 BUSCH, A. W.: "Aerobic Biological Treatment of Wastewater," Olygodynamics Press, Houston, Tex., 1971.
5 KRAUSS, L. S.: Operating Practices for Activated Sludge Plants, *Water Pollut. Control Fed.*, vol. 37, 1965.
6 ULLRICH, A. H., and M. W. SMITH: The Biosorption Process of Sewage and Waste Treatment, *Sewage Ind. Wastes*, vol. 23, p. 1248, 1951.
7 SUBCOMMITTEE ON SEWAGE TREATMENT, COMMITTEE ON SANITARY ENGINEERING: Sewage Treatment at Military Installations, *Sewage Works J.*, vol. 18, p. 789, 1946.

APPENDIX A

NOTATION

a	Interface area, l^2
a_v	Specific surface area per unit volume, l^{-1}
A	Area, l^2
b	Maintenance-energy coefficient, t^{-1}
B	BET isotherm coefficient
C	Reactant concentration, m/l^3
d_b	Bubble diameter, l
d_p	Particle diameter, l
$\mathscr{D}$	Diffusivity, l^2/t
D	Dispersion coefficient, l^2/t
e	Efficiency of unit operation
E	Effectiveness factor
E_{O_2}	Efficiency of oxygen transfer
f	Modified effectiveness factor, $m^{-1}l^{-3}$
f'	Fanning friction factor
g	Gravitational constant, l/t^2
h	Depth of unit, l
H	Arrhenius rate coefficient, t^{-1}

k	Rate coefficient, units depend on case
k_0	Rate of removal coefficient, t^{-1}
k_g	Gas-phase transfer rate coefficient, $l^{-1}t^{-1}$
k_l	Liquid-phase transfer coefficient, l/t
K_m	Saturation coefficient, m/l^3
K_f	Freundlich coefficient
K_L	Oxygen mass transfer rate coefficient, l/t
$K_L a$	Overall oxygen mass transfer rate coefficient, t^{-1}
m	Unit mass transfer rate, $m/l^3 t$
M	Mass transfer rate, m/t
N_i	Flux of component i, $m/l^2 t$
N_p	Power number
N_{Re}	Reynolds number
N_{Sch}	Schmidt number
N_{sh}	Sherwood number
P_i	Pressure component i, m/lt
P_L	Power transfer to liquid, ml^2/t
Pr	Production rate, m/t
P_s	Fractional capacity of sorption unit
P_{tc}	Total capacity of sorption unit
r_i	Unit rate of formation of i, t^{-1}
r_0	Unit rate of removal of contaminant, t^{-1}
r_g	Unit rate of bacterial growth, t^{-1}
r_H	Hydraulic radius, l
R_i	Rate of formation of i, $m/l^3 t$
R_0	Rate of removal of limiting nutrient, $m/l^3 t$
R_g	Rate of bacterial growth, $m/l^3 t$
t	Time, t
T	Temperature, °C
v	Liquid velocity, l/t
V	Liquid volume, l^3
V_b	Bubble volume, l^3
V_p	Particle volume, l^3
w	Wetted perimeter, l
W	Weight, ml/t^2
x	Liquid-phase mole fraction
X	Cell concentration m/l^3
y	Gas-phase mole fraction
Y	True cell yield
Y_{obs}	Observed cell yield
α	Recycle ratio
δ	Film depth, l

ζ	Specific resistance
η	Conversion
Θ_C	Mean cell residence time, t
Θ_H	Hydraulic residence time, t
μ	Viscosity, m/lt
v_i	Stoichiometric coefficient
ρ	Density, m/l^3
τ	Shear stress, m/lt
ϕ	Volume fraction
Ψ	Temperature coefficient

APPENDIX B

DISSOLVED-OXYGEN SOLUBILITY DATA

Dissolved oxygen,† mg/l

Temp., °C	Chloride concentration, mg/l				
	0	5000	10,000	15,000	20,000
0	14.62	13.79	12.97	12.14	11.32
1	14.23	13.41	12.61	11.82	11.03
2	13.84	13.05	12.28	11.52	10.76
3	13.48	12.72	11.98	11.24	10.50
4	13.13	12.41	11.69	10.97	10.25
5	12.80	12.09	11.39	10.70	10.01
6	12.48	11.79	11.12	10.45	9.78
7	12.17	11.51	10.85	10.21	9.57
8	11.87	11.24	10.61	9.98	9.36
9	11.59	10.97	10.36	9.76	9.17
10	11.33	10.73	10.13	9.55	8.98
11	11.08	10.49	9.92	9.35	8.80
12	10.83	10.28	9.72	9.17	8.62
13	10.60	10.05	9.52	8.98	8.46
14	10.37	9.85	9.32	8.80	8.30
15	10.15	9.65	9.14	8.63	8.14
16	9.95	9.46	8.96	8.47	7.99
17	9.74	9.26	8.78	8.30	7.84
18	9.54	9.07	8.62	8.15	7.70
19	9.35	8.89	8.45	8.00	7.56
20	9.17	8.73	8.30	7.86	7.42
21	8.99	8.57	8.14	7.71	7.28
22	8.83	8.42	7.99	7.57	7.14
23	8.68	8.27	7.85	7.43	7.00
24	8.53	8.12	7.71	7.30	6.87
25	8.38	7.96	7.56	7.15	6.74
26	8.22	7.81	7.42	7.02	6.61
27	8.07	7.67	7.28	6.88	6.49
28	7.92	7.53	7.14	6.75	6.37
29	7.77	7.39	7.00	6.62	6.25
30	7.63	7.25	6.86	6.49	6.13

† Saturation values of dissolved oxygen in fresh and sea water exposed to dry air containing 20.90 percent oxygen under a total pressure of 760 mm Hg.

SOURCE: G. C. Whipple and M. C. Whipple, Solubility of Oxygen in Sea Water, *J. Am. Chem. Soc.*, vol. 33, p. 362, 1911. Calculated using data developed by C. J. J. Fox, On the Coefficients of Absorption of Nitrogen and Oxygen in Distilled Water and Sea Water and Atmospheric Carbonic Acid in Sea Water, *Trans. Faraday Soc.*, vol. 5, p. 68, 1909.

APPENDIX C

PROPERTIES OF WATER

Temp., °C	Density, g/cm³	Absolute viscosity, Pa-s	Vapor pressure, Pa
0	0.999841	0.001787	0.0610
5	0.999965	0.001519	0.0872
10	0.999700	0.001307	0.1228
15	0.999099	0.001139	0.1705
20	0.997992	0.001002	0.2337
25	0.997044	0.0008904	0.3167
30	0.995646	0.0007975	0.4242

SOURCE: "Handbook of Chemistry and Physics," 49th ed., Chemical Rubber Company, Cleveland, Ohio, 1969.

APPENDIX D

CONVERSION FACTORS

To convert from	To	Multiply by
	Acceleration	
foot/second^2	meter/second^2 (m/s^2)	3.048*E − 01
inch/second^2	meter/second^2 (m/s^2)	2.540*E − 02
	Area	
acre	meter^2 (m^2)	4.047*E + 03
foot^2	meter^2 (m^2)	0.290*E − 02
inch^2	meter^2 (m^2)	6.451*E − 04
	Energy	
British thermal unit (Btu)	joule (J)	1.056*E + 03
calorie	joule (J)	4.190*E + 00
foot-pound force	joule (J)	1.355*E + 00
kilowatthour	joule (J)	3.600*E + 06

To convert from	To	Multiply by
Force		
dyne	newton (N)	1.000*E − 05
kilogram-force	newton (N)	0.807*E + 00
pound force	newton (N)	4.448*E + 00
Length		
foot	meter (m)	3.048*E − 01
inch	meter (m)	2.540*E − 02
mile	meter (m)	1.609*E + 03
Mass		
grain	kilogram (kg)	6.480*E − 05
ounce-mass	kilogram (kg)	2.835*E − 02
slug	kilogram (kg)	1.459*E + 01
tonne	kilogram (kg)	1.000*E + 03
Power		
foot-pound/second	watt (W)	1.356*E + 00
horsepower	watt (W)	7.457*E + 02
newton-meter/second	watt (W)	1.000*E + 00
Pressure		
atmosphere	pascal (Pa)	1.013*E + 05
bar	pascal (Pa)	1.000*E + 05
millimeter of mercury (0°C)	pascal (Pa)	1.333*E + 02
dyne/centimeter^2	pascal (Pa)	1.000*E − 01
foot of water (4°C)	pascal (Pa)	2.989*E + 03
newton/meter^2	pascal (Pa)	1.000*E + 00
pound/inch^2	pascal (Pa)	6.895*E + 03
pound/foot^2	pascal (Pa)	4.788*E + 01
Temperature		
degree C	kelvin (K)	$T_K = T_C + 273.15$
degree F	kelvin (K)	$T_K = (T_F + 459.67)/1.8$
degree R	kelvin (K)	$T_K = T_R/1.8$
Viscosity		
poise	pascal-second (Pa-s)	1.000*E − 01
ft^2/second	meter^2/second (m^2/s)	9.290*E − 02
slug/foot-second	pascal-second (Pa-s)	4.788*E + 01
Volume		
acre-foot	meter^3 (m^3)	1.234*E + 03
barrel (oil)	meter^3 (m^3)	1.590*E − 01
foot^3	meter^3 (m^3)	2.832*E − 02
gallon (U.S. liquid)	meter^3 (m^3)	3.785*E − 03
gallon (U.K. liquid)	meter^3 (m^3)	4.546*E − 03
liter	meter^3 (m^3)	1.000*E − 03
yard^3	meter^3 (m^3)	7.646*E − 01

NAME INDEX

Adelberg, E. A., 213, 235, 242, 285, 335
Agardy, F. J., 285
Aiba, S., 109, 133
Anderson, J. J., 336
Anderson, R. E., 93
Andrews, J., 289, 292, 336
Antoine, R. L., 307, 312
Arrhenius, 201, 208
Atkinson, B., 122, 134, 289, 290, 292, 293, 311, 312

Babbitt, H., 311
Baer, G. T., 324
Ball, J. E., 260, 286
Barnard, J. L., 287
Barnhart, E. L., 109, 133
Bauchop, T., 214
Baumann, E. R., 311
Becker-Boost, E. H., 93
Berg, G. R., 209, 215
Bird, R. B., 51
Bisagni, J. J., 251, 285
Blackwood, A. C., 213
Bogan, R. H., 214
Bradfield, J. R. G., 213
Bringman, G., 216
Bryant, J. O., 36, 57
Burkhead, C. E., 186, 213
Busch, A. W., 14, 36, 51, 109, 115, 117, 134, 222, 223, 234, 235, 237, 258, 273, 284, 286, 311, 349, 352
Buswell, A. M., 322

Cairns, J., 213
Camp, T. R., 133, 145, 178
Carlson, D. A., 337
Carter, J. L., 214, 234
Carver, C. E., Jr., 133
Chang, P., 133
Chaudhuri, M., 215
Chick, H., 207
Clark, D. J., 214, 234
Clark, J. W., 336
Clark, R. H., 335
Cohen, G. N., 213
Cole, R. D., 286, 336
Conway, R., 115, 134
Cookson, J. T., 215
Cordes, E. H., 188, 213
Corey, R. B., 186, 213
Coulson, J. M., 14, 62, 93, 133, 178
Culp, G. L., 71, 93, 129, 130, 178, 215
Culp, R. L., 71, 93, 129, 178, 215

Dahling, D. R., 215
Dankwerts, P. V., 101, 102, 133
Darnell, J., 215

Davies, J., 292, 312
Dawson, R. N., 287
Dean, R. B., 215
Dick, R. I., 153, 178
Dixon, M., 188, 213
Dobbins, W., 133
Dodd, J. C., 215, 287
Dorfner, K., 93
Dornbush, J. N., 337
Dunn, C. G., 109

Eckenfelder, W. W., 104 – 109, 115, 133, 134, 178, 258, 286, 294, 297, 310, 311
Eddy, H. P., 336
Eliassen, R., 178
Ellsworth, R., 214, 234
Elsden, S. R., 214
Engberg, D. J., 274, 286
Englebrecht, R. S., 215
English, J. N., 287
Evans, F. L., 216

Faik, R., 213
Fair, G. M., 94, 167, 169, 215, 311, 335
Fenel, Z., 202, 214, 234
Flegal, T. M., 224, 234
Forney, C., 286
Friedlander, S. K., 142, 178
Friedman, A. A., 214, 279, 287, 312, 369
Fuller, W. J., 336

Galler, W. S., 310 – 312
Garrison, W. E., 285
Gaudy, A. F., 286
Gerritis, J. G., 214
Geyer, J. C., 94, 311
Gotaas, H. B., 287, 310 – 312
Gould, R. H., 286
Goulecke, C., 336
Grady, C. P. L., 202, 215, 234, 274, 285
Gram, A. L., 234
Grich, E. R., 286
Grieves, R. B., 286
Gujer, W., 285
Gunsalas, I. C., 214

Hadipetrou, L. P., 214
Hammer, M. J., 338
Handler, R. W., 213
Hansen, S., 178
Harris, H. S., 143, 178
Hart, S. A., 337
Hatch, L. P., 167, 169, 178
Herbert, D., 202, 214, 234
Higbie, R., 99, 122, 133

360

SUBJECT INDEX